International Review of

Cytology

A Survey of

Cell Biology

MOLECULAR BIOLOGY OF
RECEPTORS AND TRANSPORTERS

RECEPTORS

VOLUME 137B

International Review of **Cytology**

A Survey of **Cell Biology**

Guest Edited by

Martin Friedlander
Jules Stein Eye Institute
and Department of Physiology
UCLA School of Medicine
Los Angeles, California

Michael Mueckler
Department of Cell Biology
Washington University
School of Medicine
St. Louis, Missouri

MOLECULAR BIOLOGY OF RECEPTORS AND TRANSPORTERS

RECEPTORS

VOLUME 137B

Academic Press, Inc.
Harcourt Brace Jovanovich, Publishers
San Diego   New York   Boston   London   Sydney   Tokyo   Toronto

Copyright © 1992 by ACADEMIC PRESS, INC.
All Rights Reserved.
No part of this publication may be reproduced or transmitted in any form or
by any means, electronic or mechanical, including photocopy, recording, or
any information storage and retrieval system, without permission in writing
from the publisher.

Academic Press, Inc.
1250 Sixth Avenue, San Diego, California 92101-4311

United Kingdom Edition published by
Academic Press Limited
24–28 Oval Road, London NW1 7DX

Library of Congress Catalog Number: 52-5203

International Standard Book Number: 0-12-364538-7

PRINTED IN THE UNITED STATES OF AMERICA
92 93 94 95 96 97   BB   9 8 7 6 5 4 3 2 1

CONTENTS

Molecular and Regulatory Properties of the Adenylyl Cyclase-Coupled β-Adrenergic Receptors

Jorge Gomez and Jeffrey L. Benovic

The cAMP Receptor Family of *Dictyostelium*

Dale Hereld and Peter N. Devreotes

Rhodopsin and Phototransduction

Paul A. Hargrave and J. Hugh McDowell

Subunit Structure and Transmembrane Signaling of the Erythropoietin Receptor

Mark O. Showers and Alan D. D'Andrea

Cytokine Receptors: A New Superfamily of Receptors

Jolanda Schreurs, Daniel M. Gorman, and Atsushi Miyajima

Polymeric Immunoglobulin Receptor

Benjamin Aroeti, James Casanova, Curtis Okamoto, Michael Cardone, Anne Pollack, Kitty Tang, and Keith Mostov

Receptors for Nerve Growth Factor

Moses V. Chao, David S. Battleman, and Marta Benedetti

Asialoglycoprotein Receptor

Iris Geffen and Martin Spiess

Mannose Receptor

Suzanne E. Pontow, Vladimir Kery, and Philip D. Stahl

CONTRIBUTORS

Numbers in parentheses indicate the pages on which the authors' contributions begin.

Benjamin Aroeti (157), *Departments of Anatomy and Biochemistry, and Cardiovascular Research Institute, University of California, San Francisco, San Francisco, California 94143*

David S. Battleman (169), *Department of Cell Biology and Anatomy, Cornell University Medical College, New York, New York 10021*

Marta Benedetti (169), *Department of Cell Biology and Anatomy, Cornell University Medical College, New York, New York 10021*

Jeffrey L. Benovic (1), *Department of Pharmacology, Jefferson Cancer Institute, Thomas Jefferson University, Philadelphia, Pennsylvania 19107*

Michael Cardone (157), *Departments of Anatomy and Biochemistry, and Cardiovascular Research Institute, University of California, San Francisco, San Francisco, California 94143*

James Casanova (157), *Departments of Anatomy and Biochemistry, and Cardiovascular Research Institute, University of California, San Francisco, San Francisco, California 94143*

Moses V. Chao (169), *Department of Cell Biology and Anatomy, Cornell University Medical College, New York, New York 10021*

Alan D. D'Andrea (99), *Division of Pediatric Oncology, Dana Farber Cancer Institute, Children's Hospital, Harvard Medical School, Boston, Massachusetts 02115*

Peter N. Devreotes (35), *Department of Biological Chemistry, The Johns Hopkins University School of Medicine, Baltimore, Maryland 21205*

ix

Iris Geffen (181), *Department of Biochemistry, Biocenter, University of Basel, Basel, Switzerland*

Jorge Gomez (1), *Department of Pharmacology, Jefferson Cancer Institute, Thomas Jefferson University, Philadelphia, Pennsylvania 19107*

Daniel M. Gorman (121), *Department of Molecular Biology, DNAX Research Institute of Molecular and Cellular Biology, Palo Alto, California 94304*

Paul A. Hargrave (49), *Department of Ophthalmology and Department of Biochemistry and Molecular Biology, School of Medicine, University of Florida, Gainesville, Florida 32610*

Dale Hereld (35), *Department of Biological Chemistry, The Johns Hopkins University School of Medicine, Baltimore, Maryland 21205*

Vladimir Kery (221), *Department of Cell Biology and Physiology, Washington University School of Medicine, St. Louis, Missouri 63110*

J. Hugh McDowell (49), *Department of Ophthalmology, School of Medicine, University of Florida, Gainesville, Florida 32610*

Atsushi Miyajima (121), *Department of Molecular Biology, DNAX Research Institute of Molecular and Cellular Biology, Palo Alto, California 94304*

Keith Mostov (157), *Departments of Anatomy and Biochemistry, and Cardiovascular Research Institute, University of California, San Francisco, San Francisco, California 94143*

Curtis Okamoto (157), *Departments of Anatomy and Biochemistry, and Cardiovascular Research Institute, University of California, San Francisco, San Francisco, California 94143*

Anne Pollack (157), *Departments of Anatomy and Biochemistry, and Cardiovascular Research Institute, University of California, San Francisco, San Francisco, California 94143*

Suzanne E. Pontow (221), *Department of Cell Biology and Physiology, Washington University School of Medicine, St. Louis, Missouri 63110*

Jolanda Schreurs (121), *Department of Protein Chemistry, Chiron Corporation, Emeryville, California 94608*

Mark O. Showers (99), *Division of Pediatric Oncology, Dana Farber Cancer Institute, Children's Hospital, Harvard Medical School, Boston, Massachusetts 02115*

Martin Spiess (181), *Department of Biochemistry, Biocenter, University of Basel, Basel, Switzerland*

Philip D. Stahl (221), *Department of Cell Biology and Physiology, Washington University School of Medicine, St. Louis, Missouri 63110*

Kitty Tang (157), *Departments of Anatomy and Biochemistry, and Cardiovascular Research Institute, University of California, San Francisco, San Francisco, California 94143*

FOREWORD

The study of cells interaction, cell to cell or cell to environment, is a central area of investigation in modern cell biology. Cell to cell communication is essential for coordinate functioning of cells in a tissue, tissues in an organ, and organ systems in the organism. The regulation of synthesis, assembly, and insertion into the membrane, and functioning of these membrane domains as receptors are the subjects of this volume. While only selected proteins of the cell membrane are discussed here, as a whole they represent models for understanding many of the known biosynthetic and regulatory processes carried out by cells in response to extracellular signals. Cell surface receptors must somehow be synthesized in the appropriate quantities, make their way to the proper subcellular membrane, express themselves in a functionally appropriate fashion and, after the proper ligand or stimulus is received, transduce this information to the inside of the cell where a multitude of effector systems stand ready to transmit this information to the proper cellular machinery. Over the past decade, in great part due to the revolution in molecular biological techniques, we have begun to gain insight into these processes. Such advances were not possible before the primary structure of any of these molecules was known. Once a receptor has been "cloned" it becomes possible not only to determine its primary sequence, but, along with the application of expression systems and biophysical techniques, to analyze its secondary and tertiary structural characteristics. Site directed mutagenesis coupled to expression and functional assays permit the determination of structure–function relationships and ultimately, an understanding of the functionally important features both in the normal and diseased state. Each chapter in this volume discusses progress made not only in determining the structural characteristics of the receptors described, but functional studies that have helped us better understand this diverse group of proteins.

The first three chapters describe progress made in our understanding of members of the G protein-coupled receptor superfamily. While myocardial contractility, visual phototransduction and slime mold cell aggregation may not appear to have much in common, the receptors for their respective stimuli are similar in their general membrane topology and the class of effector molecules used to transduce the external stimuli to the inside of the cell. Remarkably, the general structure of the seven membrane helix molecules can not only transduce a stimulus in response to dramatically different stimuli (catecholamines, photons of light, or cAMP), but accomplish its signal generation via a common system, the G proteins.

Significant progress has been made in our understanding of signal transduction mediated by members of the G protein superfamily. With the advent of molecular cloning and recent advances in biophysical technology it has become possible to obtain dynamic structural information on these receptors that was previously impossible to gather. Since the hydrophobic nature of these polytopic membrane proteins currently prevents their successful crystallization, other approaches are necessary before we can obtain the kind of high resolution structural information that will be necessary before we can truly understand how these receptors function. A novel approach to obtaining such data has involved the use of site specific mutagenesis and nitroxide spin labeling to make structural determinations on appropriately mutated molecules [see Altenbach, *et al.*, (1990). *Science*, **248,** 1088]. The information obtained from such an approach can be used to refine models generated through computer-assisted techniques that take into account primary structure and known biochemical properties. The model of rhodopsin on the dust jacket of this volume illustrates the sort of modeling possible through this approach. In this case, the known coordinates of the seven transmembrane helical fragments of bacteriorhodopsin were used as a template for the seven hydrophobic domains of rhodopsin. The sequence was then manually aligned using the HOMOLOGY program of Biosym Technologies (San Diego, CA). The entire structure was energy minimized and the alignment designed to satisfy all known constraints (e.g., internal salt bridges, hydrophobic/ hydrophilic boundaries, and reactivities to hydrophobic probes). The N and C terminal domains were omitted for lack of structural information. For illustrative purposes, the point mutations found in patients with retinitis pigmentosa are labeled with gray numbers. All the mutations are single amino acid substitutions except for deletion mutants 255, 256, and 68–71. Only the vertical position of the mutations is shown and their relative position on the face of the respective helices is not indicated. The model, generated by Christian Altenbach and Wayne Hubbell of the Jules Stein Eye Institute at the UCLA School of Medicine, illustrates the usefulness of such an approach. As site directed mutatgenesis is used to replace native residues with cysteines, electron paramagnetic resonance spectroscopy can be used to detect appropriate spin labels at these cysteines. Such characteristics as segment orientation, hydrophobic/hydrophilic boundaries and other secondary and tertiary structures will be obtained through the analysis of such data. Once an accurate model is obtained, it should become possible to better understand the functioning of any receptor molecule. As progress is made in understanding structure–function relationships in rhodopsin, a paradigm may be developed that will help us to design experiments to better understand other members of the G protein-coupled receptor superfamily.

Two chapters on members of the cytokine superfamily illustrate the importance of this relatively new superfamily in regulating functions as diverse as hematopoiesis and the function and proliferation of lymphoid cells. Cytokines and their receptors represent a relatively new area of molecular immunology, but one where the interrelationship between structure and function impact significantly on the regulation of the immune system. Once again, understanding of function is limited by our lack of structural information. One advantage obtained from the study of this class of receptors (as well as some others) is that there is only a single hydrophobic membrane-spanning segment and, in general, a large hydrophilic extracellular domain. Thus, functionally important parts of these receptors are suitable for crystallization. Once again, the forces of molecular biology and biophysics have dovetailed to reveal startling information on the relationship between structure and function. While not discussed in this volume, there is another member of the immunoglobulin superfamily that has been used as a model in such an approach. The class I proteins of the major histocompatibility complex (MHC) are members of this superfamily (as is the poly Ig receptor) and serve to recognize foreign antigens in one of the early steps in activating the immune response. While there are only a very limited number of MHC molecules, the number of foreign antigens recognized by these receptors is several hundred-fold greater. Recently, molecular replacement has been used to obtain a crystal structure of antigen bound to MHC receptors [Fremont, *et al.*, (1992). *Science,* **257,** 919; Matsumura, *et al.*, (1992). *Science,* **257,** 927]. Apparently, the MHC antigen binding site is characterized by a combination of a deep central groove and one or two shallow pockets in the groove. It is the ability to use multiple combinations of deep and shallow pockets that appears to permit binding, with high affinity, to a broad range of antigens. Further analysis by mutating several different conserved residues within a peptide binding cleft confirm the importance of hydrogen bonds in this process [Latron, *et al.*, (1992). *Science,* **257,** 964]. Similar approaches should help us to better understand cytokine–receptor interactions with the goal of designing drugs that may stimulate or inhibit one or several of the cellular responses elicited by these compounds.

Chapters on the polymeric immunoglobulin receptor (poly IgR) and the asialoglycoprotein receptor (ASGPR) focus on the usefulness of these receptors as models for studying transcytosis and receptor-mediated endocytosis. Progress in our understanding of these processes has been enormous and the delineation of multiple subcellular sorting pathways has occurred as a result of this progress. Site-directed mutagenesis and deletion mutagenesis have been invaluable in determining the nature of the sorting signals and as we learn more about the structure of these receptors, we should better be able to understand the complexities of biosynthetic sorting.

Unlike the G protein-coupled receptors, growth factor receptors appear to be more heterogeneous in their mechanism of action. Chao and his colleagues discuss the nerve growth factor receptor in the context of cellular responses to growth factors. The response to mitogenic signals and neurotrophic signals may involve the same molecules and a clearer understanding of the receptors involved should help clarify this relationship. The implications for oncogenesis are apparent and the relationship here is further strengthened by the observation that one class of NGF receptors is the proto oncogene *trk*. The final chapter on the mannose receptor discusses a receptor whose true physiological role remains to be determined. Is this molecule involved in stimulating both endocytosis and phagocytic signals? Its localization to cells of the reticuloendothelial system supports the notion of its role in immunological functioning. Whether this role extends beyond passive scavenging of mannose terminal oligosaccharides remains intriguing. Furthermore, the issue of how a single receptor can recognize multiple sugars also remains unsettled.

This volume represents a heterologous collection of several members of a few superfamilies of membrane receptors. While it may seem a bit presumptuous to label a volume ''receptors'' while only containing 9 chapters on as many molecules, the problems represented in this class of proteins are of a more general nature. Each represents an area in which significant progress has been made in recent years. In most of these, it still remains to obtain sufficient structural information to make accurate statements about function. While the integration of recent advances in biophysics, biochemistry, and cell and molecular biology have brought us closer to this goal, definitive statements still elude us. As Ron Kaback pointed out in the foreword to part A of this series, the knowledge of the structure of any protein (receptors as well as transporters) will be necessary before we can elucidate the precise mechanism underlying its function. While the structure, per se, will not necessarily solve functional mysteries, it will almost certainly help us to make better educated guesses as to where to begin our experimental searches. Certainly, we will better understand the cellular processes involved in the response to receptor-mediated stimuli once such structural information is obtained.

Martin Friedlander

PREFACE

It has been over 6 years since we first considered putting together a volume devoted to the molecular biology of membrane proteins. We thought a single volume would provide sufficient space for several authors to describe the analysis, through molecular and cell biological approaches, of membrane protein structure and function. That was an exciting time since the dozen or so membrane protein receptors that had been cloned and sequenced appeared to fall into several large superfamilies that were related either through sequence homology or putative structural similarities. In the year it took us to prepare an outline for the volume's contents, the number of receptors had increased logarithmically; by 1988 the cloning and sequencing of several dozen receptors, transporters, and channels had been reported in the literature and we both retreated to our laboratories to try and keep up with the flood of information emerging from these studies.

By early 1990, we again began to think about putting together a multi-volume treatise that summarized our knowledge of membrane protein receptors, transporters, and channels to serve a useful function for both investigators already in the field as well as those "extramembranous" students and established investigators who wanted to familiarize themselves with a rapidly expanding area of membrane biology. We invited definitive reviews from active investigators in the field; we asked prospective contributors to include a summary of their knowledge of a particular membrane protein as well as to speculate on future directions. The response was very satisfying; most of the major classes and superfamilies of membrane proteins are represented and the chapters have been written by authors whose laboratories are very active in the field. Thus, the chapters are stimulating and authoritative even though we recognize that the speed with which the field is moving makes it difficult to be current by the time a volume is published. Nevertheless, we hope these volumes will provide a useful resource for those individuals interested in the field of membrane receptors, transporters, and channels.

Assembling nearly 30 chapters from as many laboratories has required the usual cajoling, pleading and, on occasion, threats. However, the credit for ultimate production is due to the authors themselves. We are grateful to the many scientists who contributed their efforts to writing the chapters in these volumes. For our part, the solicitation and editing of these volumes required extensive time that otherwise would have been spent with our families, and we are both very grateful to our wives, Sheila and Paula, and our children for their patience and understanding. Both of us would also like to thank Ron Kaback for his enthusiastic involvement in many of the chapters other than his own, as well as for his foreword to the first volume.

We would also like to thank the excellent editorial and production staff at Academic Press for their assistance with this project. In particular, Charlotte Brabants in editorial, Leslie Yarborough in book production and Cathy Reynolds in the art department have been exceptionally helpful.

Each of us has our own special thanks to extend to individuals who have helped with various aspects of assembling these volumes. Martin Friedlander would like to acknowledge the support of the Heed Ophthalmic and Heed/Knapp Foundations. My chairman, Bradley Straatsma, has been firmly supportive since my arrival at UCLA and has encouraged me through-out the course of this project. I am also grateful to Allan Kreiger, Bart Mondino, Gordon Grimes, and Joe Demer for their advice throughout the preparation of these volumes. Suraj Bhat and Dean Bok provided advice and encouragement during the early phases of this project and Eileen Fallon provided excellent editorial assistance. Bernie Gilula and Bill Beers nurtured my early interest in membrane proteins and will continue to do so. I am particularly grateful to Günter Blobel for introducing me to the field of membrane protein topogenesis and sharing his infectious enthusiasm for its study with me.

Michael Mueckler would like to acknowledge the support of the Juvenile Diabetes Foundation International. I am indebted to the past and current members of my laboratory for their patience during the preparation of these volumes and for making it a (usually) pleasant experience to come into the lab in the morning. I thank my colleagues in the Department of Cell Biology and Physiology, Robert Mercer, Edwin McCleskey, Philip Stahl, and Stephen Gluck, for their contributions to this series and for many stimulating discussions on membrane proteins over the past few years. I also thank my Ph.D. thesis advisor, Henry Pitot, and my postdoctoral mentor, Harvey Lodish, for sparking my interest in membrane proteins. Lastly, I am grateful to Alan Permutt for his continued friendship, support, and encouragement since my arrival in St. Louis.

Martin Friedlander

Michael Mueckler

Molecular and Regulatory Properties of the Adenylyl Cyclase-Coupled β-Adrenergic Receptors

Jorge Gomez and Jeffrey L. Benovic
Department of Pharmacology, Jefferson Cancer Institute, Thomas Jefferson University, Philadelphia, Pennsylvania 19107

I. Introduction

The catecholamines epinephrine and norepinephrine regulate a variety of physiological processes via their interaction with specific cell surface receptors. The catecholamine receptors were initially classified as α- and β-adrenergic based on their relative pharmacological properties (Ahlquist, 1948). Further studies by Lands *et al.* (1967) delineated two classes of β-adrenergic receptors (β_1AR and β_2AR); later studies also identified two classes of α-adrenergic receptors (α_1AR and α_2AR). Knowledge of these receptors has since increased tremendously. The ability to purify and characterize the β-adrenergic receptors and clone their respective genes has provided interesting insights into the structure, function, and regulation of this important receptor family. We now have the tools to further characterize and manipulate these receptors to provide additional information regarding structural and functional analysis. Although there is a wealth of information available on all of the adrenergic receptors (as well as a wide variety of other G protein-coupled receptors), this review will focus predominantly on the molecular and regulatory properties of the β-adrenergic receptors. Other chapters in this volume deal with other G protein-coupled receptors (Hereld and Devreotes, and Hargrave and McDowell, this volume).

1

II. Molecular Structure and Function of ß-Adrenergic Receptors

A. β-Adrenergic Receptor Subtypes

The family of β-adrenergic receptors has been classified based on its tissue distribution, pharmacological specificity, and molecular structure. Several subtypes of β-adrenergic receptors have now been identified based on their pharmacological specificity and primary amino acid sequence. The β-adrenergic receptors mediate a variety of physiological responses, including bronchodilation, myocardial contraction, and lipolysis. Lands *et al.* (1967) initially subdivided the β-adrenergic receptors into β_1 and β_2 subtypes based on the organ selectivity for a variety of β-agonists. The development of a variety of specific β-adrenergic receptor radioligands, such as [^{3}H]dihydroalprenolol and [^{125}I]iodohydroxy-benzylpindolol, enabled the direct identification of cell surface receptors that bound these ligands in a selective and saturable fashion. The use of subtype selective compounds in competitive binding assays enabled the quantitation of β_1- and β_2-adrenergic receptors in a wide variety of tissues. It was found that tissues that respond to β_1-selective agents, such as the myocardium, have predominantly the β_1AR subtype, whereas those that respond to β_2-selective agents (e.g., lung) have predominantly the β_2AR subtype. The existence of distinct β_1- and β_2-adrenergic receptors was further substantiated by the purification of the β_2AR from frog erythrocytes (Shorr *et al.*, 1981) and the β_1AR from turkey erythrocytes (Shorr *et al.*, 1982). The β_2AR has subsequently been purified from a number of mammalian tissues, including dog (Homcy *et al.*, 1983), hamster, rat, and guinea pig lung (Benovic *et al.*, 1984). The ability to purify these receptors in significant quantities has facilitated the cloning of the cDNA and gene encoding the hamster β_2AR (Dixon *et al.*, 1986) and the cDNA for the turkey β_1AR (Yarden *et al.*, 1986).

Although much of the β-adrenergic pharmacology could be explained by the existence of β_1AR and β_2AR subtypes, there were several instances in which the pharmacology of a particular system suggested the presence of additional β-adrenergic receptors. These receptors were termed atypical because their pharmacology for binding β-adrenergic ligands did not correlate with the pharmacology of either the β_1AR or β_2AR. These atypical receptors were found in a variety of tissues, including adipose, heart, liver, and skeletal muscle (Zaagsma and Nahorski, 1990; Kaumann, 1989; Challis *et al.*, 1988; Bond and Clark, 1988; Arch, 1989). The identification of a β_3-adrenergic receptor (β_3AR) subtype was definitively made by Emorine *et al.* (1989) who cloned a gene encoding a protein with high homology to the β_1 and β_2 subtypes. The amino acid sequence of this gene has 50.7% and 45.5% identity with the human β_1AR (Frielle *et al.*, 1987) and β_2AR, respectively (Figs. 1 and 2). When expressed in Chinese hamster ovary (CHO) cells, which lack endogenous β-adrenergic receptors, the β_3AR was able to stimulate cyclic AMP production with a variety of β-agonists. More recent pharmacological and cloning studies have demonstrated that the atypical βAR found in adipose tissue is indeed the β_3AR (Feve *et al.*, 1991). It remains to be determined whether there are additional β-adrenergic receptor subtypes.

B. Structure of the β-Adrenergic Receptors

The β-adrenergic receptors are members of the large family of G protein-coupled receptors (Gilman, 1987; O'Dowd *et al.*, 1989b; Dohlman *et al.*, 1991). This rapidly growing family of proteins shares some interesting structural, functional, and regulatory features, which will be discussed in this review. Depicted in Figure 1 and common to all the G protein-coupled receptors cloned to date is their proposed arrangement in the plasma membrane. Each receptor contains seven putative transmembrane domains of 20–28 hydrophobic amino acid residues (TM I–VII), three extracellular loops (E1–3), an extracellular amino terminus, three intracellular loops (I1–3), and an intracellular carboxyl terminus. The seven transmembrane domains are of sufficient length to represent membrane-spanning regions. Although this model is based predominantly on hydropathicity analysis, the structure of the analogous membrane protein bacteriorhodopsin has been elucidated using high resolution electron diffraction (Henderson and Unwin, 1975) and is consistent with this model.

Analysis of the primary amino acid sequence of the adrenergic receptor subtypes shows a high degree of conservation within the transmembrane domains with an overall 55% amino acid identity between the human β_1AR, β_2AR, and β_3AR and turkey β_1AR (Fig. 1). The third transmembrane domain is the most highly conserved (79%), whereas TM I (46%) and TM IV (33%) are the least conserved. The first (54%) and second (45%) intracellular loops are also highly conserved with many of the amino acid changes representing conservative substitutions (Figs. 1 and 2). Although the third intracellular loop has a significant level of conservation at the proximal and distal ends, the length of this loop varies significantly among the βARs (54–80 residues). The carboxyl terminus of these receptors is also poorly conserved except at the proximal end where there is 75% identity in the first 12 amino acids. There is a tremendous variability in the length of the carboxyl terminus with the longest in the turkey β_1AR (137 amino acids) and the shortest in the β_3AR (53 amino acids). The amino terminus of these receptors contains potential glycosylation sites and also varies significantly in length (34–59 residues). Whereas the extracellular domains are in general poorly conserved, the four extracellular cysteines in the βARs are totally conserved.

C. Posttranslational Modifications

An important component to the overall structure and function of the β-adrenergic receptors, as well as most of the other G protein-coupled receptors, involves

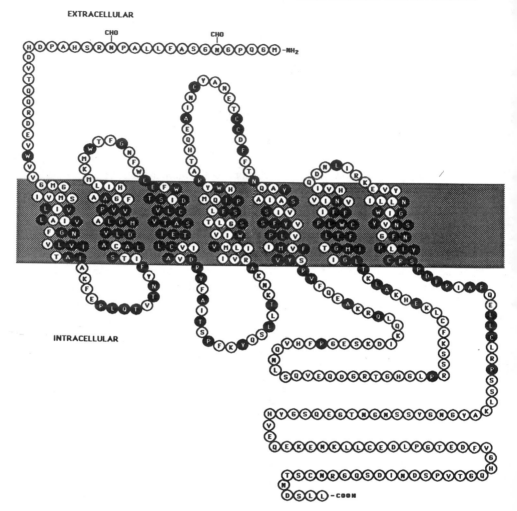

FIG. 1 Proposed topology of the human β₂-adrenergic receptor. The amino acid sequence of the human β₂AR, as it is proposed to reside in the plasma membrane, is shown. The amino acids that are identical in the human β₁-, β₂-, and β₃- and turkey β₁-adrenergic receptors are highlighted.

co-translational and posttranslational modifications. The β₂AR is one of the best characterized receptors in this respect. To date, at least four important modifications of the β₂AR have been identified including the formation of critical disulfide bonds, asparagine-linked glycosylation, fatty acid acylation (palmitylation), and phosphorylation.

The role of glycosylation in the function of the β-adrenergic receptors has not been completely elucidated. However, several lines of evidence have demon-

TABLE I

Comparison of Amino Acid Homologies between the Various Arrestin Molecules

Protein	β-arr	arr	droarr1	droarr2
β-arr	100	—	—	—
arr	58 (75)	100	—	—
droarr1	46 (67)	43 (62)	100	—
droarr2	43 (65)	42 (63)	46 (68)	100

Amino acid homologies between bovine β-arrestin (β-arr), bovine arrestin (arr), and the two *Drosophila* arrestin clones (droarr1 and droarr2) were compared using the Gap alignment program. The percent amino acid identity and percent amino acid similarity are given in parentheses.

strated that there are two asparagine-linked sites of glycosylation located in the amino-terminal domain (Asn^6 and Asn^{15}) of the β_2AR (Dohlman et al., 1987; Rands et al., 1990). In several tissues these sites appear to be heterogeneous with both high mannose, complex, and possibly mixed type chains being observed (Stiles et al., 1984a; Benovic et al., 1987b). Removal of the carbohydrate chains by endoglycosidase F treatment results in a receptor that is unchanged in its ability to bind adrenergic ligands (Stiles, 1985; Dohlman et al., 1987; Benovic et al., 1987b). Treatment of cells with tunicamycin also results in production of a βAR, which is unchanged in its ligand-binding properties (Doss et al., 1985; George et al., 1986; Boege et al., 1988; Rands et al., 1990). In contrast, there have been conflicting reports with regard to the ability of the deglycosylated βAR to stimulate adenylyl cyclase. Although there is no effect of deglycosylation on the ability of the β_2AR to couple to the G protein G_s in a reconstituted system (Benovic et al., 1987b), tunicamycin treatment of cells results in a receptor population that is either unchanged (Doss et al., 1985) or reduced (Boege et al., 1988) in its coupling to the adenylyl cyclase system. The most definitive studies on the role of glycosylation on β_2AR function have studied the expression and function of β_2AR mutants devoid of glycosylation sites at the amino terminus (Rands et al., 1990). These studies have shown that a glycosylation-deficient mutant β_2AR is not efficiently expressed or transported to the cell surface. However, receptors that do get to the cell surface are fully functional in ligand binding and in stimulating the adenylyl cyclase system. Thus, it appears that glycosylation of the β_2AR may play an important role in targeting the receptor for transport to the cell surface.

Although glycosylation of the other β-adrenergic receptors has not been extensively characterized, glycosylation of the turkey β_1AR, presumably at the one consensus site Asn^{14}, has been demonstrated (Cervantes-Olivier et al., 1985). The carbohydrate moiety on the turkey β_1AR does not appear to be important for ligand binding to the receptor. Although glycosylation of the

```
                              AMINO
HUM β₁AR    MGAGVLVLGA  SEPGNLSSAA  PLPDGAATAA  RLLVPASPPA  SLLPPASESP  EPLSQQWTAG    60
TUR β₁AR    MGDG.WLPPD  CGPHNRSGGG  G.ATAAPTGS  RQV......   ........SA  ELLSQQWEAG    43
HUM β₂AR    ..........  ..........  .MGQPGNGSA  FLLAPNRSHA  PDHDVTQQRD  EV....WVVG    35
HUM β₃AR    ..........  ..........  ..MAPWPHEN  SSLAPWPDLP  TLAPNTANTS  GLPGVPWEAA    38


                      TM I              I1                 TM II
HUM β₁AR    M.GLLMALIV  LLIVAGNVLV  IVAIAKTPRL  QTLTNLFIMS  LASADLVMGL  LVVPFGATIV   119
TUR β₁AR    M.SLLMALVV  LLIVAGNVLV  IAAIGRTQRL  QTLTNLFITS  LACADLVMGL  LVVPFGATLV   102
HUM β₂AR    M.GIVMSLIV  LAIVFGNVLV  ITAIAKFERL  QTVTNYFITS  LACADLVMGL  AVVPFGAAHI    94
HUM β₃AR    LAGALLALAV  LATVGGNLLV  IVAIAWTPRL  QTMTNVFVTS  LAAADLVMGL  LVVPPAATLA    98


              E1              TM III                 I2
HUM β₁AR    VWGRWEYGSF  FCELWTSVDV  LCVTASIETL  CVIALDRYLA  ITSPFRYQSL  LTRARARGLV   179
TUR β₁AR    VRGTWLWGSF  LCECWTSLDV  LCVTASIETL  CVIAIDRYLA  ITSPFRYQSL  MTRARAKVII   162
HUM β₂AR    LMKMWTFGNF  WCEFWTSIDV  LCVTASIETL  CVIAVDRYFA  ITSPFKYQSL  LTKNKARVII   154
HUM β₃AR    LTGHWPLGAT  GCELWTSVDV  LCVTASIETL  CALAVDRYLA  VTNPLRYGAL  VTKRCARTAV   158


                 TM IV                  E2                 TM V
HUM β₁AR    CTVWAISALV  SFLPILMHWW  R.AESDEARR  CYNDPKCCDF  VTNRAYAIAS  SVVSFYVPLC   238
TUR β₁AR    CTVWAISALV  SFLPIMMHWW  R.DEDPQALK  CYQDPGCCDF  VTNRAYAIAS  SIISFYIPLL   221
HUM β₂AR    LMVWIVSGLT  SFLPIQMHWY  R.ATHQEAIN  CYANETCCDF  FTNQAYAIAS  SIVSFYVPLV   213
HUM β₃AR    VLVWVVSAAV  SFAPIMSQWW  RVGADAEAQR  CHSNPRCCAF  ASNMPYVLLS  SSVSFYLPLL   218


                                      I3
HUM β₁AR    IMAFVYLRVF  REAQKQVKKI  DSCERRFLGG  PARPPSPSPS  PVPAPAPPPG  PPRPAAAAAT   298
TUR β₁AR    IMIFVYLRVY  REAKEQIRKI  DRCEGRFYGS  QEQPQ...PP  PLPQHQP...  ..........   265
HUM β₂AR    IMVFVYSRVF  QEAKRQLQKI  DKSEGRFHVQ  N.........  ..........  .......LS    246
HUM β₃AR    VMLFVYARVF  VVATRQLRLL  RGELGRF...  ..........  .PPEESPPAP  SRSLAPAVG    264


                     *                TM VI               E3
HUM β₁AR    APLANGRAGK  ..RRPSRLVA  LREQKALKTL  GIIMGVFTLC  WLPFFLANVV  KAF.HRELVP   355
TUR β₁AR    .ILGNGRASK  ..RKTSRVMA  MREHKALKTL  GIIMGVFTLC  WLPFFLVNIV  NVF.NRDLVP   321
HUM β₂AR    QVEQDGRTGH  GLRRSSKF.C  LKEHKALKTL  GIIMGTFTLC  WLPFFIVNIV  HVI.QDNLIR   304
HUM β₃AR    TCAPPEGVPA  CGRRPARLLP  LREHRALCTL  GLIMGTFTLC  WLPFFLANVL  RALGGPSLVP   324


                     TM VII                        *      CARBOXYL
HUM β₁AR    DRLFVFFNWL  GYANSAFNPI  IYCRSPDFRK  AFQGLLCCAR  RAARRRHATH  GDRPRASGCL   415
TUR β₁AR    DWLFVFFNWL  GYANSAFNPI  IYCRSPDFRK  AFKRLLCFPR  KADRRLHAGG  QPAPLPGGFI   381
HUM β₂AR    KEVYILLNWI  GYVNSGFNPL  IYCRSPDFRI  AFQELLCLRR  SSLK......  ...AYGNGYS   355
HUM β₃AR    GPAFLALNWL  GYANSAFNPL  IYCRSPDFRS  AFRRLLC...  RCGRRLPPEP  CAAARPALFP   381


HUM β₁AR    ARPGPP.PSP  GAASDDDDDD  VVGATPPARL  LEPWAGCNGG  AAADSDSSLD  EPCRPGFASE   474
TUR β₁AR    STLGSPEHSP  GGT.......  ..........  ...WSDCNGG  TRGGSESSLE  ERHSKTSRSE   421
HUM β₂AR    SNGNTGEQSG  YHVEQEKENK  LLCEDLPG..  TEDFVGHQGT  VPSD...NID  SQGRNCSTND   410
HUM β₃AR    SGVPAARSSP  AQPRLCQRLD  G.........  ..........  ..........  ..........   402


HUM β₁AR    SKV.......  ..........  ..........  ..........  ..........  ..........  ..  477
TUR β₁AR    SKMEREKNIL  ATTRFYCTFL  GNGDKAVFCT  VLRIVKLFED  ATCTCPHTHK  LKMKWRFKQH  QA  483
HUM β₂AR    SLL.......  ..........  ..........  ..........  ..........  ..........  ..  413
HUM β₃AR    ..........  ..........  ..........  ..........  ..........  ..........  ..
```

mammalian β_1AR and β_3AR has not been definitively shown, both of these receptors also contain consensus sequences for glycosylation at the amino terminus. The human β_1AR contains one potential glycosylation site at Asn[15] and the β_3AR contains two sites at Asn[8] and Asn[26].

The human β_2AR contains a total of 15 cysteine residues, many of which appear to exist in disulfide linkage (Dohlman et al., 1990). Four of these residues are localized in the extracellular loops, four are in transmembrane domains, and seven are in intracellular domains. A variety of mutations in specific cysteines have been made in an effort to assess the role of these residues in β_2AR function. Initial work by Dixon et al. (1987b) demonstrated that the mutagenesis of the extracellular Cys[106] and Cys[184] resulted in a β_2AR that had altered binding parameters for β-agonists while there was no effect on antagonist binding. Because the resultant effect was similar if either Cys[106] or Cys[184] was mutated, the authors suggested that these residues might be involved in disulfide linkage with each other. Dohlman et al. (1990) also observed dramatic effects on ligand binding and receptor expression when these two extracellular cysteines were mutated. Similarly, site-directed mutagenesis of the other two highly conserved extracellular cysteines (Cys[190] and Cys[191]) resulted in a marked decrease in affinity for both agonists and antagonists (Fraser, 1989; Dohlman et al., 1990). Again, these results suggest that Cys[190] and Cys[191] may be in disulfide linkage with each other. These four extracellular cysteines are highly conserved in all of the β-adrenergic receptors cloned to date (Fig. 1).

Surprisingly, cysteine residues present in the transmembrane domains of the β_2AR (Cys[77], Cys[116], Cys[125], and Cys[285]) do not appear to be important for ligand binding to the receptor. However, there is some evidence that Cys[285] may be important for G protein coupling. Moreover, Cys[116], Cys[125], and Cys[285] are highly conserved in all of the βAR subtypes, suggesting an important role for these residues in receptor structure or function. Although the intracellular cysteine residues in the β_2AR have not been systematically mutated, it has been shown that Cys[341] is not involved in ligand binding (Dixon et al., 1987b), whereas Cys[327] is not involved in ligand binding or effector coupling (Fraser, 1989).

Another important cysteine residue involved in β_2AR function is Cys[341], which is localized on the carboxyl-terminal tail of the receptor. This residue appears to be covalently modified by thioesterification with palmitic acid

FIG. 2 Alignment of the β-adrenergic receptors. The amino acid sequences from the human β_1-, β_2-, and β_3- and turkey β_1-adrenergic receptors were compared using the Gap alignment program (Genetics Computer Group). The amino acid number for each receptor is shown on the right. The putative transmembrane domains (TM I–VII) are highlighted by solid bars with the three extracellular (E1–3) and three intracellular (I1–3) loops also being denoted. The consensus serines for phosphorylation by the cAMP-dependent protein kinase are marked in bold type and the potential βARK phosphorylation sites are underlined.

(O'Dowd *et al.*, 1989a). Mutigenesis of this cysteine to a glycine results in a nonpalmitoylated form of the receptor, which is markedly reduced in its ability to stimulate adenylyl cyclase activity as well as in its ability to form a guanyl nucleotide-sensitive high-affinity state for agonists. Thus, palmitoylation may play an important role in anchoring the highly conserved proximal portion of the β_2AR carboxyl tail near the membrane to facilitate receptor–G_s coupling. Of interest is the finding that rhodopsin is also acylated by palmitic acid at two cysteine residues (Cys^{322} and Cys^{323}) in the carboxyl-terminal tail of the receptor (O'Brien and Zatz, 1984; Ovchinnikov *et al.*, 1988). Moreover, this particular cysteine is one of the most highly conserved residues among all of the G protein-coupled receptors (O'Dowd *et al.*, 1989b). This suggests that palmitoylation of this cysteine may play an important role in the structure and function of this receptor family.

Receptor phosphorylation is another posttranslational modification that has been extensively characterized. Phosphorylation of the β_2AR and many other G protein-coupled receptors plays a critical role in regulating the activity of many of these proteins. The role of phosphorylation in regulating β-adrenergic receptors will be discussed in more detail (Section III).

D. Ligand-Binding Domains

Several studies have focused on localizing the regions important for ligand binding to the β-adrenergic receptors. A number of techniques have been utilized in these studies, including sequencing sites of covalent affinity probe incorporation, site-directed mutagenesis, and production and expression of chimeric receptor cDNAs. The extracellular, intracellular, and transmembrane domains have all been extensively mutagenized and characterized in an effort to identify specific regions involved in ligand binding and G protein coupling. Mutagenesis of each of the four extracellular cysteine residues of the β_2AR results in a receptor with significantly reduced ligand binding. Examination of a variety of deletion mutants using the hamster β_2AR has revealed that most of the hydrophilic regions of the receptor are unimportant in ligand binding (Dixon *et al.*, 1987a). All of the amino terminus, most of the carboxyl terminus, and much of the third intracellular loop of the β_2AR could be removed without affecting antagonist binding (Dixon *et al.*, 1987a). However, whereas deletions in the third intracellular loop had no effect on antagonist binding, effects were observed on agonist binding (Strader *et al.*, 1987c). For example, deletions of the proximal (222–229) or distal (258–270) ends of the third intracellular loop resulted in a receptor that had only a single class of moderately high-affinity agonist-binding sites. This is in contrast to the high- and low-affinity agonist binding sites observed for the wild-type receptor. This effect appears to be due to a decreased ability of these mutant receptors to interact with G proteins. In addition, a large number of deletions in the various transmembrane domains resulted in the complete loss of receptor expression. This was also true for deletions in the first and second intracellular loops, whereas several deletions in regions near the transmembrane domains resulted in a reduced level of receptor expression.

A novel approach to elucidating the ligand binding domains involved the production of chimeric receptors constructed using the coding sequences of the α_2AR and β_2AR (Kobilka et al., 1988). Ten chimeric receptors were constructed and expressed in either COS-7 cells or Xenopus oocytes. Agonist binding was shifted to a lower affinity state as the β_2AR sequence was sequentially replaced by the α_2AR sequence, starting from the amino terminus through the fifth transmembrane domain. Interestingly, antagonist binding to the β_2AR was not affected when the first five transmembrane domains were substituted with α_2AR sequence. When the seventh transmembrane domain of the α_2AR was substituted into the β_2AR, both antagonist and agonist binding were dramatically affected. This chimeric receptor was capable of binding the α_2AR antagonist [^{3}H]yohimbine but did not bind the β_2AR antagonist [^{125}I]cyanopindolol. These studies suggest that most of the transmembrane domains play a role in agonist binding, whereas the seventh transmembrane domain is a major determinant for antagonist binding. Similar studies were performed by expression of β_1AR and β_2AR chimeras in Xenopus oocytes (Frielle et al., 1988). A pharmacological characterization of these chimeric receptors revealed that the fourth transmembrane domain is a major determinant for agonist binding, whereas the sixth and seventh transmembrane domains are important in antagonist-binding specificity. In another study, again using β_1AR–β_2AR chimeras, Marullo et al. (1990) found that the important binding determinants varied with the ligand used and concluded that various regions of all seven of the transmembrane domains are important in ligand binding.

An additional line of experimentation to identify regions important in ligand binding has utilized the covalent incorporation of affinity or photoaffinity ligands into receptor-binding sites. Photoaffinity labeling of the hamster β_2AR followed by limited proteolysis yielded two fragments of 38 and 26 kDa, representing the amino- and carboxyl-terminal domains, respectively (Dohlman et al., 1987). When separated it was found that the ligand had covalently incorporated into the 38-kDa fragment, which contains the first four transmembrane domains of the β_2AR. Additional studies involved labeling the receptor with the irreversible β-antagonist, [^{125}I]para-(bromoacetamido)-benzyl-1-iodocarazolol, followed by cleavage with V8 protease. When the resulting labeled receptor peptide was isolated and sequenced, it was found that the label had incorporated into the second transmembrane domain (Dohlman et al., 1988). Similar studies on the turkey β_1AR used the β-adrenergic photoaffinity probes [^{125}I]iodocyanopindolol-diazirine and [^{125}I]iodoazidobenzylpindolol (Wong et al., 1988). These studies revealed that both probes were incorporated into two distinct sites on the receptor. One site was identified as Trp^{330}, which is localized in the seventh transmembrane domain, whereas the second site was localized to an 8-kDa peptide, which lies in transmembrane spans III–V. Similar studies with other G protein-coupled receptors have also revealed the involvement of the transmembrane domains in ligand binding. The human platelet α_2-adrenergic receptor is specifically photolabeled in the fourth transmembrane domain (Matsui *et al.*, 1989), whereas the rat M1 muscarinic acetylcholine receptor was labeled in the third transmembrane domain using the alkylating drug propylbenzylcholine mustard (Curtis *et al.*, 1989). These studies give further confirmation of the important role of most, if not all, of the transmembrane domains in the recognition and binding of specific ligands.

The most definitive series of experiments identifying the amino acids important for ligand binding has utilized site-directed mutagenesis of specific amino acids found in the transmembrane domains. Initial studies targeted the charged residues found in the transmembrane domains because these might be expected to form a counterion for the catecholamines. When either Asp[79] (in TM II) or Asn[318] (in TM VII) were mutated a 10-fold decrease in agonist binding with no apparent effect on antagonist binding was observed (Strader *et al.*, 1987a, 1988). More detailed studies on the Asp[79] mutation yielded a 40–240-fold decrease in agonist binding to the receptor (Chung *et al.*, 1988). Initial mutagenesis studies on Asp[113], which is localized in TM III, resulted in a receptor that did not bind β-antagonists (Strader *et al.*, 1987a). More refined studies actually revealed a 10,000-fold decrease in antagonist binding and a 30–40,000-fold reduction in the ability of β-agonists to activate the Asp[113] mutant receptor (Strader *et al.*, 1988). Moreover, changing Asp[113] to a glutamic acid yields a receptor that converts antagonists to partial agonists (Strader *et al.*, 1989b).

The specific amino acids involved in hydrogen bonding to the catechol hydroxyls have also been identified (Strader *et al.*, 1989a). When Ser[204] and Ser[207] were substituted with alanine a decreased binding of isoproterenol with no effect on antagonist binding was observed. In addition, replacement of Ser[204] or Ser[207] was mimicked selectively by removal of the catechol hydroxyl moieties from the aromatic ring of the agonist. These studies suggested that interaction of catecholamine agonists with the β_2AR involves two hydrogen bonds, one between the hydroxyl side chain of Ser[204] and the *meta*-hydroxyl group of the ligand and one between Ser[207] and the *para*-hydroxyl group of the ligand. Multiple regions of the β_2AR appear critical for ligand binding. These include specific amino acid residues in the transmembrane and extracellular domains as well as proper posttranslational processing of the receptor. These regions are critical for providing the structural and specificity determinants for recognition and binding to the β-adrenergic receptor family.

E. G Protein-Coupling Domains

The β-adrenergic receptors mediate stimulation of the enzyme adenylyl cyclase via their hormone-promoted interaction with the stimulatory guanine nucleotide-regulatory protein, G_s (Gilman, 1987). This interaction leads to increases in intracellular cyclic AMP levels, activation of the cAMP-dependent protein kinase (PKA), and subsequent phosphorylation and regulation of a variety of intracellular proteins (Glass and Krebs, 1980). The mechanisms involved in the transduction of agonist binding to second messenger signaling have been extensively studied. The formation of a ternary complex composed of agonist, receptor, and G protein is one of the initial steps of activation. Formation of this complex is driven by the higher agonist affinity for the receptor–G protein complex versus the receptor alone. Formation of this complex promotes guanosine diphosphate (GDP) release and subsequent guanosine triphosphate (GTP) binding to the α-subunit of the heterotrimeric G protein. This in turn appears to induce dissociation of the α-GTP complex from the βγ-subunit. The free α_s-GTP complex is then able to directly activate adenylyl cyclase. This complex is inactivated by the hydrolysis of the bound GTP by the endogenous GTPase activity of the α-subunit (Gilman, 1987). Although most studies on β-adrenergic receptor–effector coupling have focused on the adenylyl cyclase system, there is some evidence that the βAR might also be able to regulate other effector molecules. For example, Barber et al. (1989) have provided evidence that βARs regulate Na-H exchange independent of cAMP. In addition, others have provided good evidence for a direct coupling between G_s and L type Ca^{2+} channels (Yatani and Brown, 1989) as well as an inhibition of cardiac Na^{2+} channels by βARs (Schubert et al., 1989). The mechanisms involved in β-adrenergic receptor coupling to these effector molecules have not been thoroughly characterized.

The β_2AR has served as a useful model for identifying the receptor domains involved in G protein coupling. Several lines of study have been utilized, including chimeric receptors, deletion analysis, site-specific mutagenesis, and synthetic peptides. Most of these studies have targeted the intracellular domains of the receptor, particularly the third intracellular loop, as the most likely regions involved in receptor–G protein coupling. The first evidence for the role of the third intracellular loop in β_2AR–G_s coupling was provided by Dixon et al. (1987a). They demonstrated that a deletion in the third loop of the hamster β_2AR (residues 239–272) yielded a receptor that could not stimulate cAMP production when stably transfected into L cells. Additional deletion analysis identified two major regions of the third loop, the proximal (residues 222–229) and distal (residues 258–270) portions, as important for receptor–G protein coupling (Strader et al., 1987c). The use of a more subtle deletion from the proximal portion of the third intracellular loop of the human β_2AR (residues 267–273) revealed that although the ability of this receptor to stimulate adenylyl cyclase was reduced >50%, there was no effect on high-affinity agonist binding to the mutant receptor (Hausdorff et al., 1990a). This suggests that the formation of a high-affinity agonist-receptor-G_s complex is not sufficient to fully activate the G protein.

The use of chimeric receptors has also provided significant insight into the domains involved in receptor–G protein coupling. One human α_2AR construct that contained portions of the fifth and sixth transmembrane domains and the connecting third intracellular loop from the human β_2AR (residues 214–295) stimulated adenylyl cyclase when incubated with α-agonists (Kobilka *et al.*, 1988). Additional studies that involved 12–22 amino acid insertions of the α_2AR into the human β_2AR revealed a critical role of the distal portion of the third intracellular loop of the β_2AR in G_s coupling (Liggett *et al.*, 1991). Moreover, a β_2AR containing the proximal and distal portions of I3 and proximal portion of the carboxyl-terminal tail of the α_2AR appears to couple preferentially to G_i. Similar studies utilizing a chimeric M1 muscarinic acetylcholine receptor, which contained either the I3 or a 12-amino acid proximal I3 segment from the turkey β_1AR, resulted in a receptor that gave a twofold to fourfold increase in cAMP production and normal inositol phosphate production upon stimulation with the muscarinic agonist carbachol (Wong *et al.*, 1990). Although substitution of I2 from the β_1AR had no effect on adenylyl cyclase activity, the combined replacement of I2 and the dodecapeptide in I3 stimulated adenylyl cyclase fully while only stimulating inositol phosphate release ~25%. These results suggest that the interaction of both the I2 and I3 loops is important for the specificity of receptor–G protein coupling.

Additional studies have further delineated the important role of the carboxyl-terminal region of the third intracellular loop in β_2AR–G_s coupling (O'Dowd *et al.*, 1988). Deletions of residues 263–273 in the human β_2AR decreased the ability of the receptor to stimulate adenylyl cyclase activity by 50% but depicted normal ligand binding activity for isoproterenol. Similarly, deletions or substitutions in the proximal end of the carboxyl-terminal tail of the β_2AR led to a marked decrease in the stimulation of adenylyl cyclase activity. Mutations in the second intracellular loop, in particular the substitution of threonine for Pro[138], resulted in a shift of up to sixfold in the EC_{50} for activation and a 0–30% reduction in maximal adenylyl cyclase activity. In addition, all of the mutations involving the first intracellular loop of the β_2AR resulted in a substantially reduced level of receptor expression.

Synthetic peptide studies have also helped to elucidate the domains involved in receptor–G protein coupling. These studies largely stemmed from the initial observations that the peptide mastoparan, a 14-amino acid wasp venom peptide, could directly activate G proteins (Higashijima *et al.*, 1988). Thus, in one study two peptides, comprising the proximal and distal 15 amino acids of the β_2AR third intracellular loop, were able to specifically activate the GTPase activity of G_S about threefold (Cheung *et al.*, 1991). Okamoto *et al.* (1991) also demonstrated that the peptide RRSSKFCLKEHKALK, which comprises the distal portion of the third intracellular loop of the human β_2AR, specifically stimulated GTPγS binding to G_s about threefold. Moreover, this peptide was also capable of activating adenylyl cyclase in wild-type S49 lymphoma cell membranes, whereas no effect was seen with unc⁻ mutant membranes, which have an uncoupled receptor and G_s. Thus, peptides from the third intracellular loop of the β_2AR are capable of specifically activating G_s.

Clearly, there are multiple domains involved in receptor–G protein interaction. These regions include the second intracellular loop, the proximal and distal portions of the third intracellular loop, and the amino-terminal portion of the carboxyl-terminal tail. The requirement for multiple sites of interaction plays an important role in not only providing specificity for G protein binding but also for initiating G protein activation.

F. Gene Organization of the β-Adrenergic Receptors

An interesting feature of the genes encoding the three β-adrenergic receptor subtypes is the complete lack of introns in their coding regions. This appears to be a common feature among many members of this G protein-coupled receptor family and suggests evolution from a common ancestor. Sequences have been reported for the hamster and human β_2AR (Dixon et al., 1986; Kobilka et al., 1987a; Schofield et al., 1987; Emorine et al., 1987) and the rat β_1AR (Machida et al., 1990; Shimomura and Terada, 1990). In addition, the β_3AR gene has also been cloned; however, only the amino acid sequence has been reported (Emorine et al., 1989). Characterization of the human β_2AR gene reveals a start site of transcription at nucleotide position -219 relative to the start of translation (Kobilka et al., 1987a). TATA-like sequences are found at positions -238 and -252, and a reverse complement CAAT box is found at nucleotide -293. The analogous sequences are also found at similar positions in the hamster β_2AR gene (Kobilka et al., 1987a). It has been reported that the β_3AR gene also contains an A/T-rich region reminiscent of a TATA box, a reverse sequence CAAT box analogous to the β_2AR, and a second CAAT box (Emorine et al., 1991). The promoter region of the rat β_1AR gene has not been thoroughly analyzed, although potential CAAT boxes have been identified at positions -436 and -453 relative to the proposed start of translation (Shimomura and Terada, 1990). Another interesting feature of all of the βAR genes is the presence of clusters rich in guanine and cytosine. This may play an important role in regulating the low levels of expression that are observed for these receptors (Maniatus et al., 1987).

The expression of the β_2AR is regulated by a variety of agents, including glucocorticoids and thyroid hormones (Stiles et al., 1984b; Collins et al., 1989a). Both the human and hamster β_2AR genes contain consensus sequences for transcriptional regulation by glucocorticoids (Kobilka et al., 1987a; Chung et al., 1987; Hadcock and Malbon, 1988a). In vitro and in vivo studies have demonstrated that the density of the β_2AR increases in response to steroids, in particular corticosteroids (Collins et al., 1988). In addition, Northern blot analysis has demonstrated that the regulation of the β_2AR by glucocorticoids is at the transcriptional level because β_2AR mRNA levels double after a 60-minute exposure to these steroids. This increase of mRNA levels was found to be specific for the

β_2AR and was blocked by actinomycin D, an inhibitor of RNA polymerase (Hadcock and Malbon, 1988a). This increase in β_2AR mRNA levels following glucocorticoid treatment appears to be due to a fourfold increase in the rate of transcription with no apparent effect on β_2AR message half-life (Collins *et al.*, 1988; Hadcock and Malbon, 1988a).

Transcription of the β_2AR gene also appears to be regulated by cyclic adenosine monophosphate (cAMP) (Collins *et al.*, 1989b). Short-term (30 minute) exposure of hamster DDT$_1$ MF-2 cells to a β-agonist or to cAMP analogs stimulates the rate of β_2AR gene transcription leading to a transient threefold to fourfold increase in steady-state mRNA levels. This appears to be due to an increase in the rate of transcription because the half-life of the β_2AR mRNA is unchanged in treated cells. This increase in the β_2AR gene transcription rate has been shown to be due to the presence of a cAMP-responsive element in the 5' flanking region of the human β_2AR gene (Collins *et al.*, 1989b, 1990).

Potential regulatory regions in the β_1AR and β_3AR promoters have not been well characterized. However, it has been reported that the β_3AR promoter region contains some sequence homology with the promoter for the adipocyte P2 lipid-binding protein (Emorine *et al.*, 1991). Although the significance of this finding has yet to be established, it has been demonstrated that differentiation of 3T3-F442A preadipocyte cells results in about a fivefold increase in β_1AR and β_3AR mRNA levels (Feve *et al.*, 1990). Moreover, treatment of these cells with dexamethasone, which results in increased β_2AR mRNA levels, leads to decreased mRNA levels for the β_1AR and β_3AR. These results suggest that the regulation of the human β_1AR and β_3AR genes is clearly distinct from that of the human β_2AR gene.

III. Molecular Mechanisms of Receptor Desensitization

The ability of an organism to regulate the intensity of a response in the presence of a continuous stimulus plays an important role in cell function. This phenomenon, often termed desensitization, is well recognized in biological regulatory and sensory systems (Koshland *et al.*, 1982). The βAR-coupled adenylyl cyclase system has provided an excellent model for studying the mechanisms involved in desensitization. The mechanisms involved in regulating β_2AR responsiveness are complex and include rapid alterations, such as receptor phosphorylation, uncoupling, and sequestration, which occur within minutes of agonist activation (Stadel *et al.*, 1983a,b; Clark, 1986; Benovic *et al.*, 1988; Hausdorff *et al.*, 1990b), as well as slower changes that may involve increases in receptor degradation and decreases in receptor synthesis (Hadcock and Malbon, 1988b; Collins *et al.*, 1989b). Moreover, although the receptor itself may serve as the primary locus for desensitization, there is also evidence that G proteins (Rich *et al.*, 1984; Katada *et al.*, 1985; Carlson *et al.*, 1989) and effector enzymes, such as adenylyl cyclase (Yoshimasa *et al.*, 1987, 1988), may also be directly regulated.

A. Classification of Desensitization

Desensitization has predominantly been characterized as either homologous or heterologous. Homologous desensitization is an agonist-specific phenomenon whereby receptor stimulation with a given agonist only reduces that particular agonist-stimulated response. In contrast, heterologous desensitization is an agonist-nonspecific phenomenon in which treatment with a variety of agents (e.g., cAMP analogs, prostaglandins, phorbol esters), which do not directly bind to βARs, leads to a reduced responsiveness to β-agonists. In addition, there also appear to be both rapid and much slower forms of both homologous and heterologous desensitization. Thus, there are a wide variety of mechanisms that play a role in desensitizing βAR responsiveness. In this section we will give an overview of the mechanisms currently thought to play a major role in βAR desensitization. It is important to realize, however, that much of the work on this system has been done either *in vitro* or in isolated cells. Thus, at present it is not clear which desensitization pathways are physiologically important. However, one might postulate that rapid desensitization might be important where rapid changes in agonist concentrations occur. In contrast, slower forms of desensitization may play a larger role in more chronic pathological conditions.

B. Involvement of a Receptor-Specific Protein Kinase

The mechanisms involved in rapidly regulating receptor responsiveness have been extensively studied (Sibley *et al.*, 1987; Benovic *et al.*, 1988; Hausdorff *et al.*, 1990b). Agonist activation of the β_2AR can lead to rapid receptor desensitization via multiple pathways. Green and Clark (1981) and Green *et al.* (1981) demonstrated that an agonist-promoted desensitization of the β_2AR is observed in the cyc- mutant of S49 lymphoma cells, which lacks G_s. This work was later extended by the observation that an agonist-promoted phosphorylation of the β_2AR also occurred during desensitization in cyc- and kin- (cAMP-dependent protein kinase-deficient) S49 lymphoma mutant cells (Strasser *et al.*, 1986a). These results suggested that the protein kinase involved in the cAMP-independent receptor phosphorylation either was stimulated by a second messenger other than cAMP or was able to preferentially phosphorylate the agonist-occupied form of the receptor. These studies also precipitated the search for the protein kinase involved and led to the identification of an enzyme that does specifically phosphorylate the agonist-occupied form of the receptor (Benovic *et al.*, 1986a). This kinase, termed

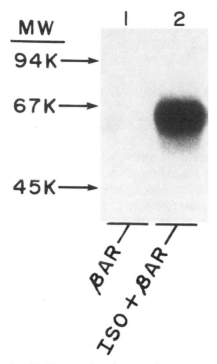

FIG. 3 Phosphorylation of purified hamster lung β_2AR by βARK. Reconstituted β_2AR was incubated with a βARK preparation for 30 minutes at 30°C in the presence (lane 2) or absence (lane 1) of 10 μM ($-$) isoproterenol. The phosphorylated receptor was resolved on a 10% polyacrylamide gel. The molecular weight standards are shown times 10^{-3}.

the β-adrenergic receptor kinase (βARK), has been purified from bovine brain, consists of a single subunit of 80 kDa, and is able to specifically phosphorylate the agonist-occupied form of the β_2AR to a stoichiometry of ~8 mol Pi/mol receptor (Benovic *et al.*, 1987a; Fig. 3). The kinetics of the phosphorylation reaction (K_m = 0.25 μM for β_2AR and 35 μM for ATP) suggest a high-affinity interaction between the receptor and kinase.

The sites of βARK phosphorylation on the β_2AR appear to be localized to the carboxyl-terminal tail of the receptor. Dohlman *et al.* (1987) demonstrated that most, if not all, of the sites on βARK-phosphorylated β_2AR are removed by carboxypeptidase treatment, which removes the carboxyl tail of the receptor. Mutagenesis studies also suggest the carboxyl terminus of the β_2AR is the major locus of βARK phosphorylation (Bouvier *et al.*, 1988; Hausdorff *et al.*, 1989). The use of synthetic peptides from the β_2AR also target the carboxyl tail as the major phosphorylation domain (Benovic *et al.*, 1990). Moreover, the kinetics of peptide phosphorylation by βARK (K_m ~3–4 mM) as well as the ability of some peptides to inhibit β_2AR phosphorylation by βARK suggest that βARK likely interacts with multiple intracellular regions of the β_2AR. It has been determined that βARK will only phosphorylate peptides that have negatively charged amino acids on the amino-terminal side of a serine or threonine residue (Onorato et al., 1991). This requirement may be served by glutamic or aspartic acid or possibly by phosphoserine (C. Y. Chen and J. L. Benovic, unpublished observation). Moreover, the acidic residue may be up to five amino acids amino-terminal to the serine or threonine. This finding enables speculation as to the potential βARK phosphorylation sites in each of the β-adrenergic receptors (Fig. 2). The human β_2AR has four putative sites in the carboxyl tail (six sites in the hamster β_2AR) and two potential sites in the third intracellular loop. Although it has not been demonstrated that βARK phosphorylates the β_2AR in I3, the α_2AR does appear to be phosphorylated by βARK in this region of the receptor (Benovic et al., 1987d; Onorato et al., 1991). The human β_1AR has five potential sites in the carboxyl tail and one site in I3, whereas the turkey β_1AR has 15 potential sites in the carboxyl tail and none in I3. Furthermore, whereas the human β_3AR has no sites in the carboxyl tail, there are three potential sites in I3. Although phosphorylation of the β_1AR and β_3AR by βARK has not been directly assessed, we might predict that both the human and turkey β_1ARs will serve as good substrates for βARK. It is not clear whether the β_3AR will be phosphorylated by βARK.

Several lines of evidence support a role for βARK in mediating rapid agonist-specific desensitization of the β_2AR. A number of studies have utilized cells expressing mutant β_2ARs that lack the proposed serine and threonine βARK sites in the carboxyl terminus. These mutant receptors are phosphorylated at a reduced level and also show a delayed onset of desensitization following agonist stimulation (Bouvier et al., 1988; Hausdorff et al., 1989; Liggett et al., 1989). However, these receptors show a normal pattern of desensitization following a prolonged agonist exposure (Strader et al., 1987b; Bouvier et al., 1988). A second line of evidence for βARK involvement in desensitization comes from the use of protein kinase inhibitors. Polyanions, such as heparin and dextran sulfate, are potent inhibitors of βARK with K_i of 10–50 nM (Benovic et al., 1989b). The ability of heparin to potently inhibit βARK activity has been used in a permeabilized cell system to assess the role of βARK in desensitization (Lohse et al., 1989, 1990a). These studies demonstrate that inhibitors of βARK are able to markedly reduce agonist-induced phosphorylation and desensitization of the β_2AR. Using a similar strategy the rate of β_2AR phosphorylation and desensitization induced by βARK was also assessed in permeabilized A431 cells (Roth et al., 1991). Agonist-induced phosphorylation of the β_2AR by βARK had a $\tau_{1/2}$ <20 seconds, whereas βARK-mediated desensitization proceeded with a $\tau_{1/2}$ <15 seconds. The effects of βARK were much more rapid than those induced by PKA or sequestration in these cells. Studies in human SK-N-MC neurotumor cells, which express

β_1AR and D1 dopamine receptors but not β_2AR, suggest that heparin does not block agonist-induced desensitization of the β_1AR but does block desensitization of D1 dopamine receptors (Zhou and Fishman, 1991). These studies suggest that the human β_1AR may not be regulated by βARK phosphorylation even though there are a large number of potential phosphorylation sites on the β_1AR (Fig. 2). A final line of evidence for the role of βARK in mediating β_2AR desensitization is provided by *in vitro* studies using purified proteins. These studies are described in detail in Section III,C.

Several lines of evidence suggest that βARK may serve a role as a general agonist-dependent receptor kinase. These include the finding that multiple agonists, such as β-agonists, prostaglandin E_1 (Strasser *et al.*, 1986b), and somatostatin (Mayor *et al.*, 1987), are each capable of inducing a translocation of the βARK activity from the cytosol to the plasma membrane. The time course of this translocation event correlates well with the rate of agonist-induced desensitization, suggesting that the prostaglandin E_1 and somatostatin receptors may also be regulated by βARK phosphorylation. Additional lines of study have demonstrated that both the human platelet α_2-adrenergic (Benovic *et al.*, 1987d) and chick heart M2 muscarinic acetylcholine (Kwatra *et al.*, 1989) receptors can serve as substrates for βARK *in vitro*. In both cases the phosphorylation was agonist-dependent and a stoichiometry comparable to the hamster β_2AR was observed, typically 4–8 mol phosphate/mol receptor. βARK is also capable of phosphorylating rhodopsin in a light-dependent manner, albeit to a significantly lower stoichiometry (Benovic *et al.*, 1986b, 1987a). Given the requirement of βARK for acidic and hydroxy amino acids, it is interesting to note that many of the other cloned G protein-coupled receptors have acidic- and serine-rich domains localized in their carboxyl terminus or third intracellular loops (O'Dowd *et al.*, 1989b).

Information on the structure of βARK has been obtained by isolation of a cDNA encoding the bovine enzyme (Benovic *et al.*, 1989a). This was accomplished by isolating and sequencing CNBr fragments from the purified kinase. Based on the amino acid sequences obtained, two synthetic oligonucleotide probes were synthesized and used to screen a bovine brain cDNA library. Two of the clones isolated had an open reading frame that encoded a protein of 689 amino acids (79.7 kDa). The overall topology of βARK suggests an amino-terminal domain of ~197 amino acids, a central protein kinase catalytic domain of ~239 amino acids, and a carboxyl-terminal domain of ~253 amino acids. To verify that this isolated cDNA indeed encoded βARK, a mammalian expression plasmid was constructed and used to transiently transfect COS-7 cells (Benovic *et al.*, 1989a). Both the agonist-occupied β_2AR and light-bleached rhodopsin served as good substrates for the expressed kinase, with the β_2AR being the preferred substrate. Northern blot analysis suggests that βARK mRNA levels are higher in tissues that have a high degree of sympathetic innervation (brain, spleen > heart, lung > liver, muscle). The human (Benovic *et al.*, 1991a) and mouse (Benovic *et al.*, 1991b) βARK genes have been localized on chromosomes 11q13 and 19, respectively. However, at present, little is known about the βARK gene or about potential mechanisms of transcriptional regulation.

Several lines of evidence suggest that βARK is a member of a multigene family. When bovine genomic DNA blots were probed with a 720-bp piece of the βARK cDNA, five hybridizing bands were observed under conditions in which only one band should be observed for βARK (Benovic *et al.*, 1989a). This suggests that the additional bands may be due to βARK-related genes. In addition, when a bovine brain cDNA library was screened using the labeled βARK cDNA as a probe, one additional class of cDNA clones was isolated (Benovic *et al.*, 1991b). These clones contain an open reading frame, which encodes a protein of 688 amino acids with an overall 84% amino acid identity with βARK. The substrate specificity of this novel kinase, termed βARK2, is currently being elucidated. However, preliminary studies using transient expression in COS-7 cells suggest that βARK2 is less active than βARK at phosphorylating both rhodopsin and the $\beta_2 AR$ (Benovic *et al.*, 1991b). However, more recent studies suggest there is not a significant difference between the ability of overexpressed purified βARK and βARK2 to phosphorylate a number of receptor and peptide substrates *in vitro* (C. Kim and J. L. Benovic, unpublished observation). RNA analysis reveals that βARK2 mRNA is less abundant than the βARK message in most tissues. In addition, the mouse βARK2 gene has been localized on chromosome 5, distinct from the mouse βARK gene on chromosome 19.

Several other G protein-coupled receptor kinase cDNAs have also recently been isolated from other tissues and organisms. The cDNA for bovine rhodopsin kinase, the enzyme involved in the phosphorylation and desensitization of light-activated rhodopsin (Palczewski *et al.*, 1988), has recently been isolated by screening a retinal cDNA library with oligonucleotides based on amino acid sequence from the purified kinase (Lorenz *et al.*, 1991). The deduced amino acid sequence of rhodopsin kinase shows an overall 34% amino acid identity and 58% similarity to βARK. Two *Drosophila* G protein-coupled receptor kinase cDNAs have also recently been isolated using a polymerase chain reaction approach (Cassill *et al.*, 1991). One of these cDNAs, termed GPRK-1, encodes a protein that is similar to bovine βARK with an overall 64% amino acid identity (78% similarity). The other kinase, termed GPRK-2, encodes a 50-kDa protein with 40% amino acid identity and 60% similarity to βARK. The similarities in amino acid sequence and overall topology between βARK, βARK2, rhodopsin kinase, GPRK-1, and GPRK-2 is depicted in Figure 4. Although no other βARK-related kinases have been purified, Haga and Haga (1989) have identified an activity that phosphorylates the agonist-occupied M1 and M2 muscarinic acetylcholine receptors. Several lines of evidence also suggest a role for agonist-dependent receptor phosphorylation in lower eukaryotes. In *Dictyostelium* a stimulus-dependent phosphorylation of the cyclic AMP

JORGE GOMEZ AND JEFFREY L. BENOVIC

G PROTEIN-COUPLED RECEPTOR KINASE FAMILY

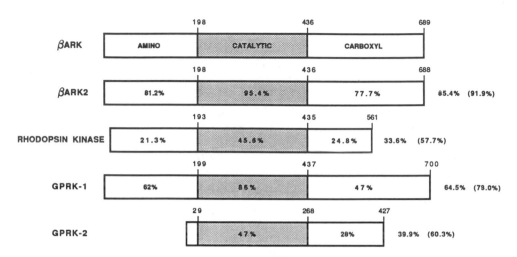

FIG. 4   Overall homology and topology of the G protein-coupled receptor kinase family. The amino acid sequences of bovine βARK (Benovic *et al.*, 1989a), bovine βARK2 (Benovic *et al.*, 1991b), bovine rhodopsin kinase (Lorenz *et al.*, 1991), and the two *Drosophila* kinases GPRK-1 and GPRK-2 (Cassill *et al.*, 1991) were compared using the GAP alignment program. The level of sequence identity with βARK is shown in each domain. The overall per cent amino identity and per cent amino acid similarity in parentheses are shown on the right.

receptor, which activates adenylyl cyclase through a G protein, plays a role in regulating chemotaxis (Vaughan and Devreotes, 1988), while in *Saccharomyces cerevisiae*, the α-mating factor receptor, which regulates an effector via interaction with a G protein, also undergoes an agonist-dependent phosphorylation, which accompanies desensitization (Reneke *et al.*, 1988).

C. Involvement of Arrestin Proteins in Desensitization

As shown in Figure 5, the current model for rapid agonist-specific desensitization of the β_2AR involves the specific binding of a cytosolic protein, termed β-arrestin, to the βARK phosphorylated receptor. The role of arrestin proteins in mediating receptor desensitization has been most extensively studied in the visual system. Retinal arrestin, also termed the 48-kDa protein or S antigen, was initially identified as a major protein, which redistributed (along with rhodopsin kinase) from the cytoplasm to the plasma membrane following light activation of rod outer segments (Kuhn, 1978). Kuhn *et al.* (1984) demonstrated that the binding of the 48-kDa protein to photoreceptor membranes was significantly

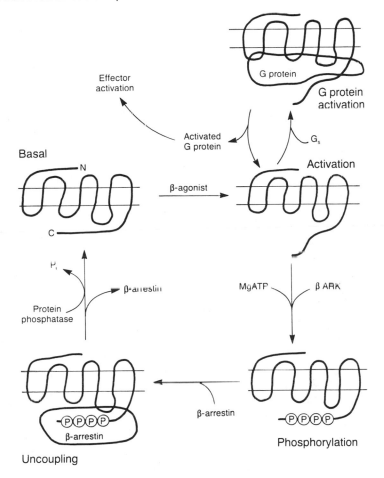

FIG. 5   Stimulus-dependent phosphorylation and desensitization of the β₂AR. Receptor activation by a β-agonist promotes interaction of the receptor with G_s, leading to G protein and effector activation. Receptor activation also promotes phosphorylation of the receptor, which is mediated by βARK. While the phosphorylated receptor appears to be partially uncoupled from the G protein, phosphorylation also promotes the interaction of β-arrestin with the β₂AR. This interaction further uncouples receptor and G protein interaction. The mechanism of β₂AR dephosphorylation remains poorly understood but may involve the sequestration of the receptor into a compartment where dephosphorylation can occur (Sibley et al., 1986).

enhanced by the phosphorylation of rhodopsin. Although phosphorylated rhodopsin has a reduced ability to interact with transducin and stimulate the cGMP phosphodiesterase, the binding of arrestin to rhodopsin suppresses the phosphodiesterase activation by ~98% (Wilden et al., 1986). The recent cloning of a bovine retinal arrestin cDNA revealed a protein of 404 amino acids (45,275 daltons), which appears to have several stretches of amino acid homology with the α-subunit of transducin (Yamaki *et al.*, 1987; Shinohara *et al.*, 1987, Wistow *et al.*, 1986).

Evidence for the involvement of an arrestin-like protein in the adenylyl cyclase system was initially suggested by studies that demonstrated that when purified reconstituted β_2AR was phosphorylated by a crude βARK preparation, the resulting receptor was largely uncoupled from G_s (Benovic *et al.*,1987c). However, if the receptor had been initially phosphorylated with a purified βARK preparation, the $\beta_2AR–G_s$ interaction remained largely intact. This suggested that there was a factor present in the crude βARK preparations that could uncouple the βARK-phosphorylated β_2AR from G_s. Moreover, when purified retinal arrestin was tested it was found to specifically impair $\beta_2AR–G_s$ coupling if the receptor had been phosphorylated by βARK (Benovic *et al.*, 1987c). Utilizing low-stringency hybridization techniques a cDNA encoding an arrestin molecule, apparently involved in β_2AR desensitization, was cloned (Lohse *et al.*, 1990b). This cDNA encodes a protein, termed β-arrestin, of 417 amino acids (47.1 kDa), which has 59% identity with retinal arrestin and appears to uncouple βARK phosphorylated β_2AR from G_s in a reconstituted system. Messenger RNA levels for β-arrestin reveal a tissue distribution that is similar to βARK.

Several lines of evidence suggest that the arrestins may also be members of a multigene family. In addition to retinal arrestin and β-arrestin, two arrestin-related genes have also recently been cloned from *Drosophila*. One gene is 3000 bp in length, contains 4 exons and 3 small introns, and encodes a protein of 364 amino acids with 43% identity with bovine arrestin (Smith *et al.*, 1990; Hyde *et al.*, 1990). A second *Drosophila* gene, composed of 3 exons and 2 introns, encodes a protein of 401 amino acids (44,972 daltons) with 42% identity to bovine arrestin (Yamada *et al.*, 1990). *In vitro* studies have demonstrated that this second *Drosophila* arrestin undergoes a Ca^{2+}-dependent phosphorylation, which is potentially mediated by a light-dependent phospholipase C (Yamada *et al.*, 1990). The current members of this arrestin gene family with their overall homology with bovine arrestin are shown in Table I. The studies to date suggest that the arrestin family of proteins play an important role in receptor desensitization via their ability to specifically interact with the phosphorylated forms of various G protein-coupled receptors.

D. Receptor Phosphorylation by the cAMP-Dependent Protein Kinase

Another pathway that is evoked by agonist appears to serve as a direct feedback mechanism of regulation. Agonist activation of the receptor leads to a rapid increase in intracellular cAMP levels, which then activates the cAMP-dependent protein kinase. This kinase plays a major role in phosphorylating and regulating the activity of a wide variety of intracellular proteins (Glass and Krebs, 1980). This pathway of desensitization has often been termed agonist-nonspecific or heterologous because any agonist that leads to increased cAMP levels will promote this form of desensitization. Cyclic AMP-dependent desensitization has been extensively studied using cAMP analogs to induce desensitization (Sibley et al., 1984b), mutant cells deficient in components of this pathway (Clark et al., 1988), cell-free desensitization systems (Nambi et al., 1984, 1985; Kunkel et al., 1989), site-directed mutagenesis (Hausdorff et al., 1989; Bouvier et al., 1989; Clark et al., 1989), specific inhibitors of the kinase (Lohse et al., 1989, 1990a; Roth et al., 1991), and purified reconstituted components of the system (Benovic et al., 1985). Moreover, it has been shown that the cAMP-mediated pathway of desensitization may play the major regulatory role under conditions of low agonist concentrations with corresponding low receptor occupancy (Clark et al., 1988; Hausdorff et al., 1989). Thus, this pathway may be the predominant mechanism for regulating peripheral receptors that respond to the low concentrations of circulating catecholamines.

As shown in Figure 2, the human β_2AR contains the consensus PKA phosphorylation sequence RRSS (Blackshear et al., 1988) in both the distal portion of the third cytoplasmic loop and proximal portion of the carboxyl terminus. While both PKA phosphorylation sites on the β_2AR can be directly phosphorylated by PKA in vitro (Benovic et al., 1985), the I3 site appears to play the major role in desensitization (Clark et al., 1989). In transfected L cells rapid heterologous desensitization is manifested as a twofold to threefold increase in the K_{act} and no change in the maximal level of agonist-stimulated adenylyl cyclase activity. When the I3 PKA site was deleted from the hamster β_2AR, this agonist-induced shift in K_{act} was abolished. In contrast, when the PKA site in the carboxyl terminus was deleted, the agonist-induced shift in K_{act} was identical to that seen for the wild-type receptor. The ability of PKA to differentially phosphorylate these two sites may in part explain these functional results. When synthetic peptides from these two domains were synthesized and used as substrates for PKA, it was found that the I3 peptide was a 10–100-fold better substrate for PKA compared with the carboxyl-terminal peptide (Blake et al., 1987; Bouvier et al., 1989). These studies suggest that the PKA site in the third intracellular loop of the β_2AR may play the major role in mediating rapid heterologous desensitization. It is interesting that both the human and turkey β_1ARs contain an analogous sequence in the same region of I3 while they do not have the consensus PKA sequence in the carboxyl terminus (Fig. 2). In contrast, the human β_3AR does not contain a good consensus PKA phosphorylation site anywhere in the molecule.

The functional consequences of β_2AR phosphorylation by PKA have also been assessed in vitro. Purified hamster β_2AR can be phosphorylated by PKA to a stoichiometry of 2 mol phosphate/mol receptor (Benovic et al., 1985). In addition, the rate of phosphorylation is stimulated twofold to threefold by ago-nist occupancy of the receptor (Benovic *et al.*, 1985; Bouvier *et al.*, 1987). When the functional consequence of PKA phosphorylation was assessed, it was found to promote an ~25% decrease in agonist-induced GTPase activity. More recent studies suggested that under more physiological Mg^{2+} concentrations (~0.5 mM) PKA phosphorylation almost totally uncouples the receptor and G_s (J. Pitcher and R. J. Lefkowitz, personal communication). Using a synthetic peptide from I3 of the human β_2AR (residues 259–273), Okamoto *et al.* (1991) demonstrated that PKA phosphorylation of this peptide resulted in a decreased ability of the peptide to stimulate G_s. Surprisingly, the phosphorylated peptide had an increased ability to couple to G_i. Although these interesting findings have not been confirmed with the intact receptor, they suggest that PKA phos-phorylation may inhibit the activation of an effector pathway by both inhibiting stimulation and stimulating inhibition.

Several lines of evidence suggest that protein kinase C phosphorylation of the β_2AR may also have a significant regulatory role. In avian erythrocytes the acti-vation of protein kinase C with phorbol esters leads to phosphorylation and de-sensitization of the β_1AR response (Kelleher *et al.*, 1984; Sibley *et al.*, 1984a). However, the effects of phorbol esters are cell-dependent and can lead to either increases or decreases in cyclase responsiveness (Katada *et al.*, 1985; Dixon *et al.*, 1988; Newman *et al.*, 1989). Although *in vitro* studies have demonstrated that the mammalian β_2AR can be directly phosphorylated by protein kinase C (Bouvier *et al.*, 1987), the functional consequences of this phosphorylation have not been directly assessed. Using site-directed mutagenesis Johnson *et al.* (1990) have more fully characterized the phorbol ester-induced desensitization of the hamster β_2AR. Phorbol ester treatment of stably transfected L cells results in three discernable effects: (1) a twofold to threefold increase in the K_{act} for β-agonist stimulation; (2) a twofold to threefold increase in the maximal level of stimulation by β-agonists; and (3) a decrease in the G_i-mediated inhibition of forskolin stimulation. When the major site of PKC phosphorylation (residues 259–262 in I3) on the β_2AR was deleted, the phorbol ester-induced increase in K_{act} was eliminated without affecting the maximal level of stimulation or the de-crease in G_i-mediated inhibition. Deletion of the other consensus PKC site of phosphorylation (residues 343–348 in the carboxyl tail) had no effect on the K_{act}. These results suggest that the PKC phosphorylation site on the third intra-cellular loop of the β_2AR is predominantly responsible for the phorbol ester-mediated increase in K_{act}. The effects on the maximal level of stimulation and decreased G_i-mediated inhibition appear to be due to a direct effect on G_i.

E. Receptor Sequestration

The rapid redistribution of cell surface βARs may also play an important role in receptor desensitization. Early studies by Chuang and Costa demonstrated that β-agonist treatment of frog erythrocytes led to a decrease in βARs from a plasma membrane fraction with a concomitant increase in cytosolic receptors (Chuang and Costa, 1979; Chuang et al., 1980). These "cytosolic" receptors could be isolated in a light vesicle fraction by high-speed centrifugation (Stadel et al., 1983b). Similar studies in astrocytoma and C6 glioma cells provided additional evidence for βAR redistribution. In these studies cells were treated with a β-agonist and lysed and cellular fractions were separated on a sucrose gradient. β-Agonist treatment was found to cause a loss of plasma membrane-associated receptors and an increase in receptors sedimenting at lower sucrose densities (Harden et al., 1980; Waldo et al., 1983; Frederich et al., 1983). In addition, the development and use of the hydrophilic β-antagonist [^{3}H]CGP-12177 enabled a direct demonstration of the loss of cell surface βARs following agonist treatment (Staehelin and Hertel, 1983; Hertel et al., 1983a, b).

The functional role of this rapid sequestration of cell surface receptors remains elusive. A number of groups have demonstrated that the functional uncoupling of the βAR and adenylyl cyclase system precedes the sequestration of the receptor (Harden et al., 1980; Waldo et al., 1983; Toews et al., 1984). In addition, the blockade of agonist-induced sequestration using either low temperature (Homburger et al., 1980), pretreatment with concanavilin A (Waldo et al., 1983; Wakshull et al., 1985), or treatment with phenylarsine oxide (Hertel et al., 1985) does not affect the rapid receptor uncoupling from adenylyl cyclase. Moreover, sequestered receptors isolated from either frog erythrocytes (Strulovici et al., 1983) or S49 lymphoma cells (Clark et al., 1985) were fully functional when characterized.

Several studies have attempted to further dissect the mechanisms involved in mediating receptor sequestration. There appears to be a direct correlation between agonist occupancy of the receptor and sequestration (Lohse et al., 1990a). However, the sites for β_2AR phosphorylation do not appear to be important for sequestration (Strader et al., 1987b; Bouvier et al., 1988; Hausdorff et al., 1989). Thus, sequestration appears to be an agonist-specific phenomenon that does not require receptor phosphorylation. The carboxyl-terminal tail of the β_2AR truncated at either position 354 (Strader et al., 1987b) or position 365 (Bouvier et al., 1988) is also not required for sequestration. When truncated at position 365 the receptor sequesters to a higher extent than the wild-type receptor (40% vs. 20%). However, one additional study suggested that when domains proximal to position 354 are truncated receptor sequestration is impaired (Cheung et al., 1989). Several studies have also demonstrated that β_2AR sequestration does not require G protein coupling (Cheung et al., 1990; Campbell et al., 1991; Hausdorff et al., 1991). The rate of β_2AR sequestration has also been compared with the rates of receptor phosphorylation in A431 cells using specific inhibitors of βARK, PKA, and receptor sequestration (Roth et al., 1991). These studies demonstrated that βARK ($\tau_{1/2}$ <20 seconds)- and PKA ($\tau_{1/2}$ ~3.5 minutes)-mediated phosphorylation of the β_2AR are substantially faster than receptor sequestration ($\tau_{1/2}$ ~10 minutes).

It is clear that sequestration of the β_2AR is an agonist-specific effect that does not require receptor phosphorylation or G protein coupling. However, because sequestration does not appear to be required for rapid desensitization to occur, its role remains poorly defined. One study that shed some light on the potential role of receptor sequestration involved the characterization of the phosphorylation state of sequestered versus plasma membrane-associated receptors (Sibley et al., 1986). In these studies frog erythrocytes, incubated with [^{32}P]inorganic phosphate to label the intracellular ATP pool, were desensitized with isoproterenol and the light vesicle fraction was then isolated. It was observed that the total cellular β_2AR pool was phosphorylated to a stoichiometry of 2.1 mol phosphate/mol receptor. In contrast, the sequestered receptors were phosphorylated to a stoichiometry of only 0.75 mol/mol, a level similar to that observed under basal conditions (no agonist). When the light vesicle fraction was characterized, it was shown to contain a potent protein phosphatase activity, which could dephosphorylate purified βARK-phosphorylated β_2AR. Thus, receptor sequestration may play a role in receptor dephosphorylation such that when recycled back to the plasma membrane the receptor is fully functional.

F. Receptor Down-Regulation

In addition to the initial rapid uncoupling and sequestration events, a much slower mechanism of desensitization involving receptor down-regulation also occurs. The down-regulation of βARs following prolonged agonist treatment has been detected in a variety of cells (Shear et al., 1976; Su et al., 1979, 1980; Homburger et al., 1980; Wang et al., 1990). In contrast to the rapid mechanisms involved in βAR–G protein uncoupling, down-regulation does not appear to require phosphorylation of the receptor. Thus, studies that have utilized mutant receptors containing altered phosphorylation sites have normal patterns of down-regulation (Strader et al., 1987b; Bouvier et al., 1988; Hausdorff et al., 1989; Cheung et al., 1989). When a 10-amino acid segment from the carboxyl tail of the β_2AR was substituted (residues 355–364), the receptor did not undergo any agonist-induced phosphorylation, sequestration, or rapid desensitization (Hausdorff et al., 1991) but displayed a time course of down-regulation identical to the wild-type receptor. Although receptor phosphorylation is not required for down-regulation, one study has shown that mutagenesis of the PKA sites of phosphorylation decreases the overall rate and extent of agonist-induced desensitization (Bouvier et al., 1989).

One enlightening finding was the demonstration that two tyrosine residues (Tyr^{350} and Tyr^{354}) localized in the carboxyl tail of the human β_2AR are important for down-regulation (Valiquette et al., 1990). Substitution of these two tyrosines with alanine led to a reduction in both the rate ($\tau_{1/2}$ ~7.5 hours for mutant, $\tau_{1/2}$ ~4.5 hours for wild-type) and extent (52% βAR reduction for mutant,

81% for wild-type) of down-regulation. Intracellularly localized tyrosine residues also appear to be important for the internalization of a number of other receptors, including the low density lipoprotein (Davis *et al.*, 1986) and mannose-6-phosphate (Lobel *et al.*, 1989) receptors. These results suggest that β_2ARs may be internalized via clathrin-coated pits, although this has yet to be directly demonstrated.

Although the loss of cell surface receptors is one mechanism for down-regulation, a second important event appears to involve a reduction in the synthesis of new receptors. Following cAMP or β-agonist treatment there is a rapid transient increase in β_2AR mRNA levels in the cell (Collins *et al.*, 1989b). However, following this initial rapid increase there is then a steady decline in β_2AR mRNA levels such that 24 hours after treatment the steady-state mRNA levels have decreased to ~50% of basal levels (Hadcock and Malbon, 1988b; Bouvier *et al.*, 1989; Collins *et al.*, 1989b; Hadcock *et al.*, 1989a). β-Agonists appear to induce a greater decrease in mRNA levels compared with cAMP analogs (Hadcock and Malbon, 1988b; Collins *et al.*, 1989b; Hadcock *et al.*, 1989a). The mechanism for the decreased mRNA levels appears to be due to destabilization of the β_2AR message because the mRNA half-life in agonist-treated cells is ~5 hours versus ~12 hours in control cells (Hadcock *et al.*, 1989b). β_2AR down-regulation has also been explored in a number of mutant S49 lymphoma cells (Hadcock *et al.*, 1989b). In these studies it was demonstrated that activation of the cAMP-dependent protein kinase directly induced a down-regulation of β_2AR mRNA levels. Agonist occupancy or receptor–G_s coupling was not required for the PKA-mediated down-regulation. However, H21a mutant S49 lymphoma cells, which are G_s–adenylyl cyclase uncoupled, displayed an agonist-induced mRNA down-regulation despite no increase in cAMP levels. The authors speculated that receptor–G_s coupling to an effector molecule distinct from adenylyl cyclase might also be able to promote β_2AR mRNA down-regulation.

Down-regulation of the β_2AR plays an important role in long-term regulation of receptor responsiveness. An agonist-induced increase in receptor internalization and degradation is one major mechanism, and long-term decreases in receptor synthesis also play a major role. Together, these mechanisms may serve to reduce cell surface β_2AR levels to <20% of normal levels.

IV. Conclusions

The β-adrenergic receptor has proven to be an excellent model system for understanding the structural, functional, and regulatory properties of G protein-coupled receptors. This receptor family shares many common structural features, including seven transmembrane domains, three intracellular and extracellular loops, an extracellular amino terminus, an intracellular carboxyl terminus, and multiple posttranslational modifications. These structural features give the receptor the unique ability to bind specific extracellular ligands and translate this binding into specific G protein activation. Posttranslational modifications also play a critical role in regulating receptor activity. Multiple protein kinases, including βARK, the cAMP-dependent protein kinase, and protein kinase C, play a role in rapidly phosphorylating and desensitizing the β_2AR, whereas additional mechanisms to internalize and degrade the receptor are important for long-term regulation of receptor activity. Future research in this area will focus on elucidating additional structural motifs of the receptor, which are important for the functional and regulatory properties of this interesting receptor family.

Acknowledgments

The authors thank Dr. Ray Penn for critical review of the manuscript. This work was supported in part by National Institutes of Health grants GM44944 and HL45964.

References

Ahlquist, R. P. (1948). *Am. J. Physiol.* **153**, 586–600.

Arch, J. R. S. (1989). *Proc. Nutr. Soc.* **48**, 215–223.

Barber, D. L., McGuire, M. E., and Ganz, M. B. (1989). *J. Biol. Chem.* **264**, 21038–21042.

Benovic, J. L., Shorr, R. G. L., Caron, M. G., and Lefkowitz, R. J. (1984). *Biochemistry* **23**, 4510–4518.

Benovic, J. L., Pike, L. J., Cerione, R. A., Staniszewski, C., Yoshimasa, T., Codina, J., Caron, M. G., and Lefkowitz, R. J. (1985). *J. Biol. Chem.* **260**, 7094–7101.

Benovic, J. L., Strasser, R. H., Caron, M. G., and Lefkowitz, R. J. (1986a). *Proc. Natl. Acad. Sci. U.S.A.* **83**, 2797–2801.

Benovic, J. L., Mayor, F., Jr., Somers, R. L., Caron, M. G., and Lefkowitz, R. J. (1986b). *Nature* **322**, 869–872.

Benovic, J. L., Mayor, F., Jr., Staniszewski, C., Lefkowitz, R. J., and Caron, M. G. (1987a). *J. Biol. Chem.* **262**, 9026–9032.

Benovic, J. L., Staniszewski, C., Cerione, R. A., Codina, J., Lefkowitz, R. J., and Caron, M. G. (1987b). *J. Recept. Res.* **7**, 257–281.

Benovic, J. L., Kuhn, H., Weyand, I., Codina, J., Caron, M. G., and Lefkowitz, R. J. (1987c). *Proc. Natl. Acad. Sci. U.S.A.* **84**, 8879–8882.

Benovic, J. L., Regan, J. W., Matsui, H., Mayor, F., Jr., Cotecchia, S., Leeb-Lundberg, F. L. M., Caron, M. G., and Lefkowitz, R. J. (1987d). *J. Biol. Chem.* **262**, 17251–17253.

Benovic, J. L., Bouvier, M., Caron, M. G., and Lefkowitz, R. J. (1988). *Annu. Rev. Cell Biol.* **4**, 405–428.

Benovic, J. L., DeBlasi, A., Stone, W. C., Caron, M. G., and Lefkowitz, R. J. (1989a). *Science* **246**, 235–240.

Benovic, J. L., Stone, W. C., Caron, M. G., and Lefkowitz, R. J. (1989b). *J. Biol. Chem.* **264**, 6707–6710.

Benovic, J. L., Onorato, J., Lohse, M. J., Dohlman, H. G., Staniszewski, C., Caron, M. G., and Lefkowitz, R. J. (1990). *Br. J. Pharmacol.* **30**, 3S-15S.

Benovic, J. L., Stone, W. C., Heubner, K., Croce, C., Caron, M. G., and Lefkowitz, R. J. (1991a). *FEBS Lett.* **283**, 122–126.

Benovic, J. L., Onorato, J. J., Arriza, J. L., Stone, W. C., Lohse, M., Jenkins, N. A., Gilbert, D. J., Copeland, N. G., Caron, M. G., and Lefkowitz, R. J. (1991b). *J. Biol. Chem.* **266**, 14939–14946.

Blackshear, P. J., Nairn, A. C., and Kuo, J. F. (1988). *FASEB J.* **2**, 2957–2969.

Blake, A. D., Mumford, R. A., Strout, H. V., Slater, E. E., and Strader, C. D. (1987). *Biochem. Biophys. Res. Commun.* **147**, 168–173.

Boege, F., Ward, M., Jurss, R., Hekman, M., and Helmreich, E. J. M. (1988). *J. Biol. Chem.* **263**, 9040–9049.

Bond, R. A., and Clark, D. E. (1988). *Br. J. Pharmacol.* **95**, 723–734.

Bouvier, M., Leeb-Lundberg, L. M. F., Benovic, J. L., Caron, M. G., and Lefkowitz, R. J. (1987). *J. Biol. Chem.* **262**, 3106–3113.

Bouvier, M., Hausdorff, W. P., DeBlasi, A., O'Dowd, B. F., Kobilka, B. K., Caron, M. G., and Lefkowitz, R. J. (1988). *Nature* **333**, 370–373.

Bouvier, M., Collins, S., O'Dowd, B. F., Campbell, P. T., DeBlasi, A., Kobilka, B. K., MacGregor, C., Irons, G. P., Caron, M. G., and Lefkowitz, R. J. (1989). *J. Biol. Chem.* **264**, 16786–16792.

Campbell, P. T., Hnatowich, M., O'Dowd, B. F., Caron, M. G., Lefkowitz, R. J., and Hausdorff, W. P. (1991). *Mol. Pharmacol.* **39**, 192–198.

Carlson, K. E., Brass, L. F., and Manning, D. R. (1989). *J. Biol. Chem.* **264**, 13298–13305.

Cassill, J. A., Whitney, M., Joazeiro, C. A. P., Becker, A., and Zucker, C. S. (1991). *Proc. Natl. Acad. Sci. U.S.A.* **88**, 11067–11070.

Cervantes-Olivier, P., Durieu-Trautmann, O., Delavier-Klutchko, C., and Strosberg, A. D. (1985). *Biochemistry* **24**, 3765–3770.

Challis, R. A. J., Leighton, B., Wilson, S., Thurlby, P. L., and Arch, J. R. S. (1988). *Biochem. Pharmacol.* **37**, 947–950.

Cheung, A. H., Sigal, I. S., Dixon, R. A. F., and Strader, C. D. (1989). *Mol. Pharmacol.* **34**, 132–138.

Cheung, A. H., Dixon, R. A. F., Hill, W. S., Sigal, I. S., and Strader, C. D. (1990). *Mol. Pharmacol.* **37**, 775–779.

Cheung, A. H., Huang, R. R. C., Graziano, M. P., and Strader, C. D. (1991). *FEBS Lett.* **279**, 277–280.

Chuang, D. M., and Costa, E. (1979). *Proc. Natl. Acad. Sci. U.S.A.* **76**, 3024–3028.

Chuang, D. M., Kinnier, W. J., Farber, L., and Costa, E. (1980). *Mol. Pharmacol.* **18**, 348–355.

Chung, F. Z., Lentes, K. U., Gocayne, J., Fitzgerald, M., Robinson, D., Kerlavage, A. R., Fraser, C. M., and Venter, J. C. (1987). *FEBS Lett.* **211**, 200–206.

Chung, F. Z., Wang, C. D., Potter, P. C., Venter, J. C., and Fraser, C. M. (1988). *J. Biol. Chem.* **263**, 4052–4055.

Clark, R. B., Friedman, J., Prashad, N., and Ruoho, A. E. (1985). *J. Cyclic Nucleotide Protein Phosphor. Res.* **10**, 97–119.

Clark, R. B. (1986). *In* "Advances in Cyclic Nucleotide and Protein Phosphorylation Research," (P. Greengard and G. A. Robison, eds.), pp. 155–209. Raven, New York.

Clark, R. B., Kunkel, M. W., Friedman, J., Goka, T. J., and Johnson, J. A. (1988). *Proc. Natl. Acad. Sci. U.S.A.* **85**, 1442–1446.

Clark, R. B., Friedman, J., Dixon, R. A. F., and Strader, C. D. (1989). *Mol. Pharmacol.* **36**, 343–348.

Collins, S., Caron, M. G., and Lefkowitz, R. J. (1988). *J. Biol. Chem.* **263**, 9067–9070.

Collins, S., Bolanowski, M. A., Caron, M. G., and Lefkowitz, R. J. (1989a). *Annu. Rev. Physiol.* **51**, 203–215.

Collins, S., Bouvier, M., Bolanowski, M. A., Caron, M. G., and Lefkowitz, R. J. (1989b). *Proc. Natl. Acad. Sci. U.S.A.* **86**, 4853–4857.

Collins, S., Altschmied, J., Herbsman, O., Caron, M. G., Mellon, P. L., and Lefkowitz, R. J. (1990). *J. Biol. Chem.* **265**, 19330–19335.

Curtis, C. A., Wheatley, M., Bansal, S., Birdsall, N. J., Eveleigh, P., Pedder, E. K., Poyner, D., and Hulme, E. C. (1989). *J. Biol. Chem.* **264**, 489–495.

Davis, C. G., Lehrman, M. A., Russel, P. W., Anderson, R. G. W., Brown, M. S. and Goldstein, J. L. (1986). *Cell* **45**, 15–24.

Dixon, R. A. F., Kobilka, B. K., Strader, D. J., Benovic, J. L., Dohlman, H. G., Frielle, T., Bolanowski, M. A., Bennett, C. D., Rands, E., Diehl, R. E., Mumford, R. A., Slater, E. E., Sigal, I. S., Caron, M. G., Lefkowitz, R. J., and Strader, C. D. (1986). *Nature* **321**, 75–79.

Dixon, R. A. F., Sigal, I. S., Rands, E., Register, B. R., Candelore, M. R., Blake, A. D., and Strader, C. D. (1987a). *Nature* **326**, 73–77.

Dixon, R. A. F., Sigal, I. S., Candelore, M. R., Register, R. B., Scattergood, W., Rands, E., and Strader, C. D. (1987b). *EMBO J.* **6**, 3269–3275.

Dixon, B. S., Breckon, R., Burke, C., and Anderson, R. J. (1988). *Am. J. Physiol.* **254**, C183–C191.

Dohlman, H. G., Bouvier, M., Benovic, J. L., Caron, M. G., and Lefkowitz, R. J. (1987). *J. Biol. Chem.* **262**, 14282–14288.

Dohlman, H. G., Caron, M. G., Strader, C. D., Amlaiky, N., and Lefkowitz, R. J. (1988). *Biochemistry* **27**, 1813–1817.

Dohlman, H. G., Caron, M. G., DeBlasi, A., Frielle, T., and Lefkowitz, R. J. (1990). *Biochemistry* **29**, 2335–2342.

Dohlman, H. G., Thorner, J., Caron, M. G., and Lefkowitz, R. J. (1991). *Annu. Rev. Biochem.* **60**, 653–688.

Doss, R. C., Kramarcy, N. R., Harden, T. K., and Perkins, J. P. (1985). *Mol. Pharmacol.* **27**, 507–516.

Emorine, L. J., Marullo, S., Delavier-Klutchko, C., Kaveri, S. V., Durieu-Trautmann, O., and Strosberg, A. D. (1987). *Proc. Natl. Acad. Sci. U.S.A.* **84**, 6995–6999.

Emorine, L. J., Marullo, S., Briend-Sutren, M. M., Patey, G., Tate, K., Delavier-Klutchko, C., and Strosberg, A. D. (1989). *Science* **245**, 1118–1121.

Emorine, L. J., Feve, B., Pairault, J., Briend-Sutren, M. M., Marullo, S., Delavier-Klutchko, C., and Strosberg, D. A. (1991). *Biochem. Pharmacol.* **41**, 853–859.

Feve, B., Emorine, L. J., Briend-Sutren, M. M., Lasnier, F., Strosberg, A. D., and Pairault, J. (1990). *J. Biol. Chem.* **265**, 16343–16349.

Feve, B., Emorine, L. J., Lasnier, F., Blin, N., Baude, B., Nahmias, C., Strosberg, A. D., and Pairault, J. (1991). *J. Biol. Chem.* **266**, 20329–20336.

Fraser, C. M. (1989). *J. Biol. Chem.* **264**, 9266–9270.

Frederich, R. C., Jr., Waldo, G. L., Harden, T. K., and Perkins, J. P. (1983). *J. Cyclic Nucleotide Protein Phosphor. Res.* **9**, 103–118.

Frielle, T., Collins, S., Daniel K. W., Caron, M. G., Lefkowitz, R. J., and Kobilka, B. K. (1987). *Proc. Natl. Acad. Sci. U.S.A.* **84**, 7920–7924.

Frielle, T., Daniel, K. W., Caron, M. G., and Lefkowitz, R. J. (1988). *Proc. Natl. Acad. Sci. U.S.A.* **85**, 9494–9498.

George, S. T., Ruoho, A. E., and Malbon, C. C. (1986). *J. Biol. Chem.* **261**, 16559–16564.

Gilman, A. G. (1987). *Annu. Rev. Biochem.* **56**, 615–649.

Glass, D. B., and Krebs, E. G. (1980). *Annu. Rev. Pharmacol. Toxicol.* **20**, 363–388.

Green, D. A., and Clark, R. B. (1981). *J. Biol. Chem.* **256**, 2105–2108.

Green, D. A., Friedman, J., and Clark, R. B. (1981). *J. Cyclic Nucleotide Protein Phosphor. Res.* **7**, 161–172.

Hadcock, J. R., and Malbon, C. C. (1988a). *Proc. Natl. Acad. Sci. U.S.A.* **85**, 8415–8419.

Hadcock, J. R., and Malbon, C. C. (1988b). *Proc. Natl. Acad. Sci. U.S.A.* **85**, 5021–5025.

Hadcock, J. R., Wang, H. Y., and Malbon, C. C. (1989a). *J. Biol. Chem.* **264**, 19928–19933.

Hadcock, J. R., Ros, M., and Malbon, C. C. (1989b). *J. Biol. Chem.* **264**, 13956–13961.

Haga, K., and Haga, T. (1989). *Biomed. Res.* **10**, 293–299.

Harden, T. K., Cotton, C. U., Waldo, G. L., Lutton, J. K., and Perkins, J. P. (1980). *Science* **210**, 441–443.

Hausdorff, W. P., Bouvier, M., O'Dowd, B. F., Irons, G. P., Caron, M. G., and Lefkowitz, R. J. (1989). *J. Biol. Chem.* **264**, 12657–12665.

Hausdorff, W. P., Hnatowich, M., O'Dowd, B. F., Caron, M. G., and Lefkowitz, R. J. (1990a). *J. Biol. Chem.* **265**, 1388–1393.

Hausdorff, W. P., Caron, M. G., and Lefkowitz, R. J. (1990b). *FASEB J.* **4**, 2881–2889.

Hausdorff, W. P., Campbell, P. T., Ostrowski, J., Yu, S. S., Caron, M. G., and Lefkowitz, R. J. (1991). *Proc. Natl. Acad. Sci. U.S.A.* **88**, 2979–2983.

Henderson, R., and Unwin, P. N. T. (1975). *Nature* **257**, 28–32.

Hertel, C., Muller, P., Portenier, M., and Staehelin, M. (1983a). *Biochem. J.* **216**, 669–674.

Hertel, C., Staehelin, M., and Perkins, J. P. (1983b). *J. Cyclic Nucleotide Protein Phosphor. Res.* **16**, 245–259.

Hertel, C., Coulter, S. J., and Perkins, J. P. (1985). *J. Biol. Chem.* **260**, 12547–12553.

Higashijima, T., Uzu, S., Nakajima, T., and Ross, E. M. (1988). *J. Biol. Chem.* **263**, 6491–6494.

Homburger, V., Lucas, M., Cantau, B., Perit, J., and Bockaert, J. (1980). *J. Biol. Chem.* **255**, 10436–10444.

Homcy, C. J., Rockson, S. G., Countaway, J., and Egan, D. A. (1983). *Biochemistry* **22**, 660–668.

Hyde, D. R., Mecklenburg, K. L., Pollock, J. A., Vihtelic, T. S., and Benzer, S. (1990). *Proc. Natl. Acad. Sci. U.S.A.* **87**, 1008–1012.

Johnson, J. A., Clark, R. B., Friedman, J., Dixon, R. A. F., and Strader, C. D. (1990). *Mol. Pharmacol.* **38**, 289–293.

Katada, T., Gilman, A. G., Watanabe, Y., Bauer, S., and Jakobs, K. H. (1985). *Eur. J. Biochem.* **151**, 431–437.

Kaumann, A. J. (1989). *Trends Pharmacol. Sci.* **10**, 316–320.

Kelleher, D. J., Pressin, J. E., Ruoho, A. E., and Johnson, G. L. (1984). *Proc. Natl. Acad. Sci. U.S.A.* **81**, 4316–4320.

Kobilka, B. K., Frielle, T., Dohlman, H. G., Bolanowski, M. A., Dixon, R. A. F., Keller, P., Caron, M. G., and Lefkowitz, R. J. (1987a). *J. Biol. Chem.* **262**, 7321–7327.

Kobilka, B. K., Kobilka, T. S., Daniel, K., Regan, J. W., Caron, M. G., and Lefkowitz, R. J. (1988). *Science* **240**, 1310–1316.

Koshland, D. E., Jr., Goldbeter, A., and Stock, J. B. (1982). *Science* **217**, 220–225.

Kuhn, H. (1978). *Biochemistry* **17**, 4389–4395.

Kuhn, H., Hall, S. W., and Wilden, U. (1984). *FEBS Lett.* **176**, 473–478.

Kunkel, M. W., Friedman, J., Shenolikar, S., and Clark, R. B. (1989). *FASEB J.* **3**, 2067–2074.

Kwatra, M. M., Benovic, J. L., Caron, M. G., Lefkowitz, R. J., and Hosey, M. M. (1989). *Biochemistry* **28**, 4543–4547.

Lands, A. M., Arnold, A., McAuliff, J. Pl., Cuduena, F. P., and Brown, T. G. (1967). *Nature* **214**, 597–598.

Liggett, S. B., Bouvier, M., Hausdorff, W. P., O'Dowd, B., Caron, M. G., and Lefkowitz, R. J. (1989). *Mol. Pharmacol.* **36**, 641–646.

Liggett, S. B., Caron, M. G., Lefkowitz, R. J., and Hnatowich, M. (1991). *J. Biol. Chem.* **266**, 4816–4821.

Lobel, P., Fujimoto, K., Ye, R. D., Griffiths, G., and Kornfeld, S. (1989). *Cell* **57**, 787–796.

Lohse, M. J., Lefkowitz, R. J., Caron, M. G., and Benovic, J. L. (1989). *Proc. Natl. Acad. Sci. U.S.A.* **86**, 3011–3015.

Lohse, M. J., Benovic, J. L., Caron, M. G., and Lefkowitz, R. J. (1990a). *J. Biol. Chem.* **265**, 3202–3209.

Lohse, M. J., Benovic, J. L., Codina, J., Caron, M. G., and Lefkowitz, R. J. (1990b). *Science* **248**, 1547–1550.

Lorenz, W., Inglese, J., Palczewski, K., Onorato, J. J., Caron, M. G., and Lefkowitz, R. J. (1991). *Proc. Natl. Acad. Sci. U.S.A.* **88**, 8715–8719.

Machida, C. A., Bunzow, J. R., Searles, R. P., Van Tol, H., Tester, B., Neve, K. A., Teal, P., Nipper, V., and Civelli, O. (1990). *J. Biol. Chem.* **265**, 12960–12965.

Maniatus, T., Goodbourn, S., and Fischer, J. A. (1987). *Science* **236**, 1237–1245.

Marullo, S., Emorine, L. J., Strosberg, A. D., and Delavier-Klutchko, C. (1990). *EMBO J.* **9**, 1471–1476.

Matsui, H., Lefkowitz, R. J., Caron, M. G., and Regan, J. W. (1989). *Biochemistry* **28**, 4125–4130.

Mayor, F., Benovic, J. L., Caron, M. G., and Lefkowitz, R. J. (1987). *J. Biol. Chem.* **262**, 6468–6471.

Nambi, P., Sibley, D. R., Stadel, J. M., Michel, T., Peters, J. R., and Lefkowitz, R. J. (1984). *J. Biol. Chem.* **259**, 4629–4633.

Nambi, P., Peters, J. R., Sibley, D. R., and Lefkowitz, R. J. (1985). *J. Biol. Chem.* **260**, 2165–2171.

Newman, K. B., Michael, J. R., and Feldman, A. M. (1989). *Am. J. Respir. Cell Mol. Biol.* **1**, 517–523.

O'Brien, P. J., and Zatz, M. (1984). *J. Biol. Chem.* **259**, 5054–5057.

O'Dowd, B. F., Hnatowich, M., Regan, J. W., Leader, W. M., Caron, M. G., and Lefkowitz, R. J. (1988). *J. Biol. Chem.* **263**, 15985–15992.

O'Dowd, B. F., Hnatowich, M., Caron, M. G., Lefkowitz, R. J., and Bouvier, M. (1989a). *J. Biol. Chem.* **264**, 7564–7569.

O'Dowd, B. F., Lefkowitz, R. J., and Caron, M. G. (1989b). *Annu. Rev. Neurosci.* **12**, 67–83.

Okamoto, T., Murayama, Y., Hayashi, Y., Inagaki, M., Ogata, E., and Nishimoto, I. (1991). *Cell* **67**, 723–730.

Onorato, J. J., Palczewski, K., Regan, J. W., Caron, M. G., Lefkowitz, R. J., and Benovic, J. L. (1991). *Biochemistry* **30**, 5118–5125.

Ovchinnikov, Y. A., Abdulaev, N. G., and Bogachuk, A. S. (1988). *FEBS Lett.* **230**, 1–5.

Palczewski, K., McDowell, J. H., and Hargrave, P. A. (1988). *J. Biol. Chem.* **263**, 14067–14073.

Rands, E., Candelore, M. R., Cheung, A. H., Hill, W. S., Strader, C. D., and Dixon, R. A. F. (1990). *J. Biol. Chem.* **265**, 10759–10764.

Reneke, J. E., Blumer, K. J., Courchesne, W. E., and Thorner, J. (1988). *Cell* **55**, 221–228.

Rich, K. A., Codina, J., Floyd, G., Sekura, R., Hildebrandt, J. D., and Iyengar, R. (1984). *J. Biol. Chem.* **259**, 7893–7901.

Roth, N. S., Campbell, P. T., Caron, M. G., Lefkowitz, R. J., and Lohse, M. J. (1991). *Proc. Natl. Acad. Sci. U.S.A.* **88**, 6201–6204.

Schofield, P. R., Rhee, L. M., and Peralta, E. G. (1987). *Nucleic Acids Res.* **15**, 3636.

Schubert, B., VanDongen, A. M. J., Kirsch, G. E., and Brown, A. M. (1989). *Science* **245**, 516–519.

Shear, M., Insel, P. A., Melmon, K. L., and Coffino, P. (1976). *J. Biol. Chem.* **251**, 7572–7576.

Shimomura, H., and Terada, A. (1990). *Nucleic Acids Res.* **18**, 4591.

Shinohara, T., Dietzschold, B., Craft, C. M., Wistow, G., Early, J. J., Donoso, L. A., Horwitz, J., and Tao, R. (1987). *Proc. Natl. Acad. Sci. U.S.A.* **84**, 6974–6979.

Shorr, R. G. L., Lefkowitz, R. J., and Caron, M. G. (1981). *J. Biol. Chem.* **256**, 5820–5826.

Shorr, R. G. L., Strohsacker, M. W., Lavin, T. N., Lefkowitz, R. J., and Caron, M. G. (1982). *J. Biol. Chem.* **257**, 12341–12350.

Sibley, D. R., Nambi, P., Peters, J. R., and Lefkowitz, R. J. (1984a). *Biochem. Biophys. Res. Commun.* **121**, 973–979.

Sibley, D. R., Peters, J. R., Nambi, P., Caron, M. G., and Lefkowitz, R. J. (1984b). *J. Biol. Chem.* **259**, 9742–9749.

Sibley, D. R., Strasser, R. H., Benovic, J. L., Daniel, K., and Lefkowitz, R. J. (1986). *Proc. Natl. Acad. Sci. U.S.A.* **83**, 9408–9412.

Sibley, D. R., Benovic, J. L., Caron, M. G., and Lefkowitz, R. J. (1987). *Cell* **48**, 913–922.

Smith, D. P., Shieh, B. H., and Zuker, C. S. (1990). *Proc. Natl. Acad. Sci. U.S.A.* **87**, 1003–1007.

Stadel, J. M., Nambi, P., Shorr, R. G. L., Sawyer, D. F., Caron, M. G., and Lefkowitz, R. J. (1983a). *Proc. Natl. Acad. Sci. U.S.A.* **80**, 3173–3177.

Stadel, J. M., Strulovici, B., Nambi, P., Lavin, T. N., Briggs, M. M., Caron, M. G., and Lefkowitz, R. J. (1983b). *J. Biol. Chem.* **258**, 3032–3038.

Stachelin, M., and Hertel, C. (1983). *J. Recep. Res.* **3**, 35–43.

Stiles, G. L. (1985). *Arch. Biochem. Biophys.* **237**, 65–71.

Stiles, G. L., Benovic, J. L., Caron, M. G., and Lefkowitz, R. J. (1984a). *J. Biol. Chem.* **259**, 8655–8663.

Stiles, G. L., Caron, M. G., and Lefkowitz, R. J. (1984b). *Physiol. Rev.* **64**, 661–743.

Strader, C. D., Sigal, I. S., Register, R. B., Candelore, M. R., Rands, E., and Dixon, R. A. F. (1987a). *Proc. Natl. Acad. Sci U.S.A.* **84**, 4384–4388.

Strader, C. D., Sigal, I. S., Blake, A. D., Cheung, A. H., Register, R. B., Rands, E., Zemcik, B. A., Candelore, M. R., and Dixon, R. A. F. (1987b). *Cell* **49**, 855–863.

Strader, C. D., Dixon, R. A. F., Cheung, A. H., Candelore, M. R., Blake, A. D., and Sigal, I. S. (1987c). *J. Biol. Chem.* **262**, 16439–16443.

Strader, C. D., Sigal, I. S., Candelore, M. R., Rands, E., Hill, W. S., and Dixon, R. A. F. (1988). *J. Biol. Chem.* **263**, 10267–10271.

Strader, C. D., Candelore, M. R., Hill, W. S., Sigal, I. S., and Dixon, R. A. F. (1989a). *J. Biol. Chem.* **264**, 13572–13578.

Strader, C. D., Candelore, M. R., Hill, W. S., Dixon, R. A. F., and Sigal, I. S. (1989b). *J. Biol. Chem.* **264**, 16470–16477.

Strasser, R. H., Sibley, D. R., and Lefkowitz, R. J. (1986a). *Biochemistry* **25**, 1371–1377.

Strasser, R. H., Benovic, J. L., Caron, M. G., and Lefkowitz, R. J. (1986b). *Proc. Natl. Acad. Sci. U.S.A.* **83**, 6362–6366.

Strulovici, B., Stadel, J. M., and Lefkowitz, R. J. (1983). *J. Biol. Chem.* **258**, 6410–6414.

Su, Y. F., Harden, T. K., and Perkins, J. P. (1979). *J. Biol. Chem.* **254**, 38–41.

Su, Y. F., Harden, T. K., and Perkins, J. P. (1980). *J. Biol. Chem.* **255**, 7410–7419.

Toews, M. L., Waldo, G. L., Harden, T. K., and Perkins, J. P. (1984). *J. Biol. Chem.* **259**, 11844–11850.

Valiquette, M., Bonin, H., Hnatowich, M., Caron, M. G., Lefkowitz, R. J., and Bouvier, M. (1990). *Proc. Natl. Acad. Sci. U.S.A.* **87**, 5089–5093.

Vaughan, R. A., and Devreotes, P. N. (1988). *J. Biol. Chem.* **263**, 14538–14543.

Wakshull, E., Hertel, C., O'Keefe, E. J., and Perkins, J. P. (1985). *J. Cell Biol.* **29**, 127–141.

Waldo, G. L., Northup, J. K., Perkins, J. P., and Harden, T. K. (1983). *J. Biol. Chem.* **258**, 13900–13908.

Wang, H. Y., Hadcock, J. R., and Malbon, C. C. (1990). *Receptor* **1**, 13–32.

Wilden, U., Hall, S. W., and Kuhn, H. (1986). *Proc. Natl. Acad. Sci. U.S.A.* **83**, 1174–1178.

Wistow, G. J., Katial, A., Craft, C., and Shinohara, T. (1986). *FEBS Lett.* **196**, 23–28.

Wong, S. K. F., Slaughter, C., Ruoho, A. E., and Ross, E. M. (1988). *J. Biol. Chem.* **263**, 7925–7928.

Wong, S. K. F., Parker, E. M., and Ross, E. M. (1990). *J. Biol. Chem.* **265**, 6219–6224.

Yamada, T., Takeuchi, Y., Komori, N., Kobayashi, H., Sakai, Y., Hotta, Y., and Matsumoto, H. (1990). *Science* **248**, 483–486.

Yamaki, K., Takahashi, Y., Sakuragi, S., and Matsubara, K. (1987). *Biochem. Biophys. Res. Commun.* **142**, 904–910.

Yarden, Y., Rodriguez, H., Wong, S. K. F., Brandt, D. R., May, D. C., Burnier, J., Harkins, R. N., Chen, E. Y., Ramachandran, J., Ullrich, A., and Ross, E. M. (1986). *Proc. Natl. Acad. Sci. U.S.A.* **83** 6795–6799.

Yatani, A., and Brown, A. M. (1989). *Science* **245**, 71–74.

Yoshimasa, T., Sibley, D. R., Bouvier, M., Lefkowitz, R. J., and Caron, M. G. (1987). *Nature* **327**, 67–70.

Yoshimasa, T., Bouvier, M., Benovic, J. L., Amlaiky, N., Lefkowitz, R. J., and Caron, M. G. (1988). *In* "Molecular Biology of Brain and Endocrine Peptidergic Systems" (K. W. McKerns and M. Chretien, eds.), pp. 123–139. Plenum, New York.

Zaagsma, J., and Nahorski, S. R. (1990). *Trends Pharmacol. Sci.* **11**, 3–7.

Zhou, X. M., and Fishman, P. H. (1991). *J. Biol. Chem.* **266**, 7462–7468.

The cAMP Receptor Family of *Dictyostelium*

Dale Hereld and Peter N. Devreotes

Department of Biological Chemistry, The Johns Hopkins University School of
Medicine, Baltimore, Maryland 21205

I. Introduction

Dictyostelium is a cellular slime mold that feeds on bacteria in soil. This simple
eucaryotic organism survives periods of starvation by undergoing a developmental process in which up to 1 million amebas aggregate to form a visible
multicellular structure consisting of a mass of spores supported by a fibrous
stalk. These spores can endure long periods of deprivation but quickly germi-
nate to yield amebas when nutrients return to the environment.

Figure 1 illustrates the stages of the developmental program. The aggregation
of the individual amebas into a central mound is highly organized. After a short
delay the mounds transform into elongated, migrating slugs. Under appropriate
conditions the migrations cease and cells in the anterior of the structure are
elaborated into a stalk. As the growing stalk rises up from the surface, cells in
the posterior portion of the organism, destined to become the spores, are lifted
to the top of the stalk. The entire process is completed in about 24 hours.

This developmental program serves as a model for numerous complex
processes, including chemotaxis, cell-to-cell communication, morphogenesis,
and cellular differentiation. These complex problems can be approached with
comparative ease due to the practical time scale on which these events take
place. Because growth and development are independent phases of the life
cycle, mutations that arrest development can be conveniently selected and prop-
agated. Methods of targeted gene disruption and DNA-mediated mutant com-
plementation are facilitated by the small, 40,000-kb genome size. It is hoped
that studies of *Dictyostelium* will shed light on mechanisms underlying related
processes in higher organisms (Devreotes, 1989).

These new approaches are beginning to reveal and permit analysis of the
components of signal transduction systems required for development. Eight

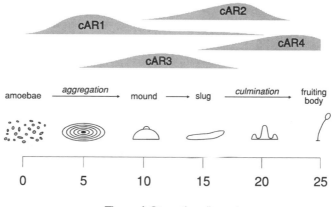

FIG. 1 *Dictyostelium* development and developmental pattern of cAMP receptor expression. Characteristic morphologies observed during development are illustrated at times when they typically can be seen. The amoebae at 0 hours are depicted at higher magnification than the subsequent multicellular forms. The shaded regions indicate the relative abundances of mRNAs encoding the four cAMP receptors throughout development (Saxe *et al.*, 1991b; A. Kimmel, personal communication).

α-subunits and a β-subunit of G proteins, two adenylyl cyclases, and a phospholipase C have been cloned (Van Haastert and Devreotes, in press; Pitt *et al.*, 1992; P. Van Haastert, personal communication). In addition, there is a family of surface cAMP receptors, expressed at distinct developmental stages, which sense extracellular signals that are critical to the developmental process (Klein *et al.*, 1988; Saxe *et al.*, 1991a, b). In this review, we will summarize our current understanding of cell surface cyclic adenosine 3′, 5′-monophosphate (cAMP) receptors in *Dictyostelium*. Some recent reviews that go beyond the scope of this review include those by Caterina and Devreotes (1991), Firtel (1991), Gerisch (1987), and Van Haastert and Devreotes (in press).

II. Role of cAMP in *Dictyostelium* Development

A role for extracellular cAMP was initially suggested when it was identified as a chemoattractant for aggregation-stage cells (Konijn *et al.*, 1967). As had been inferred from the coordinated cellular movements (Shaffer, 1957), it was determined that concentric waves of cAMP, emanating from aggregation centers at 6-minute intervals, serve to recruit surrounding amebas into central mounds (Tomchik and Devreotes, 1981). cAMP waves are generated by the signaling response; when stimulated, cells respond by making and secreting more cAMP, thereby relaying the cAMP signal to more distant cells (Shaffer, 1975; Devreotes and Steck, 1979). As a wave passes, cAMP concentrations decline because the signaling response adapts, halting further secretion of cAMP, and cAMP is readily hydrolyzed by secreted and membrane-associated phosphodiesterases (Malchow *et al.*, 1972; Orlow *et al.*, 1981).

When approached by a wave, cells move up the cAMP gradient in the direction of the aggregation center. As the wave crest passes, directed movement ceases; adaptation of the chemotactic response prevents cells from reversing their direction and following the receding wave (Tomchik and Devreotes, 1981). During the interval between waves, cells regain sensitivity, enabling them to respond repeatedly. Approximately 30 waves are needed to complete the aggregation process.

Extracellular cAMP also modulates the expression of a number of genes at distinct stages in development (Firtel, 1991). The expression of early genes requires the intermittent stimuli provided by the periodic cAMP waves. Expression of the late genes appears to require the expression of certain early genes followed by continuous cAMP application for several hours. These empirically determined schedules of cAMP addition, which elicit appropriate gene expression in cell suspensions, presumably reflect the prevailing condition in the multicellular organism (Firtel *et al.*, 1989).

Extracellular cAMP, acting through a surface receptor, elicits, within a few seconds to minutes of its application, a variety of physiological changes (Caterina and Devreotes, 1991; Devreotes and Zigmond, 1988; Janssens and Van Haastert, 1987). In addition to the activation of adenylyl cyclase, the stimulus elicits rapid activations of guanylyl cyclase and phosphatidylinositol (PI)-specific phospholipase C (PI-PLC) and increases in the rates of proton and K^+ efflux and Ca^{2+} influx. The cAMP stimulus induces alterations in cytoskeletal components, which may function in the chemotactic response. Phosphorylation of myosin heavy and light chains and alterations in amounts of actin and myosin associated with the cytoskeleton occur. Many of these cAMP-induced responses are transient even when the stimulus is persistently applied. These can be divided into two kinetic classes, suggesting that at least two distinct mechanisms of adaptation are at work. The fast responses peak at 10 seconds after cAMP stimulation and subside within 30 seconds; these include PI-PLC and guanylyl cyclase activation. The slower responses include adenylyl cyclase activation and myosin light chain phosphorylation, which peak in 1 minute and subside within 5 minutes.

Adaptation, which accounts for the transient nature of the cAMP-signaling response, has been investigated in considerable detail. The properties of adaptation are illustrated in Figure 2A. When the stimulus is increased and held constant, cells immediately respond but the response subsides within a few minutes. Cells that have adapted to one level of cAMP can respond to a further increment.

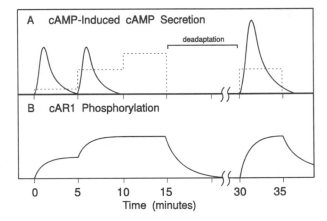

FIG. 2   Kinetics of adaptation of cAMP-induced adenylyl cyclase activation and cAR1 phosphorylation. (A) The rate of [³H]cAMP secretion is shown (——), which results from the treatment of cells
with the indicated cAMP stimuli (----). At time zero a stimulus that occupies half of the cAMP receptors was presented, resulting in a submaximal response. A similar response was elicited when
cAMP was increased to a saturating level at 5 minutes. An additional increase in the stimulus (at 10
minutes) could not trigger additional cAMP secretion. Removal of the stimulus at 15 minutes initiated deadaptation. During deadaptation saturating stimuli can evoke only partial responses (not
shown). Deadaptation is complete by 30 minutes when a maximal response could again be elicited.
(B) The extent of cAR1 phosphorylation, resulting from the schedule of cAMP stimulation illustrated
in A, is shown. The capacity of receptors to be phosphorylated parallels the system's responsiveness.

The magnitude of any response is proportional to the fractional increase in receptor occupancy (Dinauer *et al.*, 1980a). When cells have adapted to a saturating dose of a cAMP, no further response can be elicited. Adaptation is fully
reversible; a deadaptation process is initiated by removal of the stimulus. When
cAMP is removed, cells gradually recover the ability to respond. The half-time
of the recovery, about 3–4 minutes, is independent of the initial level of adaptation. Deadaptation is complete within 15 minutes (Dinauer *et al.*, 1980b).

Regulation of the signaling response can be viewed as a balance between a
rapid excitation and a slower adaptation process. The balance can be perturbed
by increasing receptor occupancy. When the stimulus is applied a response is
initiated by a rapid increase in excitation. Adaptation slowly increases to offset
the excitation signal and reestablishes the balance. Negative feedback models in
which a second messenger, such as intracellular cAMP, triggers inhibition, perhaps via protein kinase A, are unlikely because adaptation occurs despite blocking adenylyl cyclase activation (Theibert and Devreotes, 1983) and most
intracellular responses by deletion of an essential G protein α-subunit (Kesbeke
et al., 1988; Snaar-Jagalska *et al.*, 1988b).

III. Major cAMP Receptor of Early Development

Living cells express surface cAMP binding activity, which increases dramatically in response to starvation and peaks after about 5 hours, coincident with the acquisition of chemotactic sensitivity (Green and Newell, 1975). A major cAMP-binding protein on the surface of these aggregation-stage cells was first identified using the photoaffinity label, [^{32}P]8-azido-cAMP (Juliani and Klein, 1981; Theibert *et al.*, 1984). The putative receptor was a polypeptide of 40 kD, which exhibited a stimulus-induced electrophoretic mobility shift on sodium dodecyl sulfate (SDS) gels, indicative of a covalent modification. Removal of cells from cAMP initiated a reversal of the mobility shift. Metabolic labeling later established that the cAMP-induced modification was due to phosphorylation. The ^{32}P-labeled protein was purified approximately 6000-fold by SDS solubilization, hydroxyapatite chromatography, and gel electrophoresis. Purification data indicate that there are 2×10^5 copies of the polypeptide per cell, in agreement with the number of cell surface cAMP-binding sites (Klein *et al.*, 1987a). This putative cell surface receptor is hereafter referred to as cAR1.

cDNAs were isolated from λgt11 expression libraries by screening with antiserum elicited by the purified receptor (Klein *et al.*, 1988). The validity of the clones obtained was established by several different criteria. First, two fusion proteins encoded by nonoverlapping, partial cDNAs cross-reacted immunologically with cAR1 but not with one another. Second, as previously determined by *in vitro* translation and immunoprecipitation, the receptor mRNA encoded a 37-kDa polypeptide, was around 2 kb in size, and peaked in abundance roughly 4 hours into development (Klein *et al.*, 1987b). Subsequently, the authenticity of the cDNA was confirmed with the demonstration of cAMP binding by cells expressing the exogenous DNA (Klein *et al.*, 1988; Johnson *et al.*, 1991) and the loss of cAMP binding and responsiveness in cells unable to express the corresponding gene (Klein *et al.*, 1988; Sun *et al.*, 1990; Sun and Devreotes, 1991).

Sequence of the cDNAs revealed seven clusters of hydrophobic amino acids and a weak homology with mammalian rhodopsin and β-adrenergic receptor, indicating that the cAMP receptor belongs to the family of G protein-linked receptors, which are proposed to pass through the plasma membrane seven times (Dohlman *et al.*, 1991). A model for cAR1 (Fig. 3A), based on hydropathy data and homology with bovine rhodopsin, was proposed (Klein *et al.*, 1988). Its features include six highly hydrophobic transmembrane segments and a seventh (nearest the C terminus) with the potential to form an amphipathic α-helix. These range in length between 18 and 25 residues and presumably assume α-helical structures. Short loops of 11–24 residues interconnect the membrane-spanning segments on the intracellular and extracellular faces of the plasma membrane. A short N-terminal region faces the extracellular milieu. At

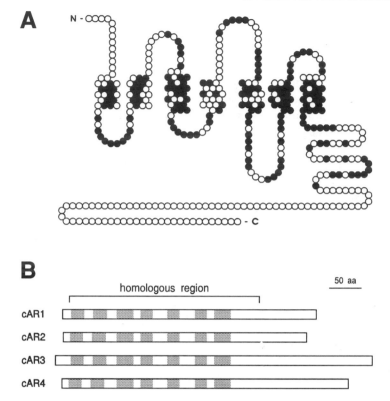

FIG. 3   Structural comparison of the cAMP receptors. (A) The topological model for cAR1 of Klein *et al.* (1988) is shown. Transmembrane segments are drawn compressed (as though in α-helical configuration) and aligned in the plane of the plasma membrane. Above and below this plane are extracellular and intracellular, respectively. Those residues conserved in all four cARs are indicated (filled circles). (B) Linear representations of the four cARs are shown aligned. The region of homology and seven transmembrane segments (shaded) are depicted. The C termini are at the right. The scale shows the length corresponding to 50 amino acid residues.

the C terminus roughly one third of the polypeptide (132 residues) constitutes a serine-rich, intracellular domain.

Biochemical and genetic evidence suggests that cAMP receptors are coupled to G proteins. In isolated membranes cAMP stimulates GTPase activity and guanosine 5′-[γ-thio]triphosphate (GTPγS) binding (Snaar-Jagalska *et al.*, 1988a, c), and GTP induces a change in the affinity of cAMP binding (Van Haastert, 1984). Also, *frigid* A mutants, which are deficient in a G protein α-subunit, designated Gα2, are blocked in multiple cAMP-mediated responses (Kesbeke *et al.*, 1988).

cAR1 is a rare protein, but it appears to be a major phosphoprotein in crude membranes. Upon stimulation of cells with cAMP, incorporation of ^{32}P into cAR1 increases roughly 10–20-fold. Serine residues in the C-terminal cytoplasmic domain appear to be the major target of the modification (unpublished results). The cAMP-induced shift in the electrophoretic mobility of cAR1 appears to be a consequence of phosphorylation. The two processes have nearly identical kinetics and cAMP dose dependence (Vaughan and Devreotes, 1988) and the shift can be reversed *in vitro* by phosphatase treatment of cAR1 (R. Vaughan, personal communication). The number of phosphate modifications was determined from measurements of the cAR1-associated ^{32}P, the specific radioactivity of the γ-phosphoryl of intracellular ATP, and, by immunoblotting, the amount of cAR1 protein (Klein *et al.*, 1987b). In the absence of cAMP, cAR1 is phosphorylated substoichiometrically (0.2 mol phosphate/mol receptor). Upon stimulation cAR1 phosphorylation increases to roughly 4 mol/mol, indicating that there are at least four sites of phosphorylation.

Although its functional significance is not known, phosphorylation of the receptor is well correlated with adaptation of adenylyl cyclase activation and may be a major mechanism of receptor regulation (Devreotes and Sherring, 1985). As illustrated in Figure 2B, when a subsaturating stimulus is applied and held constant, concomitant with the activation of adenylyl cyclase, there is an increase in the level of cAR1 phosphorylation. As the extent of modification reaches a plateau, specified by the new level of receptor occupancy, the physiological response subsides. A further increase in stimulus triggers a fresh response and a further increase in phosphorylation, which again plateaus as the response ceases. When the stimulus is removed, the extent of phosphorylation declines and the cells regain sensitivity. Thus, adaptation seemingly rises only to the extent needed to offset the excitatory signal sent to the effector in a manner that closely parallels the extent of cAR1 phosphorylation. It is possible that phosphorylated cAR1 transmits an inhibitory signal to adenylyl cyclase.

A role for phosphorylation in the desensitization of other G protein-linked receptors has been substantiated. For the β_2-adrenergic receptor (Gomez and Benovic, this volume) and rhodopsin (Hargrove and McDowell, this volume), the evidence suggests that receptor phosphorylation results in their uncoupling from G_s and transducin, respectively (Dohlman *et al.*, 1991). Kinases that preferentially modify ligand-occupied (or light-bleached) receptors have been purified and their cDNAs have been cloned (Benovic *et al.*, 1989; Lorenz *et al.*, 1991). β-Adrenergic receptor kinase (βARK) and rhodopsin kinase, as they are called, are highly homologous. Additional proteins, β-arrestin and arrestin, have been identified which bind to the respective phosphoreceptors and interfere with receptor–G protein interactions. In the case of β-adrenergic receptor, an additional mechanism believed to play a leading role at low agonist concentrations involves the modification of receptors, regardless of occupancy, by protein kinase A. This latter mechanism is unlikely in the case of cAR1 because it lacks protein kinase A consensus sites, and caffeine, which blocks adenylyl cyclase activation, does not interfere with its phosphorylation or adaptation. The existence of *Dictyostelium* homologs of βARK and arrestin is an attractive possibility.

IV. A Family of cAMP Receptors

At low stringency cAR1 cDNA hybridized to several novel genomic DNA fragments, suggesting that there is a family of several highly homologous genes.
Probes derived from different portions of the cAR1 cDNA predicted that these homologous genes resemble cAR1 within its membrane-spanning portions but not
within its cytoplasmic tail (Saxe *et al.*, 1991a). Genomic fragments and cDNAs
encoding three additional receptors were isolated and the complete amino acid sequence of each was obtained. Based on their high degree of similarity to cAR1,
they were designated cAR2 (Saxe *et al.*, in preparation), cAR3 (Johnson *et al.*, in
preparation), and cAR4 (A. Kimmel, personal communication).

Each of the receptors exhibits seven potential transmembrane segments (TMS
I–VII). Their N-terminal regions vary in length from 10 to 23 amino acid residues. More strikingly, the lengths of C-terminal cytoplasmic domains range from
116 (cAR2) to 219 (cAR3) residues. Between the N-terminal and C-terminal
tails, the only length differences are insertions of 1–3 residues (relative to cAR1)
in the extracellular loops connecting TMS II and III of cARs 2, 3, and 4. The
four cAMP receptors are highly homologous over a contiguous central region of
about 290 amino acids, which starts within TMS I and extends about 40 residues
beyond TMS VII (Fig. 3B). No homology is discernable in the short N-terminal
regions or the distal portions of the cytoplasmic tails. Within the homologous region, any two cARs are 70% identical; 56% of the residues in this region are
common to all four cARs (Fig. 3A). These identities reside largely in the TMS
and intracellular portions (connecting loops and the proximal part of the cytoplasmic tail), and to a lesser extent in the extracellular loops.

A. Time of Expression

As illustrated in Figure 1, the expression of each of the receptors is developmentally regulated. Each appears at a characteristic time after the onset of starvation
in the following order: cAR1, cAR3, cAR2, and cAR4. In a parallel series of
RNA blots, the major mRNAs for cAR1, cAR3, cAR2, and cAR4 were maximal
at 5, 10, 15, and 20 hours of development, respectively (Saxe *et al.*, 1991b; A.
Kimmel, personal communication). These times marked the aggregation, mound,
slug, and culmination stages, respectively. For cAR1 and cAR3, the amount of
receptor protein has been determined by immunoblot and closely parallels the
abundance of the major mRNA, indicating that expression is controlled primarily at the transcriptional level. The level of each protein decreased a few hours
after its mRNA subsided (Klein *et al.*, 1987b; Johnson *et al.*, in preparation).

Beyond the period of time when cAR1 is expressed, relatively lower cAMP
binding can be measured, suggesting that cAR3, cAR2, and cAR4 are expressed

in lower numbers than is cAR1. The relative amounts of cAR1 and cAR3 protein have been determined from immunoblots using antisera specific to each and known amounts of cAR1 and cAR3 as standards (Johnson *et al.*, in preparation). The maximal amount of cAR3 (5×10^3 sites/cell) was roughly one tenth that of cAR1 (7×10^4 sites/cell). Protein levels of cAR2 and cAR4 have not yet been determined.

B. Binding Properties

To verify that the cARs were cAMP-binding proteins, cDNAs of cAR1, cAR2, and cAR3 were expressed in transformed cells using a constitutive actin promoter (Johnson *et al.*, 1992). Stable cell lines were obtained which expressed substantial quantities of these proteins in undeveloped cells, which ordinarily express little cAMP binding. Each expressed receptor was specifically labeled by the photoactivatible cAMP analog [^{32}P]8-azido-cAMP. The photolabeled bands corresponding to cAR1, cAR2, and cAR3 were 40, 38, and 64 kDa, respectively.

Binding of cAMP and its analogs have been studied in developed wild-type cells by several groups of investigators. These sites are primarily cAR1. Equilibrium binding studies are consistent with the existence of two affinity states with dissociation constants of about 10 nM and 150 nM, the latter being the predominant form (Janssens and Van Haastert, 1987). It is technically difficult to study the binding properties of cARs 2–4 in wild-type cells because there is overlap in their expression during development and they are expressed in small amounts compared with cAR1. The stably expressing cell lines obviate these difficulties.

The results of [^{3}H]cAMP binding studies on the cAR1, cAR2, and cAR3 expressing cell lines are shown in Table I. In phosphate buffer, which represents physiological conditions, binding by cAR1 cells was composed of two affinity states of around 20 nM and 200 nM (K_d values), in reasonable agreement with the binding characteristics of aggregation-stage wild-type cells. Like cAR1, cAR3 displayed two affinity states but of somewhat lower affinity and a greater proportion of binding sites of the low-affinity type; some have a high affinity (K_d of 10–50 nM), whereas most (94%) are lower in affinity (K_d around 500 nM). In contrast, cAR2 exhibited no measurable binding, indicating that its affinity for cAMP is low in phosphate buffer ($K_d > \sim 2$ μM).

Ammonium sulfate shifts cAMP binding to aggregation-stage cells to a single high-affinity form, due largely to a reduction of the dissociation rate (Van Haastert and Kien, 1983; Van Haastert, 1985). Similar results, shown in Table I, were obtained with cAR1 cells (Johnson *et al.*, 1991). cAR2, although immeasurable in phosphate buffer, was comparable to cAR1 in ammonium sulfate, exhibiting a single high-affinity state ($K_d \sim 11$ nM). In contrast to both cAR1 and

TABLE I

cAMP Binding Properties of cAR1, cAR2, and cAR3

	Affinities (nM)							
	Phosphate buffer				Ammonium sulfate			
cAR subtype	High	(%)	Low	(%)	High	(%)	Low	(%)
Control	40^a	(23^a)	350^a	(77^a)	1.8	(100)	—	—
cAR1	25^a	(22^a)	230	(78^a)	3.5	(100)	—	—
cAR2	—		—		11	(100)	—	—
cAR3	47^a	(7^a)	680^a	(93^a)	4.6	(6^a)	220^a	(94^a)

Binding of [^{3}H]cAMP by undifferentiated cell lines; each receptor subtype was evaluated in both phosphate and ammonium sulfate binding conditions. Control cells, which were transformed with only the vector used to express cARs in the other cell lines, were starved 4 hours and analyzed. Tabulated are the K_d values (nanomolars) of the affinity states detected. Adjacent to these in parentheses is the percent of total sites with the corresponding affinity. Total sites per cell ($\times 10^{-3}$) in ammonium sulfate were 96, 370, 210, and ~600 for control, cAR1, cAR2, and cAR3 cells, respectively. Values are derived from Johnson et al. (1992). Properties of cAR4 have not yet been determined.

^{a}Standard error exceeds 20%.

cAR2, cAR3 persisted in exhibiting two affinity states in ammonium sulfate. The majority of cAR3 sites are of the lower affinity type in ammonium sulfate as well as in phosphate buffer.

The binding of cAMP analogs to receptors expressed on aggregation-stage cells has been studied in detail (Van Haastert and Kien, 1983) and led to the proposals that (1) cAMP is bound by hydrogen bonds at N^6H_2 of the adenine moiety and $O^{3'}$ of the ribose moiety; (2) the adenine moiety is held by hydrophobic interactions; and (3) a stereospecific interaction with the phosphate moiety is required for receptor activation. Points (1) and (2) have been supported by recent studies with cAR1 cells; point (3) was not addressed (Johnson et al., 1992). Although cAR2 and cAR3 also appear to bind cAMP by hydrogen bonding with the N^6H_2 and $O^{3'}$ positions of the ligand, hydrophobic interactions with the adenine moiety appear to contribute less to cAMP binding by cAR2 than by either cAR1 or cAR3. In addition, it was demonstrated that substitution of the axial exocyclic oxygen with sulfur in the analog, (S_p)-cAMPS, greatly reduces the cAR1 affinity (30–100-fold lower than for cAMP); however, the impact was far less substantial for both cAR2 and cAR3. Johnson et al. (1992) suggested that the S_p analog may encounter steric interference in the cAR1-binding site, whereas the cAR2 and cAR3 sites might accommodate the larger sulfur atom more easily.

V. Roles of cAMP Receptors in Development

We are beginning to delineate the individual function of each receptor. cAR1-deficient cells are blocked in early development. These cells were created initially by an antisense approach in which a complementary RNA, expressed from a transforming DNA, hybridizes to and neutralizes the cAR1 mRNA (Klein *et al.*, 1988; Sun *et al.*, 1990). In later experiments the cAR1 gene was disrupted by homologous recombination with a transforming DNA (Sun and Devreotes, 1991). Both methods yielded similar results. Phenotypically, cAR1-deficient cells remained as a monolayer upon starvation. They were impaired in cAMP chemotaxis and cAMP binding. Biochemical characterization of antisense cells revealed that exogenous cAMP did not trigger activation of either adenylyl or guanylyl cyclase as occurs in wild-type cells. In the gene-disrupted cells, expression of the cAMP phosphodiesterase gene was delayed, cAR3 gene expression was both diminished and delayed, and expression of the prespore and prestalk cell-specific genes, D19 and Dd63, was absent. Thus, cAR1 appears to be necessary for the major effects of extracellular cAMP during aggregation, namely, chemotaxis, cAMP signaling, and regulation of gene expression. Without cAR1, cells are incapable of aggregating and differentiating as they must to produce spores. However, the possibility that cAR1 is needed for the proper expression or function of other mediators of these processes cannot be excluded.

Less is understood regarding the roles of cAR2, cAR3, and cAR4. Preliminary analysis of similar gene disruptions of cAR2, cAR3, and cAR4 suggest that, as expected from the time course of expression, each receptor plays an important role at a specific stage in development consistent with their patterns of expression. Although it is not known which signal transduction pathways are utilized by these receptors, they might mediate cell movements and late gene expression in multicellular stages. cAR2, preferentially expressed in predecessors of stalk cells (Saxe *et al.*, 1991b), may mediate specific functions of these differentiated cells. The apparent requirement for constant extracellular cAMP to elicit late gene expression suggests that the responsible receptor is not subject to adaptation. If phosphorylation of the C-terminal cytoplasmic domain of cAR1 is necessary for adaptation, it is possible that the other cAMP receptors, with their markedly different C termini, may have distinct adaptation properties. Like cAR1, evidence indicates that cAR2 and cAR3 can also be phosphorylated in response to a cAMP stimulus (R. Johnson, personal communication). Future studies must focus on characterizing these sites, determining the kinetics of their phosphorylation, and attempting to ascertain the adaptation properties of these receptors.

In summary, extracellular cAMP participates in a wide variety of physiological responses during development in *Dictyostelium*. It serves as a chemoattractant, a cell to cell signaling molecule, and an inducer of cell type-specific gene expression. During development these cells serially express a family of four cell surface cAMP receptors. All of the receptors display the characteristic seven-transmembrane domain structure and, thus, are likely to transduce extracellular signals by activating G proteins. Eight distinct G protein α-subunit genes have been cloned and sequenced. Although similar in structure, each cAMP receptor has a characteristic affinity and pharmacological specificity and may have a unique pattern of regulation by covalent modification. The challenges for the future are to learn how the biochemical features of each receptor suit it for its specific role, whether each receptor is coupled to a distinct G protein or signal transduction pathway, and how these pathways interact.

References

Benovic, J. L., DeBlasi, A., Stone, W. C., Caron, M. G., and Lefkowitz, R. J. (1989). *Science* **246**, 235–240.

Caterina, M. J., and Devreotes, P. N. (1991). *FASEB J.* **5**, 3078–3085.

Devreotes, P. (1989). *Science* **245**, 1054–1058.

Devreotes, P. N., and Sherring, J. A. (1985). *J. Biol. Chem.* **260**, 6378–6384.

Devreotes, P. N., and Steck, T. L. (1979). *J. Cell. Biol.* **80**, 300–309.

Devreotes, P. N., and Zigmond, S. H. (1988). *Annu. Rev. Cell. Biol.* **4**, 649–686.

Dinauer, M. C., Steck, T. L., and Devreotes, P. N. (1980a). *J. Cell. Biol.* **86**, 554–561.

Dinauer, M. C., Steck, T. L., and Devreotes, P. N. (1980b). *J. Cell. Biol.* **86**, 545–553.

Dohlman, H. G., Thorner, J., Caron, M. G., and Lefkowitz, R. J. (1991). *Annu. Rev. Biochem.* **60**, 653–688.

Firtel, R. A. (1991). *Trends Genet.* **7**, 381–388.

Firtel, R. A., Van Haastert, P. J. M., Kimmel, A., and Devreotes, P. N. (1989). *Cell* **58**, 235–239.

Gerisch, G. (1987). *Annu. Rev. Biochem.* **56**, 853–879.

Green, A. A., and Newell, P. C. (1975). *Cell* **6**, 129–136.

Janssens, P. M. W., and Van Haastert, P. J. M. (1987). *Microbiol. Rev.* **51**, 396–418.

Johnson, R. L., Vaughan, R. A., Caterina, M. J., Van Haastert, P. J. M., and Devreotes, P. N. (1991). *Biochemistry* **30**, 6982–6986.

Johnson, R. L., Van Haastert, P. J. M., Kimmel, A. R., Saxe, C. L., Jastorff, B., and Devreotes, P. N. (1992). *J. Biol. Chem.* **267**, 4600–4607.

Juliani, M. H., and Klein, C. (1981). *J. Biol. Chem.* **256**, 613–619.

Kesbeke, F., Snaar-Jagalska, B. E., and Van Haastert, P. J. M. (1988). *J. Cell. Biol.* **107**, 521–528.

Klein, P., Knox, B., Borleis, J., and Devreotes, P. (1987a). *J. Biol. Chem.* **262**, 352–357.

Klein, P., Vaughan, R., Borleis, J., and Devreotes, P. N. (1987b). *J. Biol. Chem.* **262**, 358–364.

Klein, P. S., Sun, T. J., Saxe, C. L., Kimmel, A. R., Johnson, R. L., and Devreotes, P. N. (1988). *Science* **241**, 1467–1472.

Konijn, T. M., Van de Meene, J. G. C., Bonner, J. T., and Barkley, D. S. (1967). *Proc. Natl. Acad. Sci. U.S.A.* **58**, 1152–1154.

Lorenz, W., Inglese, J., Palczewski, K., Onorato, J. J., Caron, M. G., and Lefkowitz, R. J. (1991). *Proc. Natl. Acad. Sci. U.S.A.* **88**, 8715–8719.

Malchow, D., Nagele, B., Schwarz, H., and Gerisch, G. (1972). *Eur. J. Biochem.* **28**, 136–142.

Orlow, S. J., Shapiro, R. I., Jakob, F., and Kessin, R. H. (1981). *J. Biol. Chem.* **256**, 7620–7627.

Pitt, G. S., Milona, N., Borleis, J., Lin, K. C., Reed, R. R., and Devreotes, P. N. (1992). *Cell.* **69**, 305–315.

Saxe, C. L., Johnson, R. L., Devreotes, P. N., and Kimmel, A. R. (1991a). *Genes Dev.* **5**, 1–8.
Saxe, C. L., Johnson, R., Devreotes, P. N., and Kimmel, A. R. (1991b). *Dev. Gen.* **12**, 6–13.
Shaffer, B. M. (1957). *Am. Naturalist* **91**, 19–35.
Shaffer, B. M. (1975). *Nature* **255**, 549–552.
Snaar-Jagalska, B. E., Jakobs, K. H., and Van Haastert, P. J. M. (1988a). *FEBS Lett.* **236**, 139–144.
Snaar-Jagalska, B. E., Kesbeke, F., Pupillo, M., and Van Haastert, P. J. M. (1988b). *Biochem. Biophys. Res. Commun.* **156**, 757–761.
Snaar-Jagalska, B. E., De Wit, R. J. W., and Van Haastert, P. J. M. (1988c). *FEBS Lett.* **232**, 148–152.
Sun, T. J., Van Haastert, P. J. M., and Devreotes, P. N. (1990). *J. Cell. Biol.* **110**, 1549–1554.
Sun, T. J., and Devreotes, P. N. (1991). *Genes Dev.* **5**, 572–582.
Theibert, A. and Devreotes, P. N. (1983). *J. Cell. Biol.* **97**, 173–177.
Theibert, A., Klein, P., and Devreotes, P. N. (1984). *J. Biol. Chem.* **259**, 12318–12321.
Tomchik, K. J., and Devreotes, P. N. (1981). *Science* **212**, 443–446.
Van Haastert, P. J. M. (1984). *Biochem. Biophys. Res. Commun.* **124**, 597–604.
Van Haastert, P. J. M. (1985). *Biochem. Biophys. Acta* **845**, 254–260.
Van Haastert, P. J. M., and Devreotes, P. N. (in press). *In* "Sensory Transduction in Genetically Tractable Organisms" (J. Kurjan, ed.). Academic Press, San Diego.
Van Haastert, P. J. M., and Klein, F. (1983). *J. Biol. Chem.* **258**, 9636–9642.
Vaughan, R. A., and Devreotes, P. N. (1988). *J. Biol. Chem.* **263**, 14538–14543.

Rhodopsin and Phototransduction

Paul A. Hargrave[*†] and J. Hugh McDowell[*]

*Department of Ophthalmology and †Department of Biochemistry and Molecular Biology, School of Medicine, University of Florida, Gainesville, Florida 32610

I. Introduction

When light strikes a rod photoreceptor cell in the retina, rhodopsin molecules become photoexcited and a series of biochemical events rapidly follows (Fig. 1). Photoexcited rhodopsin activates a G protein, transducin, which in turn activates a cyclic guanosine monophosphate (cGMP)-phosphodiesterase (PDE). The resulting hydrolysis of cGMP leads to closure of cation channels in the plasma membrane, which triggers a neural signaling event. In this chapter we will concentrate on what is currently known about rhodopsin and how it functions as a photoreceptor protein. This is of intrinsic importance because of our understanding of the molecular basis of the visual transduction process. It is of more general significance because rhodopsin is a well-studied member of the class of G protein-linked receptors. Thus, whatever we learn about the function of rhodopsin in signal transduction may be of importance in understanding the behavior of the entire class of receptors.

II. The Rod Cell and Its Protein

A. Structure, Function, and Composition of the Rod Cell

Two types of photoreceptor cells (rods and cones) are present in the vertebrate retina. Rod cells operate optimally at low light intensities and contain the photosensitive pigment rhodopsin. Cones operate at higher light intensities and facilitate color vision in many animals. Humans have three different types of cone cells, one each containing red-, green-, and blue-sensitive visual pigments, or opsins. Rod cells are more numerous in most species, and their outer segments have been relatively easy to isolate and study. For that reason most of our

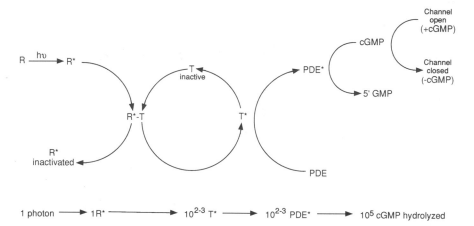

FIG. 1   The phototransduction scheme. Rhodopsin (R) is activated (to R*), which in turn activates transducin (T). Activated transducin (T*) activates cGMP-PDE, which hydrolyzes cGMP, causing closure of the cGMP-gated channel in the plasma membrane. The degree of amplification, shown below, emphasizes that the absorption of one photon leads to hydrolysis of 10^5 cGMP. The rhodopsin cycle and transducin cycle are presented in more detail in Figures 11 and 13. Adapted from Chabre and Deterre (1989) with permission.

information concerning photoreceptor cell function comes from the study of the more accessible rod cells. This chapter will address certain aspects of the molecular properties of rhodopsin and the proteins with which it interacts in rod cells. Other reviews have dealt with these and related issues (Liebman *et al.*, 1987; Chabre and Deterre, 1989; Applebury, 1991; Stryer, 1991; Palczewski and Benovic, 1991; Williams, 1991; Khorana, 1992; Hargrave and McDowell, 1992).

Rod cells are highly specialized neurons, elongated and compartmentalized to serve specific cellular functions. Their role is to transduce light into an electrical signal. Their outer segments contain disk-shaped membranes, densely packed with the photoreceptor protein rhodopsin. Additionally, there are structural proteins and cytoplasmic proteins important in visual transduction. After the absorption of a photon by rhodopsin and the subsequent biochemical events (Fig. 1), closure of cation channels in the outer segment plasma membrane leads to neural signaling of second-order neurons at the synapse at the base of the rod cell. Although we can describe features characteristic of rod cells generally (Fig. 2), there is considerable variation between species. Frog rod cells are approximately 7 μm in diameter and 43–64 μm in length, depending upon the species of frog (Rosenkranz, 1977). The isolated mouse rod cell shown in Figure 3 measures 84 μm in length. Positions of the nucleus and overall dimensions vary even within the same species (Lolley *et al.*, 1986).

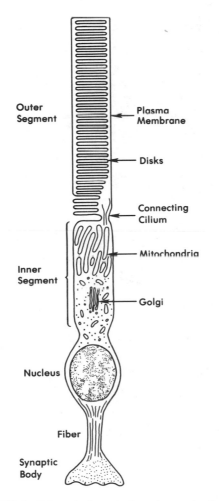

FIG. 2   Schematic diagram of vertebrate rod cell.

The outer segment, which contains stacks of disk membranes, is a modified ciliary process. The inner segment contains the major metabolic machinery for the cell, including the endoplasmic reticulum and Golgi apparatus. At the apex of the inner segment are densely packed mitochondria, which supply the energy of adenosine triphosphate (ATP) for the major metabolic functions of the rod cell: macromolecular synthesis and ion pumping. Vesicles containing rhodopsin are also found in this region. These vesicles are in transit from their site of origin in the Golgi to deliver their contents to the outer segment. They are thought to fuse with the ciliary plasma membrane, thus adding new rhodopsin molecules

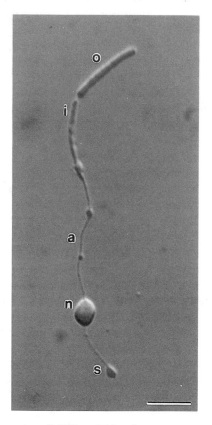

FIG. 3   Mouse rod photoreceptor cell. Differential interference contrast micrograph of intact disso-
ciated mouse rod cell. O, outer segment; i, inner segment; a, axon; n, nucleus; s, synapse. Bar, 10
μm. Reproduced from Lolley *et al.* (1986), with permission.

and other molecules to the outer segment (Bird *et al.*, 1988). Rod cells have a
specialized system for membrane turnover, adding new disks at the base of the
outer segment and periodically shedding disk stacks from the apex of the outer
segment for phagocytosis by enveloping processes of the adjacent pigment epithelium cells (Besharse, 1986).

The rod outer segment (ROS) cilium is a nonmotile cilium typical for sensory
cells that has the usual nine microtubule doublets but lacks center microtubules.
In the periciliary region, actin and myosin have been located, and it is believed
that this contractile system is involved in initiating new disk formation at the
base of the outer segment (Williams, 1991).

Although disks at the base of the outer segment are initially continuous with
the plasma membrane, they eventually are pinched off and become morphologically separate (Steinberg *et al.*, 1980). Through sorting mechanisms that are not

yet understood, disk membranes achieve a distinct and different protein composition compared with the plasma membrane from which they are formed (Section II,C). The disks were originally thought to be free-floating but can be shown to have some structural integrity as disk stacks when the plasma membranes are carefully removed (Cohen, 1972). Thin filaments can be seen between disks when tissue is prepared by rapid-freezing techniques (Usukura and Yamada, 1981). Such filaments have also been observed in freeze-fracture studies of intact retina, located at 14-nm intervals along the disk edge, slightly recessed from the edge (Roof and Heuser, 1982).

Disks also maintain some mode of attachment to the plasma membrane and can be seen to be attached by their rim after hypotonic lysis of the outer segment (Molday and Molday, 1987a). Fibrous elements can be observed to mediate this connection (Usukura and Yamada, 1981; Roof and Heuser, 1982), which may include a spectrin-like cytoskeletal protein (Wong and Molday, 1986). A speculative organizational scheme for these membrane and cytoskeletal components has been developed by Molday (1989) and is presented in Figure 4. A brief description of our current knowledge of these proteins follows in Sections II,D–F.

B. Isolation of Rod Cell Outer Segments

Whole intact mature rod cells can be prepared, but they are fragile and can be obtained only in limited quantities. However, for most studies complete intact rod cells free of other retinal cell types are not required. The type of preparation needed depends upon the purposes of the experiment. When detergent extracts containing rhodopsin were needed for spectroscopic studies, crude preparations of rod outer segments sufficed. To perform more exacting biochemical studies it became necessary to prepare rod outer segments free of other retinal cells and organelles. This was achieved after shaking or homogenizing of the retina followed by differential centrifugation and density gradient centrifugation (Papermaster and Dreyer, 1974; Hargrave, 1982).

To perform various types of metabolic experiments and to ensure that there has been little or no loss of rod cell constituents, methods have been developed for preparation of sealed outer segments. These can be prepared from bovine retinas by centrifugation on sucrose-Ficoll gradients (Schnetkamp et al., 1979) or from rat (Shuster and Farber, 1984) or frog (Hamm and Bownds, 1986) retinas by Percoll gradient centrifugation. Frog rod outer segments linked to inner segments containing mitochondria have been useful for metabolic experiments (Cote et al., 1989). Electropermeabilized frog rod outer segments have been prepared that contain pores suitable for the introduction of small molecules while retaining the bulk of higher molecular weight components (Binder et al., 1989).

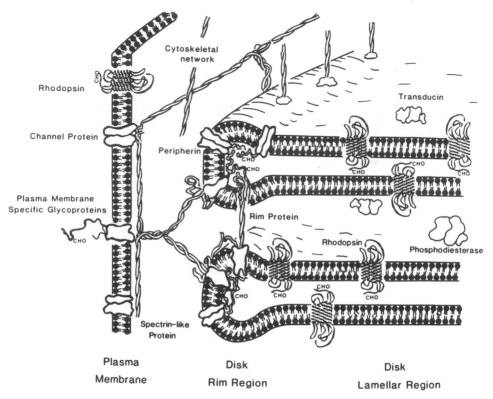

FIG. 4   Speculative model for the organization of proteins in ROS plasma and disk membranes. Reproduced from Molday (1989), with permission.

C. Separation of Disk Membranes from the Plasma Membrane

The bulk of the membranes in the bovine rod outer segment (94–95%) are disk membranes, with the remaining 5–6% being plasma membranes. Because of the similarity of physical properties of these membranes and the lack of information about membrane-specific markers, separation of disk and plasma membranes has been greatly impeded. In one study some separation of these membrane populations was obtained by binding intact rods to concanavalin A-containing beads and washing away (disk) membranes that did not bind (Kamps *et al.*, 1982). Plasma membrane-specific markers have been identified by labeling intact rods with lactoperoxidase (Clark and Hall, 1982; Witt and Bounds, 1987) or by labeling neuraminidase-treated rods with ricin-gold-dextran conjugates (Molday and Molday, 1987b).

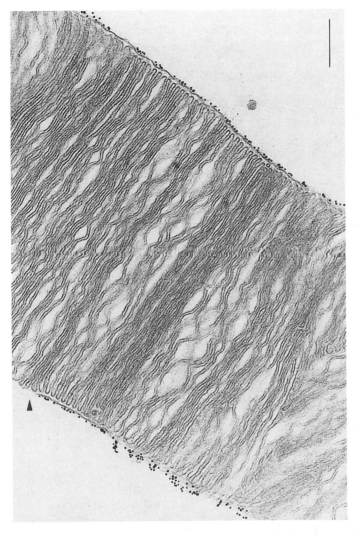

FIG. 5 Bovine rod cell labeled with ricin-gold-dextran. The transmission electron micrograph shows gold label (black dots) on the plasma membrane, marking glycoproteins that have been treated with neuraminidase followed by ricin-gold-dextran. Arrow shows an area where the plasma membrane has been broken and the disks are accessible but not labeled. Bar, 0.5 mm. Reproduced from Molday and Molday (1987b), with permission.

A convincing disk and plasma membrane separation was achieved by Molday and Molday (1987a). Purified rod outer segments are treated with neuraminidase to expose galactose residues on cell surface sialoglycoproteins. These glycoproteins are then labeled with ricin-gold-dextran particles (Fig. 5).

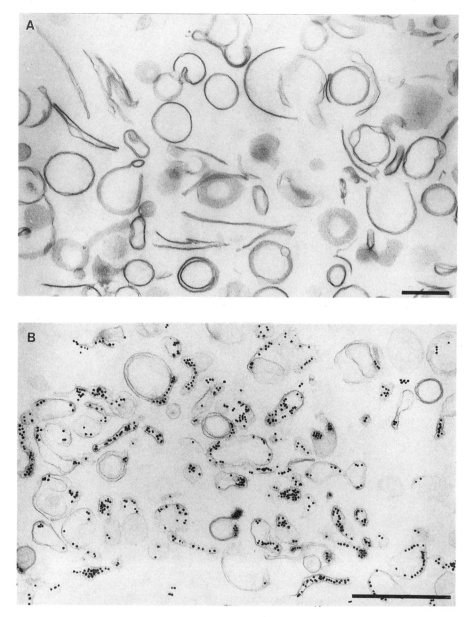

FIG. 6   Sucrose-density gradient fractions of ROS membranes. Bovine rod cells were treated with
neuraminidase and labeled with ricin-gold-dextran (Fig. 5). After lysis and treatment with trypsin to
free membranes from fibrous protein attachments, membranes were fractionated by sucrose-density
centrifugation. (A) Unlabeled intact disks and vesicles. (B) Vesicles derived from the plasma membrane and containing ricin-gold-dextran label (black dots). Reproduced from Molday and Molday
(1987a), with permission.

Such labeling both increases the density of the plasma membrane fraction and provides a tag allowing monitoring of the purification by electron microscopy. The labeled rod outer segments are then hypotonically lysed and briefly exposed to trypsin to cleave filaments connecting disks to the plasma membrane. Separation of the denser plasma membrane-gold-dextran fraction from disks is achieved by density gradient centrifugation. The plasma membrane fraction contains the gold–dextran label (Fig. 6A) as well as the proteins identified as being plasma membrane markers. The disk membrane fraction is completely free of these marker proteins (Fig. 6B). Proteins identified in the various parts of the rod cell are listed in Table I, and more detail is given concerning their characteristics in the following sections.

D. Proteins of the Plasma Membrane

1. cGMP-Gated Cation Channel

The light-sensitive cation channel of ROS is directly and cooperatively activated by cGMP (Fesenko *et al.*, 1985). When cGMP is bound to the channel it allows Na^+, and to a lesser extent Ca^{2+}, to enter the cell (Kaupp *et al.*, 1988). Despite previous reports to the contrary, the channel appears to be localized exclusively in the plasma membrane (Cook *et al.*, 1989). Channel activity resides in a 63-kDa glycoprotein, which is probably active as an oligomer. When isolated by detergent extraction and column chromatography, the 63-kDa polypeptide is present as a complex with a 240-kDa cytoskeletal protein that has been shown to be rod cell spectrin (Molday *et al.*, 1990). This and other data suggest that the channel is present in the plasma membrane in close association with the cytoskeleton. The complete primary structure of the channel has been determined (Kaupp *et al.*, 1989). It has only limited amino acid sequence similarity to other ion channels and appears to represent a new family of ion channels with a simpler structure (Kaupp, 1991). The cDNA codes for a 79-kDa polypeptide, suggesting that in rod cells a larger precursor is synthesized and proteolytically processed to form the mature form of the channel. Expression of cDNA for the channel in *Xenopus* oocytes produces the expected cGMP-gated channel activity (Molday *et al.*, 1991).

2. Sodium–Calcium Exchanger

The Na^+–Ca^{2+} exchanger of bovine rod cells is a 220-kDa glycoprotein localized exclusively in the outer segment plasma membrane (Reid *et al.*, 1990). Its function is to extrude Ca^{2+} that enters as part of the dark current through the cGMP-gated cation channel. K^+ is co-transported with Ca^{2+}, allowing Na^+ to enter the cell (Friedel *et al.*, 1991; Schnetkamp, 1989). The exchanger may

TABLE I

Rod Outer Segment Proteins

Protein name	M_r (kDa)[a]	Properties[b]	Reference
Plasma membrane			
Rhodopsin	34–36	CB, SS, Con A, WGA, MAb, iodination	Molday and Molday, 1987a Polans et al., 1986 Witt and Bownds, 1987 Kamps et al., 1982
G3PD	38	CB, SS	Hsu and Molday, 1990
GLUT-1 glucose transporter	45	MAb	Hsu and Molday, 1991
cGMP-gated cation channel	52,63	CB, SS, iodination	Molday and Molday, 1987a Witt and Bownds, 1987 Clark and Hall, 1982 Kamps et al., 1982
	110	Ricin and Con A binding; SS, NMA-sensitive	Molday and Molday, 1987a Clark and Hall, 1982 Kamps et al., 1982
	160	Ricin and Con A binding; CB, NMA-sensitive iodination	Molday and Molday, 1987a Kamps et al., 1982
Na^+–Ca^{2+} exchanger	220–230	Ricin binding; NMA, trypsin-sensitive	Molday and Molday, 1987a Witt and Bownds, 1987 Clark and Hall, 1982
	270	CB; trypsin-sensitive	Molday and Molday, 1987a
Disk membrane			
Peripherin–rds	33–35	MAb	Molday and Molday, 1987a Connell and Molday, 1990
Rhodopsin	34–36	MAb, CB protease-sensitive	Molday and Molday, 1987a Papermaster and Dreyer, 1974 Hargrave, 1982
	43	CB, SS, iodination	Molday and Molday, 1987a Witt and Bownds, 1987
	57–63	SS	Molday and Molday, 1987a
Rim protein	220	CB, SS, Con A, WGA, trypsin-sensitive; iodination	Molday and Molday, 1987a Papermaster et al., 1978 Witt and Bownds, 1987
Other proteins			
Υ-subunit transducin	6–10	CB	Kühn, 1980
Υ-subunit PDE	13	CB	Baehr et al., 1979
Guanylate kinase	22	CB, enzyme activity	Hall and Kühn, 1986
Recoverin	26	CB, MAb	Dizhoor et al., 1991 Polans et al., 1991
Small G proteins	20–30	MAb	Wieland et al., 1990a
Phosducin	33	CB, MAb	Lee et al., 1987
β-subunit transducin	35–38	CB, MAb	Kühn, 1980

(continued)

TABLE I (*continued*)

Protein Name	M_r (kDa)[a]	Properties[b]	Reference
Retinol dehydrogenase	37	CB, MAb	Ishiguro *et al.*, 1991
α-subunit transducin	37–40	CB, MAb	Kühn, 1980
Arrestin (S antigen)	48	CB, MAb	Kühn, 1978
Rhodopsin kinase	63,67	CB, radioactivity, enzyme activity	Palczewski *et al.*, 1988a Kelleher and Johnson, 1990 Lorenz *et al.*, 1991
PDE β-subunit	84	CB	Baehr *et al.*, 1979
Protein kinase C	85	CB, radioactivity	Wolbring and Cook, 1991
PDE α-subunit	88	CB	Baehr *et al.*, 1979
Guanylate cyclase	110	CB	Hayashi and Yamazaki, 1991
Spectrin	240	CB, MAb	Wong and Molday, 1986

[a]Apparent molecular weights vary between different references, mainly because of the use of different polyacrylamide gel systems.

[b]CB, Coomassie blue staining; SS, silver staining; Con A, concanavalin A binding; WGA, wheat germ agglutinin binding; MAb, detected with monoclonal antibody; NMA, neuraminidase.

account for ~4% of total plasma membrane protein. It is heavily glycosylated, decreasing in molecular weight by 15 kDa after treatment with neuraminidase (Reid *et al.*, 1990). The exchanger binds to concanavalin A (Cook and Kaupp, 1988), to wheat germ agglutinin (Nicoll and Applebury, 1989), and to *Ricinis communis* agglutinin (Reid *et al.*, 1990) after neuraminidase treatment, which can be used to advantage in its purification.

3. Rhodopsin

Rhodopsin is the dominant protein of the plasma membrane, comprising as much as 60% of its protein content. Its presence there may simply be as a supply needed for the formation of disk membranes. In the plasma membrane rhodopsin is not oriented properly to be able to efficiently capture light. In addition, the high cholesterol lipid composition of the plasma membrane appears to offer too rigid an environment to allow rhodopsin to activate transducin even if it were photolyzed (Boesze-Battaglia and Albert, 1990).

4. Glyceraldehyde-3-Phosphate Dehydrogenase

Another major protein constituent isolating with the plasma membrane and comprising more than 11% of its protein content is an extrinsic membrane protein, the glycolytic enzyme glyceraldehyde-3-phosphate dehydrogenase (G3PD) (Hsu and Molday, 1990). It is associated only with the plasma membrane and is extracted from it with 0.15 M salt. Rod cell G3PD specifically and reversibly associates with the rod cell plasma membrane in an analogous fashion to the behavior of red blood cell G3PD and its plasma membrane. The presence of such a comparatively large quantity of a glycolytic enzyme in rod outer segments suggests that the rod outer segment may obtain some of its energy needs for transduction through the glycolytic pathway (Hsu and Molday, 1990).

E. Proteins of the Disk Membrane

Rhodopsin comprises ~90% of the protein of the disk membrane. Section III will be devoted to presentation of the properties of this key rod cell protein. Rhodopsin appears to be excluded from the rims of the disks, where two other proteins, the rim protein and peripherin, are found (Molday et al., 1987).

1. Rim Protein

Disk membranes contain a high molecular weight protein (~220 kDa) that is easily seen by Coomassie blue staining of sodium dodecyl sulfate (SDS) polyacrylamide gels (Molday and Molday, 1987a). Antibodies have been made to the corresponding protein from frog ROS and used to immunochemically localize it to the margins and incisures of frog disk membranes (Papermaster et al., 1978). Although there are other high molecular weight proteins present in ROS, this appears to be one of the proteins that undergoes light-dependent phosphorylation in frogs (Szuts, 1985) and to be the same as a 220-kDa transmembrane glycoprotein in cattle (Molday and Molday, 1979).

2. Peripherin

Peripherin is an intrinsic membrane glycoprotein of bovine disk membranes that has an M_r ~33 kDa by SDS-polyacrylamide gel electrophoresis (PAGE) (Molday et al., 1987). In the absence of reducing agents it migrates at M_r ~66 kDa, suggesting that it exists normally as a disulfide-linked dimer. Molday et al. (1987) obtained monoclonal antibodies (MAbs) to the protein and localized it to the disk rims, or periphery, leading them to name it "peripherin." It comprises ~5% of the protein in disk membranes, making it a relatively abundant rod cell protein. The cDNA sequence of peripherin shows it to be a 39-kDa protein containing four putative membrane-spanning regions (Connell and Molday, 1990). It has been proposed that peripherin has a role in promoting disk membrane adhesion (Connell and Molday, 1990; Travis et al., 1991).

Determination of the sequence of the bovine protein showed that it was homologous to a retina-specific gene product from mice, the *rds* gene product

(Travis *et al.*, 1989; Connell *et al.*, 1991). The retinal degeneration slow (*rds*) mutation in mice results in abnormal development of rod cell outer segments followed by their slow degeneration. This suggests that the *rds* gene product, peripherin, is a disk protein essential for normal rod outer segment morphology. Although the precise role of peripherin in rod cell function remains to be determined, defects in the gene coding for peripherin have been implicated as responsible for some cases of autosomal dominant retinitis pigmentosa (Farrar, 1991; Kajiwara, 1991).

F. Other Rod Outer Segment Proteins

The role of the important rod cell proteins transducin, cGMP-PDE, rhodopsin kinase, and arrestin will be discussed in Section IV. Other rod cell proteins of particular significance are listed in Table I and discussed in this section.

1. Rod Cell Spectrin

This 240-kDa protein was recognized by immunoreactivity toward a monoclonal antibody and shown to be distinct from the prominent 220-kDa protein (Wong and Molday, 1986). It can be partially solubilized in urea, without the aid of detergent, and does not bind concanavalin A. Its molecular weight is identical to that of the α-chain of red blood cell spectrin, a prominent cytoskeletal protein. By immunocytochemistry it has been shown to be distributed along the periphery of the rod outer segment where the disk rims are juxtaposed to the plasma membrane. It has been proposed to be an important component of a cytoskeletal network lining the inside surface of the ROS plasma membrane (Wong and Molday, 1986).

2. Small GTP-Binding Proteins

Members of a large family of more than 20 small GTP-binding proteins (M_r 20–30 kDa) have been found to occur in nearly all types of mammalian cells (Barbacid, 1987). Their exact cellular functions are unknown, but their wide distribution suggests that they are involved in basic cellular processes. Four such GTP-binding proteins, of M_r 22–27 kDa, have been found in bovine rod outer segments (Wieland *et al.*, 1990a). Two of the proteins, M_r 22 and 24 kDa, are substrates for a botulinum adenosine diphosphate (ADP)-ribosyltransferase, a finding similar to that observed for other tissues (Wieland *et al.*, 1990a). However, ADP ribosylation is greatly reduced in the presence of light, and it has been suggested that these proteins interact with photolyzed rhodopsin (Wieland *et al.*, 1990a, b). The significance of these observations is not yet clear. Three

TABLE II
Visual Pigment Sequences

AMINO TERMINUS

```
                                                  1                                      50                                                          100                                           I₁

Bovine      M...................................NGT.........EGPNFYVPFSNKTGVVRSPFEAPQYYLAEPWQ  FSMLAAYMFLLIMLGFPINFLTLYVT  VQHKKLRTPLNY
Sheep       M...................................NGT.........EGPNFYVPFSNKTGVVRSPFEAPQYYLAEPWQ  FSMLAAYMFLLIVLGFPINFLTLYVT  VQHKKLRTPLNY
Mouse       M...................................NGT.........EGPNFYVPFSNVTGVVRSPFEQPQYYLAEPWQ  FSMLAAYMFLLIVLGFPINFLTLYVT  VQHKKLRTPLNY
Human       M...................................NGT.........EGPNFYVPFSNATGVVRSPFEYPQYYLAEPWQ  FSMLAAYMFLLIVLGFPINFLTLYVT  VQHKKLRTPLNY
Chicken     M...................................NGT.........EGQDFYVPMSNKTGVVRSPFEYPQYYLAEPWK  FSALAAYMFMLILLGFPVNFLTLYVT  IQHKKLRTPLNY
Lamprey     M...................................NGT.........EGDNFYVPFSNKTGLARSPYEYPQYYLAEPWK  YSALAAYMFLILVGFPVNFLTLYVT  VQHKKLRTPLNY
Humangreen  MAQQWSLQRLAGRHPQDSYE................DST.........QSSIFTYTNSNST...RGPFEGPNYHIAPRWV  YHLTSVWMIFVVIASVFTNGLVLAAT  MRFKKLRHPLNW
Humanred    MAQQWSLQRLAGRHPQDSYE................DST.........QSSIFTYTNSNST...RGPFEGPNYHIAPRWV  YHLTSVWMIFVVIASVFTNGLVLAAT  MRFKKLRHPLNW
cf-G101     M....AAHEPVFAARHHNE.................DTT.........RESAFVTTNANNT..RDPFEGPNYHIAPRWV  YNVSSLWMIFVVIASVFTNGLVIVAT  AKFKKLRHPLNW
cf-R007     M...CDDWCDAVFAARRRGD................DTT.........RESAAFTYTNSNNT..KDPFEGPNYHIAPRWV  YHLXSCMHFFVVVAASVFTNGLVLVAS  AKFKKLRHPLNW
Chickiodop  M....AAMCAAFAARRRHEEE...............DTT.........RDSVFTTTNSNNT..RGPFEGPNYHIAPRWV  YHLTSLWMIFVVAASVFTNGLVLVAT  MKFKKLRHPLNW
cf-G103     M....AAHADEPVFAARRYNE...............ETT.........RESAFVTTNANNT..RDPFEGPNYHIAPRWV  YHLASLWMITVVIASIFTNSLVTVAT  AKFKKLRHPLNW
Humanblue   M..................RKMS.............EEE.........FYL.F.KNIS.S..V..GPWDGPQYHIAPWVA  FYLQAAFMGTVFLIGFPLRANVLVAT  LAYKKLRQPLIY
Drh1        MESFAVAAAQLG.....PHFAPLS.NGS........VVDKVFPDMAHLISPYW.NQFP.AMDPIWAKI..........  LLAYHIMGHISMCGMGVVIYIF  ATTKSLRTPANL
Blowfly     MERYS........TPLIGPSFAALT.NGS.......VTDKVFPDMAHLVHPYW.NQFPCAMEPRWAKF.........  LAAYMVLIATISMCGMGVVIYIF  STTKSLRTPANL
Drh2        MERSHLPETPFDLAHSGPRFQAQSSG.NGS......VLDNVLPDMAHLVHPYW.SRFA.PNDPMWSKI.........  LGLFTLAIMIISCCGMGVVVYIF  GGTKSLRTPANL
Drh3        MESGNVSSSLFGNVSTALRPEARLSACTRLL..GWNVPPEELAHIPEH......WLPEPEESM.NYL.........  .GTLYIFFTLNSNLGRGLVIMWF  SAAKSLRTPSAI
Drh4        MEPICRASE.........PPLRPEARSSGNGLDQFLGWNVPPDQIQYIPEH...WLTGLEPPASMHY.........  .MLGVFYIFLFCASTVCRGHVIMIF  STSKSLRTPSNM
Octopus     M..................................VESTT......LVNQTMWYNPFVDIHPFHAKFDPIPDAV  YYSVGIFIGVVGIIGILGMGVVIYLF  SKTKSLGTPANM
Squid       M..................................GRDIPDNE....TWWYNFYMDIHPHKKGFDQVPAAV  YSLGIFIAICGIIGCVGMGVVIYLF  TKTKSLGTPANM
Hummlr      M...................................NYSAPFAVSPNITVLAPGKGPWQ  VAFGITTGLLSLAAVGHLVLLISF  KVNTELKTVNNY
Humbar      M...................................GQPGNGSAFTLAPHRSHAPDHDVTQQRDEVM  VGCMGIVMSLIVLATVFGNVLVITAI  AKFERLQTVTNY
```

II

```
Bovine      ILLWLAVADLFMVFGGFTTTLYTSL
Sheep       ILLWLAVADLFMVFGGFTTTLYTSL
Mouse       ILLWLAVADLFMVFGGFTTTLYTSL
Human       ILLWLAVADLFMVLGGFTTTLYTSL
Chicken     ILLWLAVADLFMVFGGFTSTLYTSL
Lamprey     ILLWLAMARLFMVLFGCFTVTMYTSM
Humangreen  ILVWLAVADLAETVIASTISVVNQV
Humanred    ILVWLAVADLAETVIASTISVVNQV
cf-G101     ILWWLAIALDLGETVLASTISVINQI
cf-R007     ILWWLAIADLIETLLASTISVCWQF
Chickiodop  ILVWLAVADLGETVIASTISVINQI
cf-G103     ILWWLAIADLGETVLASTISVTWQV
Humanblue   ILVWVSFGGFLLCIFSVFPVFVASC
Drh1        LVIWLAISDFGI.MITNTPMMGINL
Blowfly     LVIWLAISDFGI.MITNTPMMGINL
Drh2        LVIWLAFSDFCM.MASQSPVMIINF
Drh3        LVIWLAFCDF.M.MMVKTPIFIYNS
Drh4        FVIWLAVFDL.I.MCLKAPIFIYNS
Octopus     FIIWLAMSDLSFSAIMGFPLMTISA
Squid       FIIWLAFSDFTFSLVMGFPLMTISC
Hummlr      FLLSLACDLLIGTFSMRLVTTYLL
Humbar      FITSLACADLVMGLAVFPFGAAHIL
```

```
                      e1                                                    III                           i2
                      |150
Bovine      HG.YFVFGPT    GCHLEGFFATLGGEIALWSLVVLAI    ELVVVCKPMSNFRFGENH...
Sheep       HG.YFVFGPT    GCHLEGFFATLGGEIALWSLVVLAI    ELVVVCKPMSNFRFGENH...
Mouse       HG.YFVFGPT    GCHLEGFFATLGGEIALWSLUVVLAI   ELVVVCKPMSNFRFGENH...
Human       HG.YFVFGPT    GCHLEGFFATLGGEIALWSLUVVLAI   ELVVVCKPMSNFRFGENH...
Chicken     NG.YFVGVT     GCYIEGFFATLGGEIALWSIVVLAV    EKVYICKPMGNFRFGRTH...
Lamprey     NG.YFVFGPT    MCSIEGFFATLGGEVALWSLVVLAV    EKVYICKPMGNFRFGRTH...
Humangreen  YG.YFVLGHP    MCVLEGTVVSLGGITGLWSIAIISW    EBMVVCKPFGNVRFDAKL...
Humanred    SG.YFVLGHP    MCVLEGTVVSLGGITGLWSIAIISW    EBMLVCKPFGNVRFDAKL...
cf-G101     FG.YFILGHP    MCVFEGHTVSVGGITALWSLIVISW    EBMVVCKPFGNVKFDGKM...
cf-R007     FG.YFILGHP    MCVFEGHTVSVGGITALWSLIVISW    EBMVVCKPFGNVKFDGKM...
Chickiodop  SG.YFILGHP    MCVFEGTVAZGIAGLWSLZVISW     EBMFVVCKPFGNIKFDGKL...
cf-G103     FG.YFVLGHP    MCVFEGTVSVGGITALWSLIVISM    EBMVVCKPFGNVKFDCGKW...
Humanblue   NG.YFVRGRH    VCALEGFLCTVAGLVTGWSLAFLAF    EBTVICKPFGNFRFSSKH...
Drh1        YFETWVLGPH    MCDIYAGLGSAFGCSSTWSWCHISL   DRWNTVWGHAG.RPWTIP...
Blowfly     FYETWVLGPL    MCDIYGGLGSAFGGCSSILSWCHISL  DRWNTVWGHAG.QPWTIK...
Drh2        YYETWVLGPL    MCDIYAGCGSLFGCVSIWSWCHIAF   DRWNTVWGGING.TPWTIK...
Drh3        FHQGYALGRL    GCQIFGIIGSYTGIAAGATWAFIAY   DRWNVTRPWEG.K.WTHG...
Drh4        FHRGFALGRT    WCQIFASIGSYSGIGAGHTQAAIGY   DRWNVTKFM.NRNWFFTK...
Octopus     FMKKWIGRV     ACQLYGLLGGITGFWSITWNAWTSI   DRWNVTGRPWAASKKP.SHR...
Squid       FMKYWVGRNA    ACRVYGLIGGITGLMSIWTTWNISI   DRWNVTGRPWSASKKP.SHR...
Humml r     NG.HWALGZL    ACDLWLALDVYASWASVWNLLLISF   DRYFSVTRPLSYWAARIFWA...
Humbar      MK.HWTTGHF    WCEFWTSIDVLCVTASIETLCVIAV   DRYFALTSPFKYQSLLTKWKAR...


                     IV
                     |200
Bovine      AIMG.VAFTVMVAACAAP.LVGRS...
Sheep       AIMG.VAFTVMVAACAAP.LVGRS...
Mouse       AIMG.VVFTVIWAACAAP.LVGRS...
Human       AIMG.VAFTVMVAACAAP.LAGRS...
Chicken     AIMG.VAFSWIMMACAAP.LFGRS...
Lamprey     AMMG.IAFTVVMAFSCAAP.IFGRS...
Humangreen  AIVG.IAFSWIMSAVWTAPP.IFGRS...
Humanred    AAGG.IIFSWVWAAIICMFP.IFGRS...
cf-G101     ATAG.IVFTVMSAVWCAAP.IFGRS...
cf-R007     AVAG.ILFSWLMSCAWTAPP.IFGRS...
Chickiodop  AAGG.IIFAWFWAIICWTFP.IFGRS...
cf-G103     ALTV.VLATWTIGIGVSIFP.FFGRS...
Humanblue   LALGRIAYIWFNSSIWCLAP.AFGRS...
Drh1        LAIWKIALIWFMASIWTLAP.VFGRS...
Blowfly     TSIMKILFIWMMAVFWTWP.LIGRS...
Drh2        KAIMMIFIYWYAIFWVVACTETRG...
Drh3        AVIWHRIIWLYCTPWVVPLZQF.WD...
Drh4        RAFLMIIFVWMNSIVNSVGF.VFNG...
Octopus     KAFIMIIFVWINSTINAIGP.IFGWG...
Squid       AALMIGLAWLVSFVLMAPA.ILFWQL
Humml r     VIILNVWIVSGLZSFLFIQ.MHWY.
Humbar
```

(continues)

TABLE II (continued)
Visual Pigment Sequences

```
                          e₂                              V                          I₃
                                      250                                                      300
                                       |                                                        |
Bovine      RYIPEGMQCSCGIDYYTPHEETNNES  ...FVIMFVHFIIPLIVIFFCYGQLV.  FTVKEAAAQQQ......ESATTQKAKEVTR....
Sheep       RYIPGQMQCSCGLAYTILKPEINNES  ...FVIMFVHFSIPLVIFFCYGGLV.  FTVKEAAAQQQ......ESATTQKAKEVTR....
Mouse       RYIPEGMQCSCGIDYYTLKPEVNNES  ...FVIMFVHFTIPMIVIFFCYGGLV.  FTVKEAAAQQQ......ESATTQKAKEVTR....
Human       RYIPEGLQCSCGIDYYTLKPEINNES  ...FVIMFVHFTIPMIIFFCYGGLV.  CTVKEAAAQQQ......ESATTQKAKEVTR....
Chicken     RYIPEGMQCSCGIDYYTLEPNFNNES  ...YVVMFVHFLVPFVIFFCYGRLL.  CTVKEAAAQQQ......ESASTQKAKEVTR....
Lamprey     RYMPHGLKTSCGPDVFSGSSYPCVQS  ...YMIVLKVTCCIIPLSIIVLCYLQVM.  LAIRAVAKQQK......ESESTQKAKEVTR....
Humangreen  RYMPHGLKTSCGPDVFSGSSYPCVQS  ...YMIVLKVTCCIIPLAIIMLCYLQVM.  LAIRAVAKQQK......ESASTQKAKEVTR....
Humanred    RYMPHGLKTSCGPDVFSGSEDPCVAS  ...YMIVLHITCCFIPLGIIILCYIAVM.  SAIHQVAQQQK......DSESTQKAKEVSR....
cf-G101     RYMPHGLKTSCGPDVFSGSEDPCVQS  ...YMIVLKVTCCFFPLAIILCYLQVM.  WAIRTVAQQQK......ESESTQKAKEVSR....
cf-R007     RYMPHGLKTSCGPDVFSGSSDPCVQS  ...YMVVLRVTCCFFPLAIIILCYLQVM.  LAIRAVAQQQK......ESESTQKAKEVSR....
chicklodop  RYRMPHGLRTSCGPDVFSGSEDPCVAS  ...YMVTLLLFCCLLPLSVIICYIFVW.  WAIHNVAAQQQK......ESATTQKAREDVSR....
cf-G103     RFIPEGLQCSCGPDWTVGTKYRSES  ...YTMFLFIFCFIVPLSLICFSTQLL.  RALKAVAAQQQ.....
Humanblue   RYVPEGRLTSCGIDYLRDWNPRSYL  ...IYYSIFVYIPLFLICYSYWFIAA.  VSANEKAREQQAKKDHRVKSLRSSEDAEK.SAEGKLAK...
Drh1        RYVPEGRLTSCGIDYLRDWNPRSTL  ...IYYSIFVYLPLFLICYSYWFIAA.  VSANEKAREQQAKKDHRVKSLRSSEDADK.SAEGKLAK...
Blowfly     AYVPEGRLTACSIDYMTRMNFRSYL  ...ITYSLFVYTPLFLICYSYWFIAA.  VAANEKAREGAQAKKDHVKVESLASSEDCDK.SAEGKLAK...
Drh2        RFVPEGTLSCTFDYLTDNFDTRLFV  ...ACIFFSFVCPTMIIYIYSQIVGH.  VFSHEKALREQQAKKDHVESLASNVDKNKETAEIRIAK...
Drh3        RFVPEGTYLTCSFDYLSDNFDTRLFV  ...GTIFFFSFVCPTLWILYYSQIVGH.  VFSHEKALREQQAKKDHVESLASNVDKSKETAEIRIAK...
Drh4        AYVPEGILTSCSFDYLSTDFSTRSFI  ...LCHYTCGFMLPIIIAFCYFNIVWS.  VSRHEKEHAMAAKRLRAKELRAKAQAGA..SAEMKLAK...
Octopus     AYTLEGVLCMCSFDIIRDTTTRSNI  ...LGMYIIAFMCPIVVIFFCYFRIYMS.  VSRHEKEHAMAAKRLRAKELRAKAQAGAN..AEMKLAK...
Squid       VGERTMLAGQC......YIQFISQ..  ...PIITFGTANAAFYLPVTVMCTLMRI  YRETERRADELA...ALqESIVEEKTKAAR...
Humnlr      R.ATHQEAINCY....ANETCCDFF  2MQAYALASSIVEFVPLVIHVFVYSRVF.  QEARRQLQKID.....KSakFCLKEHKALR...
Humbar
```

VI

```
Bovine      NVIBVIAFLICWLPYAGVAFYIFT.
Sheep       NVIBVIAFLICWLPYAGVAFYIFT.
Mouse       NVIBVIFFLICWLPYASVAFYIFT.
Human       NVIBVIAFLICWWYASVAFYIFT.
Chicken     NVIBVIAFLICWVTASVAFYIFT.
Lamprey     NVVLRVIGFLVCWWTASVAFYIFT.
Humangreen  NVVVMVLAFCFCWGPYAFFACFAAA.
Humanred    NVVVMILAFIVCWGPYASFATFSAV.
cf-G101     NVVVMINAYCFCWGPYTFFACFAAA.
cf-R007     NVVVMILVAYCFCWGPYTFFACFAAA.
chicklodop  NVVVMILAFILCWGPYASFATFSAL.
cf-G103     NVVVMVGSFCVCYVPYAAFAMYMVN.
Humanblue   VALVTILMFKAWTPYLVIHCKGLF.
Drh1        VALVTISLMFKAWTPYLVICYFGLF.
Blowfly     VALVTISLMFRAWTPTLVICYTGLF.
Drh2        AAIICFLFFCSWTPYGVMSLIGAF.
Drh3        ISVVITQFMLSNSPYAIIALLAQF.
Drh4        ISIVVTQFLLSNSPYAVVALLAQF.
Octopus     TLSAILLAFILTWTPYNIMVLVSFC.
Squid       TLGIIMGTFILCWLPFFIYNIVHVI.
Humnlr
Humbar
```

64

```
                      e₃                    VII                        I₄
                      350
                       |
Bovine       HQGSDFGP   ITMIPAFFAMTSAVYNRVITIMMN.    KQFRN...CMYTLCC
Sheep        HQGSDFGP   ITMIPAFFAMTSSVYNRVITIMMN.    KQFRN...CMYTLCC
Mouse        HQGSNFGP   ITMIPAFFAKSSSIYNRVITIMLN.    KQFRN...CMYTLCC
Human        HQGSDFGP   ITMIPAFFAKSSAIYNRVITIMMN.    KQFRN...CMYTLCC
Chicken      HQGSDFGP   ITMIPAFFAKSSALYNRVITIVMN.    KQFRN...CMITTLCC
Lamprey      HQGSDFGA   TTMIPAFFAKSSALYNRVITILMN.    KQFRN...CMITTLCC
Humangreen   NPGTPFHP   MMALPAFFAEASATIYNRVITVFMN.   RQFRN...CILQLF..
Humanred     NPGTAFHP   MMALPAYFAEASATIYNRVITVFMN.   RQFRN...CILQLF..
cf-G101      NPGTAMHP   LAAMPAYFAEASATIYNRIITVFMN.   RQFES...CIMQLF..
cf-R007      NPGTAFHP   LAAMPAYFAEASATIYNRIITVFMN.   RQFEV...CIMQLF..
cf-G103      NPGTAMHP   LLAALPAYFAEASATIYNRIITVFMN.  RQFES...CIMQLF..
Chickiodop   NRNHGLDL   ELVTIPSFSESACIYNRIITYCFMN.   KQFQA...CIM.RMVC
Humanblue    KFEG.LIP   LFTIWGACSAESAACYNRIVTGISH.   PKYRALAEKCP...CC
Drh1         KYEG.LIP   LFTIWGACSAESAACYNRIVTGISHC   PKYGIALKEKCP...CC
Blowfly      KIDG.LIP   FFIIWGATFAETSAVYNRIVYGISH.   PKYRIVLKEKCP...MC
Drh2         GDKTLLIP   CJIMIPACPEDRVACIDPFVTAISH.   PRYRMELQKRCPWIALNE
Drh3         GDKSLLIP   CJIMIPACEELVACIDFFVTAISH.    PRYRLELQKRCPWL..GV
Drh4         GPAENVYP   LAAELPVLPALASAINNFIVTSVSH.   PKFRELAIQGTTFPWLITCC
Octopus      GPIENVYP   LAAQLPWPELASAIHNFMITSVSH.    PKFRERIASNFPWILTCC
Squid        KDCVPETL   NE..LGYWL.PVNSTINFMCYALCN.   KAFRD....ATFRLLLL
Humnlr       QDMLIRRE   .FILLMWI-PVNSGFNPLIYC.RS.    PDFRI...AFQEILCL
```

                              CARBOXYL TERMINUS

```
                      400
                       |
Bovine       GKNPLGDDE..AST TVSKTETS....QVAPA
Sheep        GKNPLGDDE..AST TVSKTETS....QVAPA
Mouse        GKNPLGDD..ASA TASKTETS....QVAPA
Human        GKNPLGDDE..ASA TVSKTETS....QVAPA
Chicken      GKNPLGED..TSA .GKTETSSV STSQVSPA
Lamprey      GKNPLGDDESGAST .SKTEVSSV STSPVSPA
Humangreen   GKRV..DDGSELSS A.SKTEVSSV S..SVSPA
Humanred     GKRV..DDGSELSS A.SKTEVSSV S..SVSPA
cf-G101      GKRV..EDASEVSG ST..TEVSTA S
cf-R007      GKRV..DDG
cf-G103      GKK..VDDGSE.VS TSR.TEVSSV SNSSVSPA
Chickiodop   GKK..VEDASE.VS GST.TEVSTA S
Humanblue    GK.AMTDESDT.CS SQ.KTEVSTV SSTQVGPN
Drh1         VFGKV.DD...GKS SDAGSQA....T..ASEA ESEA
Blowfly      VFGKV.DD...GKA SDATSQA....T..HNES ETEA
Drh2         VFGHT.DE...PKF .DAPASD.....TETTSEA DSEA
Drh3         .KAPESSAV.AST STPQEPQQ....TTAA
Drh4         NEKSGEISS.AGS ITTQEQQQ....TTAA
Octopus      QFDEKEDD...AND AEEEVVAS....ERGGESR .DAAQMKEHM(81 residues)
Squid        QYDEKEIED..DKD AEAEIPAG....EQSGGET ADAAQMKEHM(78 residues)
Humnlr       CRWDKRAWRKIPKP PGSVHRTPSR.QC
Humber       RRSSLRAYGNGYSS NGNTGEQSGY HVEQEKEH
```

65

members of this same class of G proteins have also been detected in rhodopsin-containing vesicles involved in the transport of rhodopsin from the inner to the outer segment (Deretic and Papermaster, 1991).

3. Phosducin

Phosducin is a 33-kDa phosphoprotein that is found in the rod cell in a complex with $T_{\beta\gamma}$ (Lolley and Lee, 1990). In a reconstituted system it can compete with T_α for binding $T_{\beta\gamma}$ (Lee *et al.*, 1987). This serves to down-regulate the cycling of T_α, inhibiting activity of transducin and cGMP-PDE. If this is its cellular function, it would appear to be most effective under conditions of high light intensity when the amount of T_α might be limiting. By sequence analysis phosducin is actually 28 kDa (Lee *et al.*, 1990). It shows no sequence homology with any G proteins or any other proteins sequenced to date.

4. Guanylate Cyclase

Guanylate cyclase is the rod cell enzyme that converts guanosine triphosphate (GTP) to $3',5'$-cyclic GMP. This enzyme is of particular importance in rod cell metabolism because of the use that the rod cell makes of cGMP in regulating its outer segment cation permeability. Guanylate cyclase is found associated with the axoneme basal apparatus complexes of rod cell cilia (Fleischman and Denisevich, 1979). It can be solubilized in detergents and purified by chromatography on concanavalin A- and GTP-sepharose (Hayashi and Yamazaki, 1991; Koch, 1991). The bovine enzyme is a 110-kDa glycoprotein (Hayashi and Yamazaki, 1991; Koch, 1991), the activity of which is regulated by Ca^{2+}, increasing from fivefold to 20-fold when the calcium level is lowered from 200 nM to 50 nM (Koch and Stryer, 1988). This calcium sensitivity is due to interaction with a calcium-binding protein, recoverin.

5. Recoverin

The calcium sensitivity of guanylate cyclase is due to its interaction with a 23–26-kDa calcium-binding protein (Dizhoor *et al.*, 1991; Lambrecht and Koch, 1991). This protein was named recoverin because by promoting the resynthesis of cGMP it helps the photoreceptor recover to the dark state. It stimulates guanylate cyclase activity when cellular calcium levels decline after illumination, when the cGMP channel restricts entry of calcium but calcium is still being extruded by the Na^+–Ca^{2+} exchanger. Recoverin must be present for guanylate cyclase activation because activation does not occur when antirecoverin antibody is added. Therefore, it must be the calcium-free form of the protein that serves to activate guanylate cyclase (Dizhoor *et al.*, 1991).

III. Rhodopsin: A Protoreceptor Protein

A. Molecular Properties of Rhodopsin

Vertebrate rhodopsins are integral membrane proteins of about 40,000 molecular weight. Amino acid sequences have been determined for 20 visual pigments (Table II) and partial sequences have been determined for several others (Findlay, 1986; Neitz et al., 1991). The sequences given in Table II are as follows: bovine rhodopsin (Hargrave et al., 1983; Ovchinnikov et al., 1982); sheep rhodopsin (Findlay, 1986); mouse rhodopsin (Baehr et al., 1988); human rhodopsin (Nathans and Hogness, 1984); chicken rhodopsin (Takao et al., 1988); lamprey rhodopsin (Hisatomi et al., 1991); human green cone pigment (Nathans et al., 1986); human red cone pigment (Nathans et al., 1986); cavefish pigment (Cf-G101) (Yokoyama and Yokoyama, 1990a); cavefish pigment (Cf-R007) (Yokoyama and Yokoyama, 1990a); chicken iodopsin (Kuwata et al., 1990); cavefish pigment (Cf-G103) (Yokoyama and Yokoyama, 1990b); human blue cone pigment (Nathans et al., 1986); Drosophila pigments Drh1 (Zuker et al., 1985), Drh2 (Cowman et al., 1986), Drh3 (Zuker et al., 1987), and Drh4 (Montell et al., 1987); blowfly rhodopsin (Huber et al., 1990); octopus rhodopsin (Ovchinnikov et al., 1988); squid rhodopsin (Hall et al., 1991); human muscarinic receptor 1 (humm1r) (Peralta et al., 1987); and human β-adrenergic receptor (Humbar) (Schofield et al., 1987). Sequences in the center of i_3 of both the human β-adrenergic receptor and the human muscarinic receptor 1 have been omitted at the arrow but are available in the original references. These regions were not considered in the phylogenetic analysis. Similarly, the long carboxyl-terminal regions of squid and octopus opsin were omitted. The proteins are homologous in structure and share the same basic features. Most studies of structure and function have been conducted with bovine rhodopsin, and we believe that the knowledge gained by the study of this protein applies to most of the other vertebrate visual pigments. Bovine rhodopsin is blocked at its amino terminus with an acetyl group (Tsunasawa et al., 1980). This is a common structural feature for many soluble and membrane proteins (Persson et al., 1985). The biological significance of N-terminal acetylation is not known, but it has been suggested to be important in protein turnover (Mayer et al., 1989). Oligosaccharide chains are attached to bovine rhodopsin at two locations: asparagines 2 and 15 (Hargrave, 1977). These oligosaccharides are uniquely small by comparison with those linked to asparagines of other proteins. There is the usual heterogeneity, but the dominant component is $Man_3GlcNAc_3$ (Fukuda et al., 1979; Liang et al., 1979; Hargrave et al., 1984). Bovine rhodopsin also contains 0.1 mol galactose by chemical analysis, suggesting that 10% of the molecules may contain galactose (Fukuda et al., 1979). It has been shown in rat that galactose is transiently incorporated into rhodopsin (Smith et al., 1991). There appear to be differences in the carbohydrate content of cone visual pigments as well as species differences. Many cone

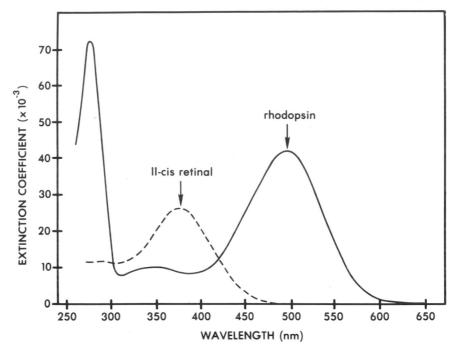

FIG. 7   UV-visible absorption spectrum of bovine rhodopsin and 11-*cis*-retinal. Adapted from Ebrey and Honig (1975), with permission.

opsins and some species of rhodopsin also contain fucose (Bunt, 1978; Bunt and Klock, 1980; Aguirre and O'Brien, 1986; Fliesler *et al.*, 1991).

To become active as a photoreceptor, the protein (opsin) must bind a molecule of 11-*cis*-retinal. The retinal forms a protonated Schiff base linkage to a specific lysine in the bovine opsin sequence, lysine 296. This forms the light-sensitive rhodopsin, which now has an absorption spectrum in the visible range, with maximal absorption at 498 nm for bovine rhodopsin (Fig. 7). Opsins with different sequences (from different photoreceptor cell types or species) yield visual pigments with different wavelengths of maximal absorption. Rhodopsin becomes palmitoylated (O'Brien and Zatz, 1984; St. Jules and O'Brien, 1986). Two cysteines, 322 and 323 in bovine rhodopsin, have been identified as the sites of palmitate incorporation (Ovchinnikov *et al.*, 1988). Other membrane proteins also become modified by palmitate, but the biological significance of this modification is not yet clear. Upon exposure to light, rhodopsin becomes phosophorylated by rhodopsin kinase. Serines and threonines on rhodopsin's cytoplasmic surface become modified. This is a reversible reaction; a protein phosphatase catalyzes the removal of the phosphates. This reaction is important in turning off the visual tranduction process (Section IV,D).

B. Rhodopsin Biosynthesis

Rhodopsin is synthesized on membrane-bound ribosomes in the endoplasmic reticulum (Papermaster and Schneider, 1982; Besharse, 1986). In contrast to many membrane proteins, rhodopsin is not synthesized with an N-terminal signal sequence, which is then proteolytically removed (Schechter *et al.*, 1979). The sequences that define rhodopsin's transmembrane segments are sufficient to direct its proper insertion into the membrane and to define the correct protein topography (Audigier *et al.*, 1987). Glycosylation, palmitoylation, possibly the binding of retinal (and undoubtedly acetylation) occur in the endoplasmic reticulum (St. Jules *et al.*, 1989; 1990). As soon as rhodopsin's amino-terminal sequence enters the lumen of the endoplasmic reticulum, glycosylation occurs at asparagines 2 and 15. A large oligosaccharide, $Glc_3Man_9GlcNAc_2$, is originally incorporated but is extensively modified as rhodopsin passes through the endoplasmic reticulum and Golgi cisternae (Kornfeld and Kornfeld, 1985). From the *trans* Golgi cisternae, rhodopsin is transported in vesicles to the base of the connecting cilium (Papermaster *et al.*, 1985). This region has been well described in frogs as the periciliary ridge complex.

Rhodopsin passes from the periciliary ridge complex at the base of the connecting cilium to the plasma membrane by a process that is poorly understood. The plasma membrane contains rhodopsin and serves as the precursor for the formation of new disks for the rod outer segment. New disks are formed by evagination of the plasma membrane in a process that requires F-actin (Vaughan and Fisher, 1989). Disk membranes are distinctly different in their protein and lipid composition compared with the plasma membrane (Molday and Molday, 1987a; Boesze-Battaglia and Albert, 1989), indicating the operation of a sorting mechanism.

C. A Model for Rhodopsin Structure and Membrane Topography

Physical studies of rod outer segment membranes have shown rhodopsin to be a transmembrane protein with approximately half of its mass embedded in the lipid bilayer (Chabre, 1985). X-ray and neutron diffraction data suggest the molecule to be prolate in shape and about 6–6.5 nm in length. As much as 50–60% of the molecule may be α-helical in structure, with these helices being oriented perpendicular to the plane of the membrane. This is consistent with a transmembrane helical bundle as a principle structural feature of the protein.

Inspection of the amino acid sequence of bovine rhodopsin shows a feature that is uncommon in globular proteins but that is now recognized as typical of integral membrane proteins; that is, the presence of uniform-length blocks of nonpolar amino acids separated by stretches of polar amino acids (Table II).

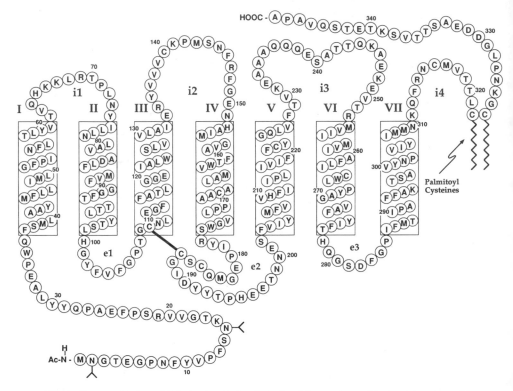

FIG. 8 Topographical model for bovine rhodopsin in the disk membrane. Rhodopsin's polypeptide chain traverses the membrane lipid bilayer seven times, shown by the transmembrane helices I–VII. Helix-connecting loops i_1–i_4 and the carboxyl-terminal region are located on the cytoplasmic surface. The N-terminal region and loops e_1–e_3 are on the extracellular surface when rhodopsin is in the plasma membrane or on the intradiskal surface in the disk. Oligosaccharide chains are attached to asparagines 2 and 15. A disulfide bridge links cysteines 110 and 187. Cysteines 322 and 323 are palmitoylated. Retinal is linked to lysine 296.

Rhodopsin's amino acid sequence contains seven stretches of predominantly hydrophobic amino acids of 21–28 amino acids in length separated by sequences of predominantly hydrophilic amino acids. This is consistent with a transmembrane protein that contains seven transmembrane segments.

There is ample evidence that rhodopsin is a transmembrane protein. Detailed information about its topography in the membrane forms the basis for the model shown in Figure 8. This model shows the rhodopsin polypeptide chain crossing the lipid bilayer of the disk membrane seven times, exposing its amino terminus to the lumen of the disk and its carboxyl terminus at the cytoplasmic surface. Rhodopsin's orientation in the plasma membrane also exposes its carboxyl terminus to the cytoplasm and places its amino terminus on the external surface of the plasma membrane. A variety of experimental observations have led to the development of this topographic model. Location of the amino terminal inside the disk is demonstrated by the fact that antibodies and lectins that bind rhodopsin's N terminus can bind rhodopsin only in disrupted disks. Anti-N-terminal antibodies also label rhodopsin on the outside surface of the rod outer segment plasma membrane. Surface assignment of the second lumenal loop is demonstrated by binding of a loop-specific antibody only to disrupted discs (Adamus *et al.*, 1991) and by binding of an antibody specific for the same loop to the external surface of the plasma membrane (Gaur *et al.*, 1988).

The carboxyl terminus is exposed to the cytoplasmic surface as shown by limited proteolysis of intact disk membranes, phosphorylation of the C-terminal sequence by rhodopsin kinase, modification of the C-terminal sequence by transglutaminase, and binding of anti-C-terminal antibodies to intact disks. Four loop regions of the polypeptide chain comprise the remainder of the cytoplasmic surface. Location of the first cytoplasmic loop is demonstrated by its modification in the disk using a nonpenetrating chemical reagent (Barclay and Findlay, 1984). The third loop is assigned to the cytoplasmic surface based upon the results of limited proteolysis (Findlay *et al.*, 1981; Hargrave *et al.*, 1982), its modification by transglutaminase (Pober *et al.*, 1978; McDowell *et al.*, 1986), its phosphorylation by rod cell kinases (McDowell *et al.*, 1985; Newton and Williams, 1991), and its binding by site-specific antibodies (Adamus *et al.*, 1991; Molday, 1989). The fourth loop is contiguous with the C terminus and is assumed to be formed by intercalation of the palmitates attached to cysteines 322 and 323 into the lipid bilayer (Ovchinnikov *et al.*, 1988). It is highly exposed on the cytoplasmic surface as shown by accessibility to an IgM antibody that binds the sequence 310–321 (Adamus *et al.*, 1991).

D. Three-Dimensional Structure of Rhodopsin

Only a few membrane proteins have yielded crystals that diffract to high resolution, allowing determination of a three-dimensional structure such as that which is available for numerous globular proteins. Not only is rhodopsin a membrane protein but it also undergoes structural changes upon exposure to light, making determination of its crystal structure an even more difficult undertaking. In the meantime a working model has been developed taking advantage of the best information available (Hargrave *et al.*, 1984; Findlay and Eliopoulos, 1990). Our model for rhodopsin shows seven helices arranged in a bundle, which serves as a container for the retinal (Fig. 9). The seven transmembrane segments comprise half the mass of the protein and must account for the 50–60% helix content. The cross-sectional area of rhodopsin is essentially the same as that determined for bacteriorhodopsin, a protein known to have seven transmembrane helices. The order of packing of the helices is unknown

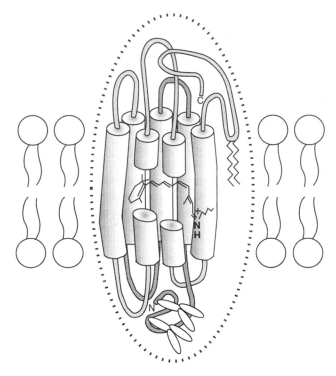

FIG. 9   Helical bundle model for rhodopsin. Rhodopsin is proposed to be comprised of a bundle of seven transmembrane helices, the interior surfaces of which form a hydrophobic pocket for 11-*cis*-retinal.

but may be modeled by analogy to what is known of the structure of another retinyl protein, bacteriorhodopsin. Here it is clear that there is a 4 + 3 circular packing arrangement of the helices (Henderson and Schertler, 1990). We assume that the physically adjacent helices are arranged in the order of their biosynthesis.

Rhodopsin's retinal chromophore has been identified as buried within the protein in the lipid bilayer (Thomas and Stryer, 1982). Its site of attachment has been shown to be via a Schiff base to lysine 296, located near the midpoint of the seventh transmembrane helix. The amino acid side chains from some of the transmembrane helices form the environment for the polyene chain of the retinal and its β-ionone ring. Differences in wavelength of maximal absorption of the various visual pigments are undoubtedly due to the difference in polarity of the amino acid environment in the retinal pocket. Details have been determined for the orientation of retinal within its binding pocket in bovine rhodopsin. A photoactivatable analog has been cross-linked to amino acids in helices III and VI, showing that the β-ionone ring orients toward these helices (Nakayama and Khorana, 1990). When tryptophan 265 in helix VI is replaced by alanine by site-specific mutagenesis, the spectrum is blue-shifted 30 nm and the ability to bind retinal is markedly reduced (Nakayama and Khorana, 1991). Similarly, re-placement of glutamic acid 122 in helix III by alanine leads to a blue shift of 20 nm (Nakayama and Khorana, 1991). Other buried amino acids, such as aspartic acid 83 in helix II, can be substituted with little effect. This suggests that Trp[265] and Glu[122], at a minimum, are located in the retinal-binding pocket and con-tribute in regulating the maximal wavelength of absorption of rhodopsin's chromophore.

Although uncharged amino acid side chains in the vicinity of the retinal have an influence on the retinal wavelength, the most important element in regulating the wavelength of the visual pigment is the counterion for the retinylidene Schiff's base. This amino acid has now been identified as glutamic acid 113 (Zhukovsky and Oprian, 1989; Sakmar et al., 1989; Nathans, 1990). Substitu-tion of this buried charged amino acid with aspartic acid has little effect, but substitution with its uncharged amide counterpart, glutamine, causes the visual pigment spectrum to shift dramatically from 498 nm to 380 nm.

E. Sequence Comparisons and Phylogenetic Analysis of Opsins

The complete amino acid sequences of 20 visual pigments are known. These se-quences were aligned (Devereux et al., 1984) to determine which amino acids are strictly conserved and for phylogenetic analysis. There are 18 residues that are conserved in the visual pigments and these are shown in bold type in Table II. Seven of these residues are also conserved in most G protein-linked receptors (Hargrave and McDowell, 1992). These include an asparagine in helix I, a pro-line in helix VI, and proline and tyrosine residues in helix VII. Two cysteine residues are conserved; one in helix III and the other in extracellular loop 2 (e_2). These are linked by a disulfide bond and are required for the structural stability of the molecule (Karnik et al., 1988). An acidic residue–arginine pair in intra-cellular loop 2 (i_2) is also conserved and is critical for the binding of transducin (Franke et al., 1990).

The residues conserved in the visual pigments are distributed almost evenly among the membrane-spanning helices (10 conserved residues) and the cyto-plasmic loops (eight residues). In the membrane-spanning region there are two conserved asparagine residues, one glycine, two prolines, two tyrosines, and a tryptophan. The functions of these conserved residues are not clear, although they are likely to be involved in creating a binding pocket for retinal. The func-tion is known for the other two residues conserved in the membrane-spanning region: the retinal-binding lysine residue in helix VII and the cysteine near the end of helix III that participates in the aforementioned disulfide bond for almost all G protein-linked receptors. A tryptophan residue in the middle of helix VI is present in all but two pigments where it is replaced by tyrosine. In bovine rhodopsin this tryptophan is involved in binding retinal (Nakayama and Khorana, 1991).

The intracellular loops (i_1, i_2, i_3, and i_4) display similar sequences, although only four residues are strictly conserved: a leucine and an asparagine in i_1 and an arginine and a valine in i_2. The role of i_1 is not known at this time; however, i_2 has been shown to be involved in binding transducin, as have i_3 and i_4 (König et al., 1989). The palmitoylation site(s) that gives rise to the putative loop i_4 appears to be missing from several of the pigments. The extracellular loops (e_1, e_2, and e_3) have three residues that are strictly conserved: a cysteine discussed herein and a glycine nearby in loop e_2 as well as a glycine in e_1. The glycine residues are conformationally less restrictive than other residues and may be required for these relatively short loops to attain a required conformation.

The amino acid sequence data in Table II were used to examine the evolutionary relationships of the visual pigments. Using these data, a phylogenetic tree was generated (Fig. 10) (Swofford, 1991). The invertebrate visual pigments branch from the vertebrates at an early evolutionary stage, as expected. On the vertebrate branch it is clear that the divergence of cone and rod pigments was also an early event. The rhodopsins show the expected phylogenetic relationships. The relationships among the cone pigments is more complex; for example, human red and green cone pigments appear to be more closely related to chicken iodopsin than to the human blue cone pigment. Clearly more cone pigment sequences are needed to best determine their relationship (Goldsmith, 1990; Fryxell and Meyerowitz, 1991; Okano et al., 1992).

F. Natural Mutations in Rhodopsin

A mutation in the rhodopsin gene is the cause of some cases of the blinding eye disease, retinitis pigmentosa (Hargrave and O'Brien, 1991). Although the disease is variable in its time course and severity, rod photoreceptor cells eventually degenerate. To date there have been 32 different mutations in rhodopsin reported to cause this disease (Table III). These mutations can be found in each of the domains of rhodopsin: the cytoplasmic surface, the transmembrane region, and the intradiskal surface. It is not yet clear why these many and varied defects in rhodopsin lead eventually to photoreceptor cell degeneration, but a few speculations can be offered.

It is easy to understand why some mutations in rhodopsin lead to a nonfunctional rhodopsin. Because Lys 296 is essential for binding of retinal, this mutant rhodopsin can never bind retinal and function in photoreception. However, it appears to be constitutively active (D. Oprian, personal communication). Other mutations in the transmembrane helices probably lead to incorrect packing of the

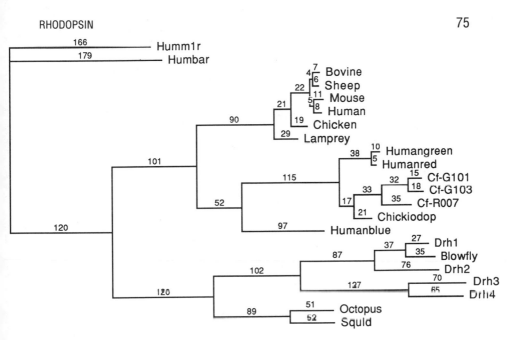

FIG. 10 Phylogenetic tree for visual pigments. This phylogenetic tree of the visual pigments was prepared from the amino acid sequence alignment shown in Table II. Gaps, indicated by dots, were treated as a new amino acid. The large i_3 loops of the adrenergic and muscarinic receptors were removed between the lower case letters at the arrow, and these regions were not used in the phylogenetic analysis. Likewise, the extended carboxyl-terminal regions of squid and octopus rhodopsins were removed from the analysis and are not shown. A general heuristic search using PAUP (Swofford, 1991), with the adrenergic and muscarinic receptors as outgroups, generated the tree shown with the branch distances given (distances are arbitrary numbers in which larger numbers indicate the greatest phylogenetic distances).

helices and also lead to a protein that cannot form a binding site for retinal. Some mutations are probably detrimental to the proper folding of rhodopsin, preventing it from inserting into the membrane and binding retinal. Pro 23-His may be a mutation of this kind. Failure to insert into the membrane of the endoplasmic reticulum and form a stable protein could lead to a buildup of material in the rod cell inner segment that could interfere with normal cell functions. It is far from clear how other mutations, such as Pro 347-Leu, lead to rhodopsins that cannot function properly in the rod cell. Presumably they interfere with functions that we do not currently understand, such as proper targeting of the protein to its appropriate location in the outer segment disk membrane. Rhodopsins carrying many of these individual mutations have been constructed, and their study in various expression systems may help to answer some of these outstanding questions (Sung *et al.*, 1991b). These natural mutations expressed in some types of the disease retinitis pigmentosa serve to challenge us further in our attempts to understand the relationship of structure and function for rhodopsin.

TABLE III

Mutations in Rhodopsin in Autosomal Dominant Retinitis Pigmentosa

Codon	Mutation	Reference
17	Thr → Met	Dryja *et al.*, 1991
		Sung *et al.*, 1991a
		Sheffield *et al.*, 1991
23	Pro → His	Sung *et al.*, 1991a
		Dryja *et al.*, 1991
		Sheffield *et al.*, 1991
23	Pro → Leu	Dryja *et al.*, 1991
45	Phe → Leu	Sung *et al.*, 1991a
51	Gly → Val	Dryja *et al.*, 1991
53	Pro → Arg	Bhattacharya *et al.*, 1991
58	Thr → Arg	Sung *et al.*, 1991a
		Dryja *et al.*, 1991
		Sheffield *et al.*, 1991
		Bhattacharya *et al.*, 1991
68–71	Deletion	Bhattacharya *et al.*, 1991
		Keen *et al.*, 1991
87	Val → Asp	Sung *et al.*, 1991a
89	Gly → Asp	Sung *et al.*, 1991a
		Dryja *et al.*, 1991
106	Gly → Trp	Sung *et al.*, 1991a
125	Leu → Arg	Dryja *et al.*, 1991
135	Arg → Leu	Sung *et al.*, 1991a
135	Arg → Trp	Sung *et al.*, 1991a
167	Cys → Arg	Dryja *et al.*, 1991
171	Pro → Leu	Dryja *et al.*, 1991
178	Tyr → Cys	Sung *et al.*, 1991a
181	Glu → Lys	Dryja *et al.*, 1991
182	Gly → Ser	Sheffield *et al.*, 1991
186	Ser → Pro	Dryja *et al.*, 1991
188	Gly → Arg	Dryja *et al.*, 1991
190	Asp → Gly	Sung *et al.*, 1991a
		Dryja *et al.*, 1991
190	Asp → Asn	Dryja *et al.*, 1991
		Bhattacharya *et al.*, 1991
		Keen *et al.*, 1991
211	His → Pro	Bhattacharya *et al.*, 1991
		Keen *et al.*, 1991
255/256	Ile deletion	Bhattacharya *et al.*, 1991
267	Pro → Leu	Sheffield *et al.*, 1991
296	Lys → Glu	Bhattacharya *et al.*, 1991
		Keen *et al.*, 1991
344	Gln → Stop	Sung *et al.*, 1991a
345	Val → Met	Dryja *et al.*, 1991
347	Pro → Leu	Sung *et al.*, 1991a
		Dryja *et al.*, 1991
		Bhattacharya *et al.*, 1991
347	Pro → Ser	Dryja *et al.*, 1991
347	Pro → Arg	Gal *et al.*, 1991

IV. Light-Dependent Biochemistry of the Rod Cell

Rhodopsin interacts with a number of rod cell proteins as it carries out its functions in the visual cycle. These events are summarized in Figure 11. This section will explore each of these transformations involving rhodopsin in the visual cycle.

A. Rhodopsin Activates Transducin

Absorption of the energy of light by rhodopsin's chromophore, 11-*cis*-retinal, is the first event in vision. It initiates a series of biochemical changes that culminates in neural signaling. After absorption of a photon isomerization of retinal to the all-*trans* isomer takes place in 200 femtoseconds, making it one of the fastest photochemical reactions ever studied (Schoenlein *et al.*, 1991). A series of additional changes occurs in the interaction of retinal and the protein opsin, and these can be followed by the sequential formation of photolysis intermediates that have different defined spectral properties (Fig. 12). Changes in the conformation of retinal are rapid at room temperature, and it is these changes that are primarily

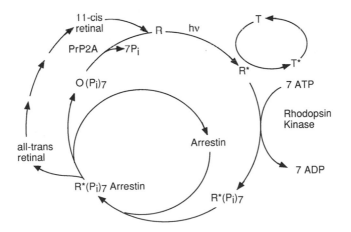

FIG. 11   The rhodopsin cycle. Rhodopsin (R) is activated (to R*) upon absorbing the energy of a photon (hυ). R* activates transducin (T). R* becomes phosphorylated at as many as seven sites [R*(P$_i$)$_7$] by rhodopsin kinase, which allows it to become inactivated by binding arrestin. The complex, [R*(P$_i$)$_7$·arrestin], loses all-*trans*-retinal and arrestin, forming phosphorylated opsin [O(P$_i$)$_7$], which becomes dephosphorylated by protein phosphatase 2A (PrP2A). All-*trans*-retinal is reduced, esterified, and reisomerized through a series of steps to reform 11-*cis*-retinal. Opsin subsequently rebinds 11-*cis*-retinal, reforming rhodopsin.

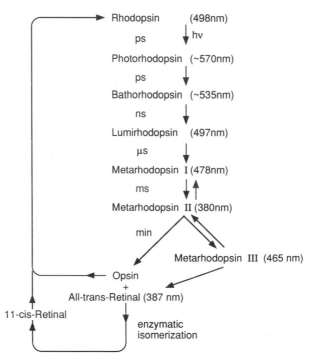

FIG. 12 Rhodopsin photolysis intermediates and the visual cycle. Following absorption of a photon by 11-*cis*-retinal, rhodopsin is converted via a series of spectroscopically identifiable intermediates to all-*trans* retinal and opsin. The wavelength of maximal absorption of each intermediate and the order of magnitude of its lifetime at 20°C are shown. Advances include determination of the 200 femtosecond time for conversion of rhodopsin to photorhodopsin (Schoenlein *et al.*, 1991), identification of an intermediate between bathorhodopsin and lumirhodopsin (Hug *et al.*, 1990) (not shown), and elucidation of many of the details of the series of enzymatic steps to convert all-*trans*-retinal back to 11-*cis*-retinal (Rando, 1991).

involved in the formation of the intermediates, bathorhodopsin, lumirhodopsin, and metarhodopsin I (Meta I). Formation of metarhodopsin II (Meta II) occurs in milliseconds and is accompanied by changes in conformation of the protein, opsin. The protein opens up as Meta II is formed (Lamola *et al.*, 1974). A tryptophan and several tyrosines assume a more hydrophilic environment (Chabre, 1985). Other amino acids change their environment as indicated by both an increase in the number of cysteines available for chemical modification (Chen and Hubbell, 1978) and by proton uptake and release (Hofmann, 1986). The previously inaccessible protonated Schiff base becomes accessible to water and other small molecules (Bownds, 1967; Cooper *et al.*, 1987). This enables retinal to be chemically reduced to its lysine binding site when water-soluble reducing agents gain access to the site at the Meta II stage of photolysis.

The cytoplasmic surface of Meta II is altered in conformation. This is detectable by the increased accessibility of the carboxyl terminus to a macromolecular probe (Kühn *et al.*, 1982). This conformational change in the protein presumably comes about as a result of movement of rhodopsin's transmembrane helices in response to a change in the geometry of the retinal bound deeply in its interior. It is by this mechanism that the information that rhodopsin has received a photon becomes communicated to the transduction machinery of the outer segment.

When rhodopsin in the rod disk membrane is bleached, several rod outer segment proteins bind strongly to the membranes (Kühn, 1978, 1984). They bind to rhodopsin in its Meta II state. Presumably the Meta II state of rhodopsin provides a cytoplasmic surface array that is sufficiently different from that of rhodopsin and its earlier photolysis intermediates to offer binding sites that were previously unavailable for these proteins. The proteins bound in such a light-dependent manner include the transducin complex, rhodopsin kinase, a 48-kDa protein (arrestin), and possibly other proteins (Wieland *et al.*, 1990b; Plouet *et al.*, 1988). The biochemical events that result from the interaction of these proteins with the Meta II rhodopsin surface are responsible for initiating the visual transduction process and for terminating it.

Transducin is bound to the surface of Meta II in preparation for being activated. The second, third, and fourth cytoplasmic loops of Meta II cooperate in the binding of transducin (König *et al.*, 1989). The importance of the third loop was first suspected due to the loss of reversible light-dependent binding when the third loop was partially excised by proteolysis (Kühn and Hargrave, 1981). Confirmation of its role has been assured by the results of site-specific mutagenesis (Franke *et al.*, 1988). Peptide competition experiments have yielded clear evidence for the participation of the three loops in transducin binding and for the lack of participation of the first loop and the carboxyl terminus (König *et al.*, 1989). Mutagenesis experiments have begun to dissect some of the individual parts of rhodopsin that are most critical for interaction with transducin. The Glu-Arg pair at the top of helix III in loop i_2 has been found to be invariant not only in rhodopsin but in all other members of the 7-helix receptor family. Reversal of this pair led to a failure of the mutant rhodopsin to bind transducin, suggesting that this invariant structural feature of the receptors was critically involved in G protein binding (Franke *et al.*, 1990).

Binding of transducin to Meta II normally leads to transducin activation, but it has been possible to construct rhodopsin mutants that will bind transducin but not activate it (Franke *et al.*, 1990). One such mutant of rhodopsin was constructed in which loop i_2 was replaced with an unrelated sequence and another mutant in which a portion of loop i_3 was deleted. In both cases the mutant rhodopsins combined with retinal to form visual pigments with normal absorption spectra. On photolysis they formed Meta II, which was stabilized as expected by binding of transducin. But when GTP was added, transducin continued to be bound; it was not released and activated. Thus, at least two sites on the rhodopsin surface [i.e., loops i_2 and i_3 are implicated in activation of transducin (Franke *et al.*, 1990)].

B. Transducin

Transducin is a member of the heterotrimeric GTP-binding protein family (Kaziro *et al.*, 1991; Birnbaumer, 1990a). These G proteins act as second messengers to transduce signals from activated members of the 7-helix receptor family. Transducin acts in vertebrate rod cells to link the photolysis of rhodopsin to the activation of the enzyme cGMP-PDE.

1. Composition and Preparation

Each member of this G protein class consists of an α-, β-, γ-subunit (Table I). The α subunits contain the site for binding guanine nucleotides (GDP and GTP) and for interacting with a receptor. The β- and γ-subunits function as an inseparable $\beta\gamma$-dimer to assist the α-subunit in binding to the receptor. Transducin subunits are characterized by their behavior on SDS-PAGE as T_α, $M_r \sim 39$ kDa; T_β, $M_r \sim 37$ kDa; and T_γ, $M_r \sim 6$–8 kDa (Table I). The functionally intact transducin trimer may be prepared by an affinity-binding procedure (Kühn, 1980). Transducin binds and remains bound to membranes containing photoactivated rhodopsin under conditions of low ionic strength, which allows other peripheral proteins to be washed free. GTP then elutes transducin from the membranes. Final purification may be achieved following chromatography on hexyl-agarose (Fung *et al.*, 1981). T_α and $T_{\beta\gamma}$ may be separated from one another by further chromatographic procedures (Fung, 1983).

2. Functional Properties

The transducin cycle is shown in Figure 13. The properties and functions of transducin have been the subject of extensive study (Fung, 1987; Chabre and Deterre, 1989). Transducin is present in approximately 1 copy for every 10 rhodopsins and is loosely associated with the rod outer segment disk membrane. This membrane association must accelerate its interaction with photoactivated rhodopsin compared with three-dimensional diffusion, thus increasing its efficiency of activation (Liebman *et al.*, 1987). Nevertheless, transducin can be eluted from the membrane in its inactive GDP form, whereas under these conditions other G proteins remain membrane-associated. This may indicate that it is less hydrophobic than other G proteins. Other G_α-subunits are myristoylated, which contributes to their membrane attachment. However, the $\beta\gamma$-dimer has

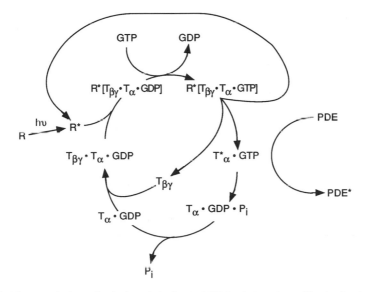

FIG. 13   The transducin cycle. Activated rhodopsin (R*) binds transducin ($T_{\alpha\beta\gamma}$), allowing T_α to bind GTP and dissociate from $T_{\beta\gamma}$. Activated T_α(T_α*·GTP) activates cGMP-PDE. Transducin is deactivated by hydrolyzing its bound GTP. Upon releasing P_i, T_α·GDP rebinds $T_{\beta\gamma}$.

been proposed to be responsible for membrane attachment (Birnbaumer, 1990a). Transducin's γ-subunit is farnesylated (Fukada *et al.*, 1990b; Lai *et al.*, 1990), which increases its hydrophobicity but does not render it insoluble.

3. Activated Transducin Has a Subsecond Lifetime

When transducin is activated with a nonhydrolyzable GTP analog, it continues to activate PDE. Only upon hydrolysis of GTP does it become inactivated. The response of a retinal rod to a dim light flash lasts less than a second, but the rate of turnoff of transducin would appear to be tens of seconds. Transducin GTPase rates measured by several techniques show that only a few GTPs are hydrolyzed per minute per transducin. However, when the hydrolysis of GTP is measured by microcalorimetric methods, the active lifetime of transducin is found to be <1 second at room temperature (Vuong and Chabre, 1991; Ting and Ho, 1991). This appears to be because there is a refractory period after the hydrolysis reaction in which GDP and P_i remain associated in the nucleotide binding site prior to dissociation. Measurement of the rate of release of P_i gives a low estimate of the rate of GTPase activity. Release of P_i returns T_α to its GDP bound form and allows it to reassociate with $T_{\beta\gamma}$.

C. cGMP-Phosphodiesterase (PDE)

1. Properties of the Enzyme

The rod outer segment enzyme that hydrolyzes cGMP to 5'-GMP is cGMP-PDE. Under physiological conditions the enzyme remains peripherally associated with rod cell membranes, but it can be removed by extraction at low ionic strength. It has been isolated and studied from frog (Miki *et al.*, 1975) and from cattle (Baehr *et al.*, 1979). The bovine enzyme has a molecular weight of 170,000 and is composed of three types of subunits: α, 88 kDa; β, 84 kDa; and γ, 13 kDa (Baehr *et al.*, 1979). There are two γ subunits present for each αβ-dimer (Deterre *et al.*, 1988; Fung *et al.*, 1990). The catalytic activity is associated with αβ- and the γ-subunit acts as an inhibitor (Hurley and Stryer, 1982). The native PDE has a low basal activity that is activated by 50- to 100-fold when the inhibitory γ-subunit is removed by treatment with trypsin (Baehr *et al.*, 1979; Hurley and Stryer, 1982). Under physiological conditions the enzyme becomes activated by the transducin α-subunit. The activated form of transducin, T_α-GTP, binds a PDE γ-subunit, preventing it from continuing to inactivate PDE αβ.

2. A Defect in PDE Leads to Retinal Degeneration

Mice homozygous for the *rd* mutation suffer retinal degeneration, showing complete loss of rod cells by day 20. The retinas show a high concentration of cGMP and a decreased level of a cGMP-PDE, suggesting a defect in the enzyme (Farber and Lolley, 1974). The defect has been traced to a mutation in the gene for the β-subunit of the enzyme (Bowes *et al.*, 1990; Pittler and Baehr, 1991).

D. Regulation of Rhodopsin Function: The Phosphorylation of Rhodopsin

1. Many Proteins Become Reversibly Phosphorylated

Reversible phosphorylation of proteins is a ubiquitous mechanism for metabolic regulation. A kinase phosphorylates amino acids in a protein sequence, giving the protein altered physical and/or enzymatic properties. A phosphatase reverses the reaction, dephosphorylating the amino acids and restoring the protein to its original state. The overwhelming majority of protein phosphorylations occur on specific serines and threonines. A large number of different kinases have been identified, but there are only three protein kinases that have been identified to date in the rod cell. These are rhodopsin kinase, cGMP-stimulated protein kinase (protein kinase A), and protein kinase C (Lee *et al.*, 1982b; Binder *et al.*,

1989). Although proteins other than rhodopsin become phosphorylated in the rod cell, we will confine our attention to the light-dependent phosphorylation of rhodopsin.

2. Light Triggers the Phosphorylation of Rhodopsin

The light-dependent phosphorylation of rhodopsin was discovered independently by three laboratories (Kühn and Dreyer, 1972; Bownds et al., 1972; Frank et al., 1973). ATP is the phosphate donor, and selected serines and threonines in rhodopsin become phosphorylated (Kühn and Dreyer, 1972; Bownds et al., 1972). The reaction has been demonstrated to occur in vivo, indicating that it is a physiologically significant reaction (Kühn, 1974). Rhodopsin phosphorylation has been studied in whole retina (Kühn and Bader, 1976), but most studies have employed isolated rod outer segments. The earliest reports indicated that approximately 1 mol phosphate was incorporated per mole of rhodopsin present (Bownds et al., 1972; Frank et al., 1973; Kühn et al., 1973) As conditions for the isolation of rod outer segments and for conducting the reaction were optimized, higher stoichiometries were achieved. When rhodopsin was subjected to chromatography or isoelectric focusing after phosphorylation, it was shown to be composed of a mixture of unphosphorylated rhodospin and rhodospins phosphorylated to varying extents (Kühn and McDowell, 1977; Shichi and Somers, 1978; Wilden and Kühn, 1982; Aton et al., 1984; Arshavsky et al., 1986). The highest level of phosphorylation of individual rhodopsin molecules has been reported to be 9 mol phosphate per mole of rhodopsin (Wilden and Kühn, 1982).

3. Phosphorylation of Opsins Is Universal

Opsins have been found to become phosphorylated in all species thus far investigated. Among the vertebrates this includes rod visual pigments from cattle (Kühn and Dreyer, 1972; Frank et al., 1973), sheep (Thompson and Findlay, 1984), frog (Bownds et al., 1972), dog (Schmidt et al., 1986), and rat (Shuster and Farber, 1984). Light-exposed cone visual pigments become phosphorylated as shown in a study of the all-cone retina of the lizard, Anolis carolinensis (Walter et al., 1986). Phosphorylation of cone cell visual pigments can also be demonstrated to occur normally in a reconstituted system using a rod cell kinase. Under these conditions both the chicken rod pigment and red-sensitive cone pigment are excellent substrates for bovine rhodopsin kinase (Fukada et al., 1990a).

Invertebrate rhodopsins become phosphorylated after light exposure. This was first shown for octopus (Paulsen and Hoppe, 1978) and has since been observed for crayfish and for the flies Drosophila (Matsumoto and Pak, 1984) and Calliphora (Paulsen and Bentrop, 1984). The universality of the phenomenon of light-stimulated photoreceptor pigment phosphorylation suggests that it is involved in an important function over a wide range of phylogeny.

4. The Phosphorylation of Rhodopsin Is Specific

One of the hallmarks of a biologically important phosphorylation reaction is its specificity. The phosphorylation of rhodopsin is specific. More than 85% of the phosphate is incorporated into amino acids in rhodopsin's carboxyl-terminal 39-amino acid cyanogen bromide (CNBr) peptide, and the remainder is located in a high molecular weight CNBr peptide (Hargrave et al., 1980). The major sites of phosphate incorporation of sheep rhodopsin have been identified as serines 334, 338, 343 and threonines 335 and 336 (Thompson and Findlay, 1984). Presumably the adjacent serines and threonines also become phosphorylated in this hydroxy-amino acid-rich block in the C terminal of various rhodopsins when the visual pigments are highly phosphorylated. An additional site of phosphorylation has been identified as Ser^{240} in the third cytoplasmic loop (McDowell et al., 1985; Newton and Williams, 1991).

5. Photoactivated Rhodopsin (R*) Is Phosphorylated at High Light Exposure

It has been generally observed that the extent of phosphorylation of rhodopsin is proportional to the fraction of rhodopsin bleached. For light exposures that bleach a large fraction of the rhodopsin present an average of 7 mol phosphate can be incorporated per mole of rhodopsin, with 9 mol incorporated maximally into any light-exposed rhodopsin molecule (Wilden and Kühn, 1982). Because rhodopsin goes through a number of spectrally identifiable intermediates after light exposure, it has been of interest to determine when in the bleaching pathway rhodopsin can become phosphorylated. It has been reported that molecular events early in the bleaching sequence set the stage for the conformational change that will allow rhodopsin to become phosphorylated (Paulsen and Bentrop, 1983). However, it appears that rhodopsin cannot be phosphorylated until Meta II is formed (Yamamoto and Shichi, 1983). Additionally, it is also necessary for the protonated Schiff base, which becomes exposed to solvent at the Meta II stage, to become deprotonated (Seckler and Rando, 1989). Presumably upon formation of Meta II and deprotonation of its Schiff base, there is a change in the conformation of rhodopsin's cytoplasmic surface that makes it capable of interacting with rhodopsin kinase. This appears reasonable inasmuch as the C-terminal region, which contains sites of phosphorylation for rhodopsin kinase, becomes more accessible in Meta II (Kühn et al., 1982). In addition, it has been shown that rhodopsin kinase binds preferentially to light-exposed rhodopsin (Kühn, 1978).

6. A High-Gain Phosphorylation Appears to Lead to Phosphorylation of Unbleached Rhodopsin

Most of the studies of rhodopsin phosphorylation, including those discussed herein, have been concerned with the reaction that occurs when the majority of the rhodopsin molecules become photoexcited. But at low levels of bleaching (<1%) Bownds *et al.* (1972) found that as many as 50 mol phosphate were incorporated per mole of rhodopsin bleached. Such a high-gain reaction would require that unbleached rhodopsin molecules also become phosphorylated. There are additional studies that would suggest that this is the case (Miller *et al.*, 1977; Sitaramayya and Liebman, 1983).

In studies with electropermeabilized frog ROS, when fewer than 1000 rhodopsins are photoexcited out of the total of 3×10^9 rhodopsins present per rod, as many as 1400 phosphates are incorporated into the rhodopsin pool for every photoexcited rhodopsin (Binder *et al.*, 1990). This number drops sharply as the percent of photoexcitation increases, eventually approaching the lower-gain reaction regularly observed under continuous bright white light illumination. The integrity of the high-gain reaction requires maintenance of rod structure; only the lower-gain phosphorylation is observed if rods are sheared to produce vesicles (Binder *et al.*, 1990). Some factor concerned with the compartmentalization of reactants in essentially intact rods must be important for production of the high-gain phosphorylation under dim light intensity. The recent finding that rhodopsin kinase is activated by photoexcited rhodopsin (Fowles *et al.*, 1988; Palczewski *et al.*, 1991b) may help to explain the phenomenon of high-gain phosphorylation.

7. What Kinase(s) Phosphorylate Rhodopsin?

Most studies to date have assumed that rhodopsin kinase is the principle kinase and perhaps the only rod cell kinase that phosphorylates rhodopsin. Rhodopsin kinase certainly does phosphorylate rhodopsin, thereby playing a most important role in rod cell visual transduction. Its properties will be discussed in detail (Section IV,D,8). Most attempts to demonstrate phosphorylation of rhodopsin by other kinases have yielded negative results (Frank and Bensinger, 1974; Binder *et al.*, 1989). It has been demonstrated that photolyzed rhodopsin can be phosphorylated in rod cells in a Ca^{2+}-dependent reaction catalyzed by protein kinase C (Kelleher and Johnson, 1986; Kapoor *et al.*, 1987; Newton and Williams, 1991). It is even possible, by properly activating protein kinase C, to demonstrate that it can phosphorylate unbleached rhodopsin (Kelleher and Johnson, 1986; Newton and Williams, 1991). Most interesting, however, is the recent observation that protein kinase C activation causes an increased level of phosphorylation of bleached rhodopsin at low light intensities (Newton and

Williams, 1991). This suggests that it could be protein kinase C that is involved in the high-gain phosphorylation observed by Binder *et al.* (1990) because the high levels of phosphorylation observed would require that some molecules of unbleached rhodopsin become phosphorylated.

8. Rhodopsin Kinase

In contrast to some of the other protein kinases, rhodopsin kinase is not activated by Ca^{2+}, lipids, or by cyclic nucleotides. It is Mg^{2+}-activated and uses Mg^{2+}-ATP as its preferred nucleotide substrate (Palczewski *et al.*, 1988b). Although GTP can substitute, it has a K_m ~10^3 greater (Shichi and Somers, 1978; Palczewski *et al.*, 1988a). The high selectivity that rhodopsin kinase shows for its nucleotide substrate not only requires the adenosine ring but requires it to be present in a specific conformation (Palczewski *et al.*, 1990).

Rhodopsin kinase shows a similarly high specificity for its protein substrate. It acts on the photoactivated form of visual pigments. It binds to opsin-containing light-exposed rod cell membranes (Kühn, 1984), a characteristic that has been exploited in the purification of the enzyme (Sitaramayya, 1986; Palczewski *et al.*, 1988a). Rhodopsin kinase has been the subject of several purification efforts and its properties have been studied by a number of laboratories (Kühn, 1984; Shichi and Somers, 1978; Sitaramayya, 1986; Palczewski *et al.*, 1988a; Kelleher and Johnson, 1990). Some of this literature has been the subject of review (Hargrave *et al.*, 1988; Palczewski and Benovic, 1991). Rhodopsin kinase from cattle will successfully phosphorylate not only the opsin from cattle but also from chicken (rods and red cones) (Fukada *et al.*, 1990a), rabbit, pig, and alligator (Palczewski *et al.*, 1988b). To a small extent it will also phosphorylate a distantly related nonvisual receptor, the β-adrenergic receptor (Benovic *et al.*, 1986).

Like many other protein kinases, it is possible to explore the substrate specificity of rhodopsin kinase by using synthetic peptides as substrates (Palczewski *et al.*, 1988b; Palczewski *et al.*, 1989). Unlike other kinases, rhodopsin kinase has a K_m for its peptide substrates that is >10^3 that for its preferred protein substrate. This relatively poor recognition of peptides appears to be due to its mechanism of action, which involves light-dependent binding to photolyzed rhodopsin prior to phosphorylating it. Although rhodopsin kinase generally requires an acidic amino acid near the serine or threonine that it will phosphorylate, it is less specific and lacks the strict consensus sequence that is required for other protein kinases (Palczewski *et al.*, 1989; Kennelly and Krebs, 1991).

Rhodopsin kinase is a member of a new class of receptor kinases that act on G protein-coupled receptors (Palczewski and Benovic, 1991). In addition to rhodopsin kinase, structurally and functionally homologous proteins include the two isozymes of the β-adrenergic receptor kinase, β-ARK (Benovic *et al.*, 1989, 1991). The amino acid sequence of rhodopsin kinase has been determined

(Lorenz *et al.*, 1991). It is a 62.9-kDa protein, the catalytic domain of 270 amino acids of which is flanked by a 185-amino acid N-terminal and a 106-amino acid C-terminal domain. It is the domains flanking the catalytic domain of protein kinases that appear to provide the binding specificity determinants for their particular substrates. There are several features of the rhodopsin kinase sequence that suggest important functional properties of the enzyme. Near the N terminus is a consensus sequence for myristoylation, and at the C terminus is a consensus sequence for isoprenylation (Lorenz *et al.*, 1991). Such lipophilic modifications, if found to occur *in vivo*, would enhance the membrane-binding capabilities of the enzyme. Rhodopsin kinase has always been difficult to completely extract from rod cell membranes, suggesting that such modifications might be used as cellular control mechanisms to promote interaction with its membrane-bound substrate.

Sites for phosphorylation by several types of protein kinases can be found within the rhodopsin kinase sequence (Lorenz *et al.*, 1991). Although rhodopsin kinase has not currently been demonstrated to be a substrate for other protein kinases, the activity of many other enzymes (including kinases) has been shown to be regulated by reversible phosphorylation. Rhodopsin kinase, like most other kinases, phosphorylates itself (Lee *et al.*, 1982a; Kelleher and Johnson, 1990; Buczylko *et al.*, 1991). It has been proposed that this autophosphorylation of rhodopsin kinase has regulatory significance (Buczylko *et al.*, 1991). The autophosphorylated enzyme shows a reduced affinity for phosphorylated rhodopsin that could promote its dissociation from it, thereby facilitating the binding of arrestin.

Rhodopsin kinase phosphorylates serine and threonine residues within the carboxyl-terminal sequence of rhodopsin. However, the enzyme can bind to photolyzed rhodopsin that lacks its carboxyl-terminal sequence (Palczewski *et al.*, 1991), indicating that this is not needed for purposes of binding. The third cytoplasmic loop connecting helices V–VI is essential for kinase binding, and other surface loops may also contribute (Palczewski *et al.*, 1991b). When rhodopsin kinase is bound to the surface of photolyzed rhodopsin, its ability to phosphorylate synthetic peptide substrates is enhanced (Fowles *et al.*, 1988; Palczewski *et al.*, 1991b). This shows that binding of the kinase to rhodopsin's surface serves to activate it.

E. Function of Arrestin in Visual Transduction

1. Arrestin Binds to Photoactivated Phosphorylated Rhodopsin

To shut off the phototransduction cascade in rod cells, R* must be inactivated. Phosophorylation of R* is required (Section IV,D) and is assisted by a regulatory protein called 48K protein, or arrestin. Arrestin binds strongly to phosphorylated R*, thereby inhibiting its ability to continue to activate transducin (Kühn et al., 1984). Knowledge about arrestin comes from a variety of sources. Kühn (1978) was the first to observe that an abundant soluble rod cell protein of ~48,000 molecular weight (by SDS-PAGE) that he called 48K protein, bound to light-exposed rod outer segment membranes. This binding was often not reproducible, and it eventually became clear that binding occurred only to the phosphorylated form of R* (Kühn et al., 1984). An affinity purification of arrestin has been developed based upon this light-dependent binding property (Wilden et al., 1986).

2. 48K Protein Is S Antigen

In an entirely different line of research it was shown that a soluble retinal protein called S antigen, induced experimental autoimmune uveitis (Wacker et al., 1977). This disease had characteristics similar to clinically observed inflammatory autoimmune diseases of the retina and uvea (Nussenblatt and Palestine, 1989). When the properties of S antigen and of 48K protein were compared, they were found to be identical in their biochemical, immunological, functional, and pathological properties (Pfister et al., 1985).

3. Physical and Chemical Properties of Arrestin

Arrestin is found widely in nature and its sequence has been determined from human, mouse, and cattle (Shinohara et al., 1989) as well as *Drosophila* (Smith et al., 1990; Yamada et al., 1990). Bovine arrestin is 404 amino acids in length (Shinohara et al., 1987) and with an N-terminal acetyl group has a molecular weight by mass spectrometry of 45,317 (Buczylko and Palczewski, in press). The purified protein is composed of different isoelectric species (Weyand and Kühn, 1990), but it does not appear that this heterogeneity is due to posttranslational modifications, such as glycosylation or phosphorylation. Arrestin readily aggregates at low ionic strength and is susceptible to limited proteolysis on storage.

The fluorescence spectrum of arrestin is dominated by contribution from tyrosine, suggesting an unusual environment for its only tryptophan residue (Kotake et al., 1991). Arrestin has essentially no α-helical structure and is composed nearly completely of β-strands and turns (Palczewski et al., 1992a). Analysis of its primary structure and the results of limited proteolysis suggest that the protein has a two-domain folded structure in which the two domains are bridged by a highly exposed connecting region (Palczewski et al., 1991a). Arrestin appears to be a flexible molecule as indicated by the gradual increase in reactivity and accessibility of its three sulfhydryl groups with increasing concentrations of denaturant (Palczewski et al., 1992a).

There have been reports that arrestin binds Ca^{2+} (Huppertz *et al.*, 1990), binds nucleotides (Glitscher and Rüppel, 1989), and has ATPase activity (Sitaramayya and Hakki, 1990; Glitscher and Rüppel, 1991). However, when highly purified arrestin was studied by several sensitive methods, no evidence for Ca^{2+} or nucleotide binding could be observed (Palczewski and Hargrave, 1991). Arrestin has a high affinity for heparin, other polyanions, and inositol hexaphosphate, and these compete for its binding to phosphorylated rhodopsin (Palczewski *et al.*, 1991c).

4. Arrestin Turns Off or Accelerates the Turn-off of PDE Activation

Binding of arrestin to rhodopsin requires that rhodopsin be phosphorylated and that it be in the Meta II conformation (Schleicher *et al.*, 1989). Such binding competitively inhibits binding and activation of transducin and hence the activation of PDE. Wilden *et al.* (1986) found that phosphorylation of rhodopsin by itself was not sufficient to turn off activation of transducin. However, Miller *et al.* (1986), working with purified phosphorylated species of rhodopsin, found that increasing levels of phosphorylation of rhodopsin produced a graded decrease in the binding affinity of transducin for R*. Greater effect was observed at low levels of rhodopsin bleaching than at high levels. Additional studies suggested that arrestin is only required at low levels of R* phosphorylation and not at high levels (Bennett and Sitaramayya, 1988). This view is further supported by electrophysiological experiments on functionally intact rod cells in the presence of inhibitors of rhodopsin kinase activity (sangivamycin) and arrestin binding (inositol hexaphosphate) (Palczewski *et al.*, 1992b). These studies lead to a view of arrestin as a protein that accelerates the turnoff of PDE activation by binding to partially phosphorylated rhodopsin.

5. Completion of the Rhodopsin Cycle

The complex of arrestin with phosphorylated rhodopsin decays rather slowly (Fig. 11). This can be monitored by the inhibition of the ability of protein phosphatase 2A to dephosphorylate phosphorylated rhodopsin (Palczewski *et al.*, 1989). Experiments in progress point to dissociation of all-*trans*-retinal from the complex as the key step that facilitates the release of arrestin (Hofmann *et al.*, 1992). Dissociation of arrestin allows protein phosphatase 2A to complete the dephosphorylation process. Dissociation of all-*trans*-retinal from rhodopsin's retinal binding site makes possible the regeneration of rhodopsin by rebinding of 11-*cis*-retinal and the reconstruction of light sensitivity to rhodopsin.

V. Rhodopsin Serves as a Model for G Protein-linked Receptors

Investigators who want to understand the function of signal transducers that function via G proteins would be well advised to learn from vertebrate photoreceptors. Through the study of this specialized system one can learn general features that help in the understanding of other analogous signaling systems (Hargrave and McDowell, 1992). Structural and functional analogies are clear between rhodopsin and receptors that bind neurotransmitters, peptides, odorants, and numerous other small molecules (Dohlman *et al.*, 1987). Rhodopsin is available in greater quantities than other receptors of its class. For this reason as well as its intrinsic importance in vision, it has been the object of many detailed investigations. It is the only member of its class to have its sequence determined by protein sequence analysis and the first to have its primary sequence determined by nucleic acid sequencing. It will probably be the first to yield three-dimensional structural information by x-ray crystallography. Posttranslational modifications (glycosylation, phosphorylation, palmitoylation) have been assigned for rhodopsin, which cannot be practically done for the other receptors. Detailed chemical and enzymatic topographical studies of rhodopsin in the disk membrane have been performed that have allowed an excellent topographical model to be developed. This has, in large part, served as the basis for construction of similar topographical models for the other related receptors.

Physical methods of study, such as x-ray and neutron diffraction, have been possible with this receptor because of availability of the large quantities of membrane in which rhodopsin is nearly the sole protein component. Study of the changes in conformation that occur with the receptor during the course of its functional cycle is uniquely possible with rhodopsin. Proteins that function in association with rhodopsin are also enriched in quantity in the rod cell. There are analogous proteins in other systems. Rhodopsin kinase is the first receptor-specific kinase of its class, and it has its β-adrenergic receptor kinase counterpart (Palczewski and Benovic, 1991). Arrestin's existence and function was discovered in the rod cell, and it has its β-arrestin counterpart (Lohse *et al.*, 1990). Transducin has been a well-studied G protein that has helped greatly in understanding G protein structure and function. It was the first G protein purified, the first shown to be an αβγ-trimer, and the first shown to dissociate upon activation and have its α-subunit be an information carrier (Birnbaumer, 1990b). Its subunits were the first to be cloned, and T_α will undoubtedly be the first to have its crystal structure determined. Thus, the rod cell provides rich sources of both the receptor, rhodopsin, and its auxiliary proteins for the study of G protein-linked receptor function.

Acknowledgments

Supported in part by research grants EY06225 and EY06226 from the National Eye Institute of the National Institutes of Health, and an unrestricted departmental award from Research to Prevent Blindness, Inc. Paul A. Hargrave is Francis N. Bullard Professor of Ophthalmology.

References

Adamus, G., Zam, Z. S., Arendt, A., Palczewski, K., McDowell, J. H., and Hargrave, P. A. (1991). *Vision Res.* **31**, 17–31.
Aguirre, G., and O'Brien, P. (1986). *Invest. Ophthalmol. Vis. Sci.* **27**, 635–655.
Applebury, M. A. (1991). *Curr. Opin. Neurobiol.* **1**, 263–269.
Arshavsky, V. Y., Antoch, M. P., and Philippov, P. P. (1986). *Biol. Membr.* **3**, 1197–1203.
Aton, B. R., Litman, B. J., and Jackson, M. L. (1984). *Biochemistry* **23**, 1737–1741.
Audigier, Y., Friedlander, M., and Blobel, G. (1987). *Proc. Natl. Acad. Sci. U.S.A.* **84**, 5783–5787.
Baehr, W., Devlin, M. J., and Applebury, M. L. (1979), *J. Biol. Chem.* **254**, 11669–11677.
Baehr, W., Falk, J. D., Bugra, K., Triantafyllos, J. T., and McGinnis, J. F. (1988) *FEBS Lett.* **238**, 253–256.
Barbacid, M. (1987). *Annu. Rev. Biochem.* **56**, 779–827.
Barclay, P. L., and Findlay, J. B. C. (1984). *Biochem. J.* **220**, 75–84.
Bennett, N., and Sitaramayya, A. (1988). *Biochemistry* **27**, 1710–1715.
Benovic, J. L., DeBlasi, A., Stone, W. C., Caron, M. G., and Lefkowitz, R. J. (1989). *Science* **246**, 235–246.
Benovic, J. L., Mayor, F., Jr., Somers, R. L., Caron, M. G., and Lefkowitz, R. J. (1986). *Nature* **321**, 869–872.
Benovic, J. L., Onorato, J. J., Arriza, J. L., Stone, W. C., Lohse, M., Jenkins, N. A., Gilbert, D. J., Copeland, N. G., Caron, M. G., and Lefkowitz, R. J. (1991). *J. Biol. Chem.* **266**, 14939–14946.
Besharse, J. C. (1986). *In* "The Retina: A Model for Cell Biology Studies" (R. Adler and D. Farber, eds.), pp. 297–352. Academic Press, San Diego, California.
Bhattacharya, S. S., Inglehearn, C. F., Keen, J., Lester, D., Bashir, R., Jay, M., and Bird, A. C. (1991). *Invest. Ophthalmol. Vis. Sci. (suppl.)* **32**, 890.
Binder, B. M., Biernbaum, M. S., and Bownds, M. D. (1990). *J. Biol. Chem.* **265**, 15333–15340.
Binder, B. M., Brewer, E., and Bownds, M. C. (1989). *J. Biol. Chem.* **264**, 8857–8864.
Bird, A. C., Flannery, J. G., and Bok, D. (1988). *Invest. Ophthalmol. Vis. Sci.* **29**, 1028–1039.
Birnhaumer, L. (1990a). *Annu. Rev. Pharmacol. Toxicol.* **30**, 675–705.
Birnbaumer, L. (1990b). *FASEB J.* **4**, 3178–3188.
Boesze-Battaglia, K., and Albert, A. D. (1989). *Exp. Eye Res.* **49**, 699–701.
Boesze-Battaglia, K., and Albert, A. D. (1990). *J. Biol. Chem.* **265**, 20727–20730.
Bowes, C., Li, T., Danciger, M., Baxter, L. C., Applebury, M. L., and Farber, D. B. (1990). *Nature* **347**, 677–680.
Bownds, D. (1967). *Nature* **216**, 1178–1181.
Bownds, D., Dawes, J., Miller, J., and Stahlman, M. (1972). *Nature* **237**, 125–127.
Buczylko, J., Gutmann, C., and Palczewski, K. (1991). *Proc. Natl. Acad. Sci. U.S.A.* **88**, 2568–2572.
Buczylko, J., and Palczewski, K. (in press). *In* "Photoreceptor Cells" (P. A. Hargrave, ed.). Academic Press, San Diego, California.
Bunt, A. H. (1978). *Invest. Ophthalmol. Vis. Sci.* **17**, 90–104.
Bunt, A. H., and Klock, I. B. (1980). *Vision Res.* **20**, 739–747.
Chabre, M. (1985). *Annu. Rev. Biophys. Biophys. Chem.* **14**, 331–360.

Chabre, M., and Deterre, P. (1989). *Euro. J. Biochem.* **179**, 255–266.

Chen, Y. S., and Hubbell, W. L. (1978). *Membr. Biochem.* **1**, 107–130.

Clark, V. M., and Hall, M. O. (1982). *Exp. Eye Res.* **34**, 847–859.

Cohen, A. L. (1972). *In* "Physiology of Photoreceptor Organs" (M. G. F. Fuortes, pp. 63–110. Springer-Verlag, Berlin.

Connell, G., Bascom, R., Molday, L., Reid, D., McInnes, R. R., and Molday, R. S. (1991). *Proc. Natl. Acad. Sci. U.S.A.* **88**, 723–726.

Connell, G. J., and Molday, R. S. (1990). *Biochemistry* **29**, 4691–4698.

Cook, N. J., and Kaupp, U. B. (1988). *J. Biol. Chem.* **263**, 11382–11388.

Cook, N. J., Molday, L. L., Reid, D., Kaupp, U. B., and Molday, R. S. (1989). *J. Biol. Chem.* **264**, 6996–6999.

Cooper, A., Dixon, S. F., Nutley, M. A., and Robb, J. L. (1987). *J. Am. Chem. Soc.* **109**, 7254–7263.

Cote, R. H., Nicol, G. D., Burke, S. A., and Bownds, M. D. (1989). *J. Biol. Chem.* **264**, 15384–15391.

Cowman, A. F., Zuker, C. S., and Rubin, G. M. (1986). *Cell* **44**, 705–710.

Deretic, D., and Papermaster, D. S. (1991). *J. Cell Biol.* **113**, 1281–1293.

Deterre, P., Bigay, J., Forquet, F., Robert, M., and Chabre, M. (1988). *Proc. Natl. Acad. Sci. U.S.A.* **85**, 2424–2428.

Devereux, J., Haeberli, P., and Smithies, O. (1984). *Nucleic Acids Res.* **12**, 387–395.

Dizhoor, A. M., Ray, S., Kumar, S., Niemi, G., Spencer, M., Brolley, D., Walsh, K. A., Philipov, P. P., Hurley, J. B., and Stryer, L. (1991). *Science* **251**, 915–918.

Dohlman, H. G., Caron, M. G., and Lefkowitz, R. J. (1987). *Biochemistry* **26**, 2657–2664.

Dryja, T. P., Hahn, L. B., Cowley, G. S., McGee, T. L., and Berson, E. L. (1991). *Proc. Natl. Acad. Sci. U.S.A.* **88**, 9370–9374.

Ebrey, T. G., and Honig, B. (1975). *Q. Rev. Biophys.* **8**, 129–184.

Farber, D. B., and Lolley, R. N. (1974). *Science* **186**, 449–451.

Farrar, G. J., Kenna, P., Redmond, R., Shiels, D., McWilliam, P., Humphries, M. M., Sharp, E. M., Jordan, S., Kumarsingh, S., and Humphries, P. (1991). *Genomics II*, 1170–1171.

Fesenko, E. E., Kolesnikov, S. S., and Lyubarsky, A. L. (1985). *Nature* **313**, 310–313.

Findlay, J. B. C. (1986). *In* "The Molecular Mechanism of Photoreception" (H. Stieve, ed.), pp. 11–30. Springer-Verlag, Berlin.

Findlay, J. B. C., Brett, M., and Pappin, D. J. C. (1981). *Nature* **293**, 314–316.

Findlay, J., and Eliopoulos, E. (1990). *TIPS* **11**, 492–499.

Fleischman, D., and Denisevich, M. (1979). *Biochemistry* **23**, 5060–5066.

Fliesler, S. J., Arakawa, H., O'Brien, P. J., Aquirre, G. D., and Acland, G. M. (1991). *Invest. Ophthalmol. Vis. Sci. (Suppl.)* **32**, 1150.

Fowles, C., Sharma, R., and Akhtar, M. (1988). *FEBS Lett.* **238**, 56–60.

Frank, R. N., and Bensinger, R. E. (1974). *Exp. Eye Res.* **18**, 271–280.

Frank, R. N., Cavanagh, H. D., and Kenyon, K. R. (1973). *J. Biol. Chem.* **248**, 596–609.

Franke, R. H., Sakmar, T. P., Oprian, D. D., and Khorana, H. G. (1988). *J. Biol. Chem.* **263**, 2119–2122.

Franke, R. R., König, B., Sakmar, T. P., Khorana, H. G., and Hofmann, K. P. (1990). *Science* **250**, 123–125.

Friedel, U., Wolbring, G., Wohlfart, P., and Cook, N. J. (1991). *Biochim. Biophys. Acta* **1061**, 247–252.

Fryxell, K. J., and Meyerowitz, E. M. (1991). *J. Mol. Evol.* **33**, 367–378.

Fukada, Y., Kokame, K., Okano, T., Shichida, Y., Yoshizawa, T., McDowell, J. H., Hargrave, P. A. and Palczewski, K. (1990a). *Biochemistry* **29**, 10102–10106.

Fukuda, M. N., Papermaster, D. S., and Hargrave, P. A. (1979). *J. Biol. Chem.* **254**, 8201–8207.

Fukada, Y., Takao, T., Ohguro, H., Yoshizawa, T., Akino, T., and Shimonishi, Y. (1990b). *Nature* **346**, 658–660.

Fung, B. K. K. (1983). *J. Biol. Chem.* **258**, 10495–10502.

Fung, B. K. K. (1987). *Prog. Ret. Res.* **6**, 151–177.

Fung, B. K. K., Hurley, J. B., and Stryer, L. (1981). *Proc. Natl. Acad. Sci. U.S.A.* **78**, 152–156.

Fung, B. K. K., Young, J. H., Yamane, H. K., and Griswold-Prenner, I. (1990). *Biochemistry* **29**, 2657–2664.

Gal, A., Artlich, A., Ludwig, M., Niemeyer, G., Olek, K., Schwinger, I., and Schinzel, A. (1991). *Genomics* **11**, 468–470.

Gaur, V. P., Adamus, G., Arendt, A., Eldred, W., Possin, D. E., McDowell, J. H., Hargrave, P. A., and Sarthy, V. (1988). *Vision Res.* **28**, 765–776.

Glitscher, W., and Rüppel, H. (1991). *FEBS Lett.* **282**, 431–435.

Glitscher, W., and Rüppel, H. (1989). *FEBS Lett.* **256**, 101–105.

Goldsmith, T. H. (1990). *Q. Rev. Biol.* **65**, 281–322.

Hall, M. D., Hoon, M. A., Ryba, N. J. P., Pottinger, J. D. D., Keen, J. N., Sabil, H. R., and Findlay, J. B. C. (1991). *Biochem. J.* **274**, 35–40.

Hall, S. W., and Kühn, H. (1986). *Eur. J. Biochem.* **161**, 551–556.

Hamm, H. E., and Bownds, M. D. (1986). *Biochemistry* **25**, 4512–4523.

Hargrave, P. A. (1977). *Biochim. Biophys. Acta* **492**, 83–94.

Hargrave, P. A. (1982). "Progress in Retinal Research" (N. N. Osbourne and G. J. Chader, eds.), pp. 1–51

Hargrave, P. A., McDowell, J. H., Curtis, D. R., Wang, J. K., Juszczak, E., Fong, S. L., Rao, J. K. M., and Argos, P. (1983). *Biophys. Struct. Mech.* **9**, 235–244.

Hargrave, P. A., Fong, S. L., McDowell, J. H., Mas, M. T., Curtis, D. R., Wang, J. K., Juszczak, E., and Smith, D. P. (1980). *Neurochem. Int.* **1**, 231–244.

Hargrave, P. A., McDowell, J. H., Feldmann, R. J., Atkinson, P. H., Rao, J. K. M., and Argos, P. (1984). *Vision Res.* **24**, 1487–1499.

Hargrave, P. A., McDowell, J. H., Siemiatkowski-Juszczak, E. C., Fong, S. L., Kühn, H., Wang, J. K., Curtis, D. R., Rao, J. K. M., Argos, P., and Feldmann, R. J. (1982). *Vis. Res.* **22**, 1429–1438.

Hargrave, P. A., and McDowell, J. M. (1992). *FASEB J.* **6**, 2323–2331.

Hargrave, P. A., and O'Brien, P. J. (1991). *In* "Retinal Degenerations" (R. E. Anderson, M. LaVail, J. Hollyfield, eds.), CRC Press, Boca Raton, Florida.

Hargrave, P. A., Palczewski, K., Arendt, A., Adamus, G., and McDowell, J. H. (1988). *In* "Molecular Biology of the Eye: Genes, Vision, and Ocular Disease" (J. Piatigorsky, P. Zelenka, and T. Shinohara, eds.), pp. 35–44. Alan R. Liss, New York.

Hayashi, F., and Yamazaki, A. (1991). *Proc. Natl. Acad. Sci. U.S.A.* **88**, 4746–4750.

Henderson, R., and Schertler, G. F. X. (1990). *Philos. Trans. R. Soc. Lond. [Biol.]* **326**, 379–389.

Hisatomi, O., Iwasa, T., Tokunaga, F., and Yasui, A. (1991). *Biochem. Biophys. Res. Commun.* **174**, 1125–1132.

Hofmann, K. P. (1986). *Photochem. Photobiophys.* **13**, 309–327.

Hofmann, K. P., Pulvermüller, A., Buczylko, J., Van Hooser, P. and Palczewski, K. (1992). *J. Biol. Chem.* **267**, 15701–15706.

Hsu, S. C., and Molday, R. S. (1990). *J. Biol. Chem.* **265**, 13308–13313.

Hsu, S. C., and Molday, R. S. (1991). *J. Biol. Chem.* **266**, 21745–21752.

Huber, A., Smith, D. P., Zuker, C. S., and Paulsen, R. (1990). *J. Biol. Chem.* **265**, 17906–17910.

Hug, S. J., Lewis, J. W., Einterz, C. M., Thorgeirsson, T. E., and Kliger, D. S. (1990). *Biochemistry* **29**, 1475–1485.

Hurley, J. B., and Stryer, L. (1982). *J. Biol. Chem.* **257**, 11094–11099.

Huppertz, B., Weyand, I., and Bauer, P. J. (1990). *J. Biol. Chem.* **265**, 9470–9475.

Ishiguro, S., Suzuki, Y., Tamai, M., and Mizuno, K. (1991). *J. Biol. Chem.* **266**, 15520–15524.

Kajiwara, K., Hahn, L. B., Mukai, S., Travis, G. H., Berson, E. L., and Dryja, T. P. (1991). *Nature* **354**, 480–483.

Kamps, K. M. P., DeGrip, W. J., and Deamen, F. J. M. (1982). *Biochim. Biophys. Acta* **687**, 296–302.

Kapoor, C. L., O'Brien, P. J., and Chader, G. J. (1987). *Exp. Eye Res.* **445**, 545–556.

Karnik, S. S., Sakmar, T. P., Chen, H. B., and Khorana, H. G. (1988). *Proc. Natl. Acad. U.S.A.* **85**, 8459–8463.

Kaupp, U. B. (1991). *Trends Neurosci.* **14**, 150–157.

Kaupp, U. B., Hanke, W., Simmoteit, R., and Luhring. H. (1988). *Cold Spring Harbor Symp. Quant. Biol.* **53**, 407–415.

Kaupp, U. B., Niidome, T., Tanabe, T., Terada, S., Bonigk, W., Stuhmer, W., Cook, N. J., Kangawa, K., Matsuo, H., Hirose, T., Miyata, T., and Numa, S. (1989). *Nature* **342**, 762–766.

Kaziro, Y., Itoh, H., Kozasa, T., Nakafuku, M., and Satoh, T. (1991). *Annu. Rev. Biochem.* **60**, 349–400.

Keen, T. J., Inglehearn, C. F., Lester, D. H., Bashir, R., Jay, M., Bird, A. C., Jay, B., and Bhattacharya, S. S. (1991). *Genomics* **11**, 199–205.

Kelleher, D. J., and Johnson, G. L. (1986). *J. Biol. Chem.* **261**, 4749–4757.

Kelleher, D. J., and Johnson, G. L. (1990). *J. Biol. Chem.* **265**, 2632–2639.

Kennelly, P. J., and Krebs, E. G. (1991). *J. Biol. Chem.* **266**, 15555–15558.

Khorana, H. G. (1992). *J. Biol Chem.* **267**, 1–4.

Koch, K. W. (1991). *J. Biol. Chem.* **266**, 8634–8637.

Koch, K. W., and Stryer, L. (1988). *Nature* **334**, 64–66.

König, B., Arendt, A., McDowell, J. H., Kahlert, M., Hargrave, P. A., and Hofmann, K. P. (1989). *Proc. Natl. Acad. Sci. U.S.A.* **86**, 6878–6882.

Kornfeld, R., and Kornfeld, S. (1985). *Annu. Rev. Biochem.* **54**, 631.

Kotake, S., Hey, P., Raghavendra, G. M., and Copeland, R. A. (1991). *Arch. Biochem. Biophys.* **285**, 126–133.

Kühn, H. (1974). *Nature* **250**, 588–590.

Kühn, H. (1978). *Biochemistry* **17**, 4389.

Kühn, H. (1980). *Nature* **283**, 587–590.

Kühn, H. (1984). *Prog. Ret. Res.* **3**, 123–156.

Kühn, H., and Bader, S. (1976). *Biochim. Biophys. Acta* **428**, 13–18.

Kühn, H., Cook, J. H., and Dreyer, W. J. (1973). *Biochemistry* **12**, 2495–2502.

Kühn, H., and Dreyer, W. J. (1972). *FEBS Lett.* **20**, 1–6.

Kühn, H., Hall, S. W., and Wilden, U. (1984). *FEBS. Lett.* **176**, 473–478.

Kühn, H., and Hargrave, P. A. (1981). *Biochemistry* **20**, 2410–2417.

Kühn, H., and McDowell, J. H. (1977). *Biophys. Struct. Mech.* **3**, 199–203.

Kühn, H., Mommertz, O., and Hargrave, P. A. (1982). *Biochim. Biophys. Acta* **679**, 95–100.

Kuwata, O., Imamoto, Y., Okano, T., Kokame, K., Kojima, D., Matsumoto, H., Morodome, A., Fukada, Y., Shichida, Y., Yasuda, K., Shimura, Y., and Yoshizawa, T. (1990). *FEBS Lett.* **272**, 128–132.

Lai, R. K., Perez-Sala, D., Cañada, F. J., and Rando, R. R. (1990). *Proc. Natl. Acad. Sci. U.S.A.* **87**, 7673–7677.

Lambrecht, H. G., and Koch, K. W. (1991). *EMBO J.* **10**, 793–798.

Lamola, A. A., Yamane, T., and Zipp, A. (1974). *Biochemistry* **13**, 738–745.

Lee, R. H., Brown, B. M., and Lolley, R. N. (1982a). *Biochemistry* **21**, 3303–3307.

Lee, R. H., Farber, D. B., and Lolley, R. N. (1982b). *Methods Enzymol* **81**, 496–506.

Lee, R. H., Fowler, A., McGinnis, J. F., Lolley, R. N., and Craft, C. M. (1990). *J. Biol. Chem.* **265**, 15867–15873.

Lee, R. H., Lieberman, B. S., and Lolley, R. S. (1987). *Biochemistry* **26**, 3983–3990.

Liang, C. J., Yamashita, K., Muellenberg, C. G., Shichi, H., and Kobata, A. (1979). *J. Biol. Chem.* **254**, 6414–6418.

Liebman, P., Parker, K. R., and Dratz, E. A. (1987). *Annu. Rev. Physiol.* **49**, 765–791.

Lohse, M. J., Benovic, J. L., Codina, J., Caron, M. G., and Lefkowitz, R. J. (1990). *Science* **248**, 1547–1550.

Lolley, R. N., and Lee, R. H. (1990). *FASEB J.* **4**, 3001–3008.

Lolley, R. N., Lee, R. H., Chase, D. G., and Racz, E. (1986). *Invest. Ophthalmol. Vis. Sci.* **27**, 285–295.

Lorenz, W., Inglese, J., Palczewski, K., Onorato, J. J., Caron, M. G., and Lefkowitz, R. J. (1991). *Proc. Natl. Acad. Sci. U.S.A.* **88**, 8715–8719.

Matsumoto, H., and Pak, W. L. (1984). *Science* **223**, 184–186.

Mayer, A., Siegel, N. R., Schwartz, A. L., and Ciechanover, A. (1989). *Science* **244**, 1480–1483.

McDowell, J. H., Curtis, D. R., Abu Bakar, U., and Hargrave, P. A. (1985). *Invest. Ophthalmol. Vis. Sci. (Suppl.)* **26**, 291.

McDowell, J. H., Ubel, A., Brown, R. A., and Hargrave, P. A. (1986). *Arch. Biochem. Biophys.* **249**, 506–514.

Miki, N., Baraban, J. M., Keirns, J. J. Boyce, J. J., and Bitensky, M. W. (1975). *J. Biol. Chem.* **250**, 6320–6327.

Miller, J. A., Paulsen, R., and Bownds, M. D. (1977). *Biochemistry* **16**, 2633–2639.

Miller, J. L., Fox, D. A., and Litman, B. J. (1986). *Biochemistry* **25**, 4983–4988.

Molday, L. L., Cook, N. J., Kaupp, U. B., and Molday, R. S. (1990). *J. Biol. Chem.* **265**, 18690–18695.

Molday, L. L., and Molday, R. S. (1987b). *Biochim. Biophys. Acta* **897**, 335–340.

Molday, R. S. (1989). *Prog. Ret. Res.* **8**, 173–209.

Molday, R. S., Hicks, D., and Molday, L. L. (1987). *Invest. Ophthalmol. Vis. Sci.* **28**, 50–61.

Molday, R. S., and Molday L. L. (1979). *J. Biol. Chem.* **254**, 4653–4660.

Molday, R. S., and Molday, L. L. (1987a). *J. Cell Biol.* **105**, 2589–2601.

Molday, R. S., Molday, L. L., Dose, A., Clark-Lewis, I., Illing, M., Cook, N. J., Eismann, E., and Kaupp, U. B. (1991). *J. Biol. Chem.* **266**, 21917–21922.

Montell, C., Jones, K., Zuker, C. S., and Rubin, G. (1987). *J. Neurosci.* **7**, 1558–1566.

Nakayama, T. A., and Khorana, H. G. (1990). *J. Biol. Chem.* **265**, 15762–15769.

Nakayama, T. A., and Khorana, H. G. (1991). *J. Biol. Chem.* **266**, 4269–4275.

Nathans, J. (1990). *Biochemistry* **29**, 937–942.

Nathans, J., and Hogness, D. S. (1984). *Proc. Natl. Acad. Sci. U.S.A.* **81**, 4851–4855.

Nathans, J., Thomas, D., and Hogness, D. S. (1986). *Science* **232**, 193–202.

Neitz, M., Neitz, J., and Jacobs, G. H. (1991). *Science* **252**, 971–974.

Newton, A. C., and Williams, D. S. (1991). *J. Biol. Chem.* **266**, 17725–17728.

Nicoll, D., and Applebury, M. L. (1989). *J. Biol. Chem.* **264**, 16207–16213.

Nussenblatt, R. B., and Palestine, A. G. (1989). "Uveitis: fundamentals and clinical practice," pp. 212–232. Year Book, Chicago.

O'Brien, P. J., and Zatz, M. (1984). *J. Biol. Chem.* **259**, 5054–5057.

Okano, T., Kojima, D., Fukada, Y., Shichida, Y., and Yoshizawa, T. (1992). *Proc. Natl. Acad. Sci. USA* **89**, 5932–5936.

Ovchinnikov, Y. A., Abdulaev, N. G., and Bogachuk, A. S. (1988). *FEBS Lett.* **230**, 1–5.

Ovchinnikov, Y. A., Abdulaev, N. G., Feigina, M. Y., Artamonov, I. D., Zolotarev, A. S., Kostina, M. B., Bogachuk, A. S., Miroshnkov, A. I., Martinov, V. I. and Kudelin, A. B. (1982). *Bioorg. Khim.* **8**, 1011–1014.

Ovchinnikov, Y. A., Abdulaev, N. G., Zolotarev, A. S., Artamonov, I. D., Bespalov, I. A., Dergachev, A. E., and Tsuda, M. (1988). *FEBS Lett.* **232**, 69–72.

Palczewski, K., Arendt, A., McDowell, J. H., and Hargrave, P. A. (1989). *Biochemistry* **28**, 8764–8770.

Palczewski, K., and Benovic, J. L. (1991). *TIBS* **16**, 387–391.

Palczewski, K., Buczylko, J., Imami, N. R., McDowell, J. H., and Hargrave, P. A. (1991a). *J. Biol. Chem.* **266**, 15334–15339.

Palczewski, K., Buczylko, J., Kaplan, M. W., Polans, A. S., and Crabb, J. W. (1991b). *J. Biol. Chem.* **266**, 12949–12955.

Palczewski, K., and Hargrave, P. A. (1991). *J. Biol. Chem.* **266**, 4201–4206.

Palczewski, K., Kahn, N., and Hargrave, P. A. (1990). *Biochemistry* **29**, 6276–6282.

Palczewski, K., McDowell, J. H., and Hargrave, P. A. (1988a). *J. Biol. Chem.* **263**, 14067–14073.

Palczewski, K., McDowell, J. H., and Hargrave, P. A. (1988b). *Biochemistry* **27**, 2306–2313.

Palczewski, K., McDowell, J. H., Jakes, S., Ingebritsen, T. S., and Hargrave, P. A. (1989). *J. Biol. Chem.* **264**, 15770–15773.

Palczewski, K., Pulvermüller, A., Buczylko, J., and Hofmann, K. P. (1991c). *J. Biol. Chem.* **266**, 18649–18654.

Palczewski, K., Raizance-Lawrence, J. H., and Johnson, W. C., Jr. (1992a). *Biochemistry* **31**, 3902–3906.

Palczewski, K., Rispoli, G., and Detwiler, P. B. (1992b). *Neuron.* **8**, 117–126.

Papermaster, D. S., and Dreyer, W. J. (1974). *Biochemistry* **13**, 2438–2444.

Papermaster, D. S., and Schneider, B. G. (1982). *In* "Cellular Aspects of the Eye" (D. S. McDevitt, ed.), pp. 475–531. Academic Press, San Diego, California.

Papermaster, D. S., Schneider, B. G., and Besharse, J. C. (1985). *Invest. Ophthalmol. Vis. Sci.* **26**, 1386.

Papermaster, D. S., Schneider, B. G., Zorn, M. A., and Kraehenbuhl, J. P. (1978). *J. Cell. Biol.* **78**, 415–425.

Paulsen, R., and Bentrop, J. (1983). *Nature* **302**, 417–419.

Paulsen, R., and Bentrop, J. (1984). *J. Comp. Physiol.* **155**, 39–45.

Paulsen, R., and Hoppe, I. (1978). *FEBS Lett.* **96**, 55–58.

Peralta, E. G., Ashkenazi, A., Winslow, J. W., Smith, D. H., Ramachandran, J., and Capon, D. J. (1987). *EMBO J.* **6**, 3923–3929.

Persson, B., Flinta, C., Von Heijne, G., and Jornvall, H. (1985). *Eur. J. Biochem.* **152**, 523–527.

Pfister, C., Chabre, M., Plouet, J., Tuyen, V. V., DeKozak, Y., Faure, J. P., and Kühn, H. (1985). *Science* **228**, 891–893.

Pittler, S. J., and Baehr, W. (1991). *Proc. Natl. Acad. Sci. U.S.A.* **88**, 8322–8326.

Plouet, J., Mascarelli, F., Loret, M. D., Faure, J. P., and Courtois, Y. (1988). *EMBO J.* **7**, 373–376.

Pober, J. S., Iwanij, V., Reich, E., and Stryer, L. (1978). *Biochemistry* **17**, 2163–2169.

Polans, A. S., Altman, L. G., and Papermaster, D. S. (1986). *J. Histochem. Cytochem.* **34**, 659–664.

Polans, A. S., Buczylko, J., Crabb, J., and Palczewski, K. (1991). *J. Cell. Biol.* **112**, 981–989.

Rando, R. R. (1991). *Biochemistry* **30**, 595–602.

Reid, D. M., Friedel, U., Molday, R. S., and Cook, N. J. (1990). *Biochemistry* **29**, 1601–1607.

Roof, D. J., and Heuser, J. E. (1982). *J. Cell Biol.* **95**, 487–500.

Rosenkranz, J. (1977). *Int. Rev. Cytol.* **50**, 26–158.

Sakmar, T. P., Franke, R. R., and Khorana, H. G. (1989). *Proc. Natl. Acad. Sci. U.S.A.* **86**, 8309–8313.

Schechter, I., Burstein, Y., Zemell, R., Ziv, E., Kantor, F., and Papermaster, D. S. (1979). *Proc. Natl. Acad. Sci. U.S.A.* **76**, 2654–2658.

Schleicher, A., Kühn, H., and Hofmann, K. P. (1989). *Biochemistry* **28**, 1770–1775.

Schmidt, S. Y., Andley, U. P., Heth, C. A., and Miller, J. (1986). *Invest. Ophthalmol. Vis. Sci.* **27**, 1551–1559.

Schnetkamp, P. P. M. (1989). *Prog. Biophys. Mol. Biol.* **54**, 1–29.

Schnetkamp, P. P. M., Klompmakers, A. A., and Daemen, F. J. M. (1979). *Biochim. Biophys. Acta* **552**, 379–389.

Schoenlein, R. W., Peteanu, L. A., Mathies, R. A., and Shank, C. V. (1991). *Science* **254**, 412–415.

Schofield, P. R., Rhee, L. M., and Peralta, E. G. (1987). *Nucleic Acids Res.* **15**, 3636.

Seckler, B., and Rando, R. R. (1989). *Biochem. J.* **264**, 489–493.

Sheffield, V. C., Fishman, G. A., Beck, J. S., Kimura, A. E., and Stone, E. M. (1991). *Am. J. Hum. Genet.* **49**, 699–706.

Shichi, H., and Somers, R. L. (1978). *J. Biol. Chem.* **253**, 7040–7046.

Shinohara, T., Dietzschold, B., Craft, C. M., Wistow, G., Early, J. J., Donoso, L. A., Horwitz, J., and Tao, R. (1987). *Proc. Natl. Acad. Sci. U.S.A.* **84**, 6975–6979.

Shinohara, T., Donoso, L., Tsuda, M., Yamaki, K., and Singh, V. K. (1989). *Prog. Ret. Res.* **8**, 51–66.

Shuster, T. A., and Farber, D. B. (1984). *Biochemistry* **23**, 515–521.

Sitaramayya, A. (1986). *Biochemistry* **25**, 5460–5468.

Sitaramayya, A., and Hakki, S. (1990). *Vis. Neurosci.* **5**, 585–589.

Sitaramayya, A., and Liebman, P. A. (1983). *J. Biol. Chem.* **258**, 12106–12109.

Smith, D., Shieh, B. H., and Zuker, C. S. (1990). *Proc. Natl. Acad. Sci. U.S.A.* **87**, 1003–1007.

Smith, S. B., St. Jules, R. S., and O'Brien, P. J. (1991). *Exp. Eye Res.* **53**, 525–537.

Steinberg, R. H., Fisher, S. K., and Anderson, D. H. (1980). *J. Comp. Neurol.* **190**, 501–508.

St. Jules, R. S., and O'Brien, P. J. (1986). *Exp. Eye Res.* **43**, 929–940.

St. Jules, R. S., Smith, S. B., and O'Brien, P. J. (1990). *Exp. Eye Res.* **51**, 427–434.

St. Jules, R. S., Wallingford, J. C., Smith, S. B., and O'Brien, P. J. (1989). *Exp. Eye Res.* **48**, 653–665.

Stryer, L. (1991). *J. Biol. Chem.* **266**, 10711.

Sung, C. H., Davenport, C. M., Hennessey, J. C., Maumenee, I. H., Jacobson, S. G., Heckenlively, J. R., Nowakowski, R., Fishman, G., Gouras, P., and Nathans, J. (1991a). *Proc. Natl. Acad. Sci. U.S.A.* **88**, 6481–6485.

Sung, C. H., Schneider, B. G., Agarwal, N., Papermaster, D. S., and Nathans, J. (1991b). *Proc. Natl. Acad. Sci. U.S.A.* **88** 8840–8844.

Swofford, D. L. (1991). "PAUP: Phylogenetic Analysis Using Parsimony. Version 3.0." Champaign, Illinois.

Szuts, E. Z. (1985). *Biochemistry* **24**, 4176–4184.

Takao, M., Yasui, A., and Tokunaga, F. (1988). *Vision Res.* **28**, 471–480.

Thomas, D. D., and Stryer, L. (1982). *J. Mol. Biol.* **154**, 145–157.

Thompson, P., and Findlay, J. B. C. (1984). *Biochem. J.* **220**, 773–780.

Ting, T. D., and Ho, Y. K. (1991). *Biochemistry* **30**, 8996–9007.

Travis, G. H., Brennan, M. B., Danielson, P. E., Kozak, C. A., and Sutcliffe, J. G. (1989). *Nature* **338**, 70–73.

Travis, G. H., Sutcliffe, J. G., and Bok, D. (1991). *Neuron* **6**, 61–70.

Tsunasawa, S., Narita, K., and Shichi, H. (1980). *Biochim. Biophys. Acta* **624**, 218–225.

Usukura, J., and Yamada, E. (1981). *Biomed. Res.* **2**, 177–193.

Vaughan, D. K., and Fisher, S. K. (1989). *Invest. Ophthalmol. Vis. Sci.* **30**, 339–342.

Vuong, T. M., and Chabre, M. (1991). *Proc. Natl. Acad. Sci. U.S.A.* **88**, 9813–9817.

Wacker, W. B., Donoso, L. A., Kalsow, C. M., Yankeelov, J. A. J., and Organisciak, D. T. (1977). *J. Immunol.* **119**, 1948–1957.

Walter, A. E., Shuster, T. A., and Farber, D. B. (1986). *Invest. Ophthalmol. Vis. Sci.* **27**, 1609–1614.

Weyand, I., and Kühn, H. (1990). *Eur. J. Biochem.* **193**, 459–467.

Wieland, T., Ulibarri, I., Aktories, K., Gierschik, P., and Jakobs, K. H. (1990a). *FEBS Lett.* **263**, 195–198.

Wieland, T., Ulibarri, I., Gierschik, P., Hall, A., Aktories, K., and Jakobs, K. H. (1990b). *FEBS Lett.* **274**, 111–114.

Wilden, U., Hall, S. W., and Kühn, H. (1986). *Proc. Natl. Acad. Sci. U.S.A.* **83**, 1174–1178.

Wilden, U., and Kühn, H. (1982). *Biochemistry* **21**, 3014–3022.

Wilden, U., Wust, E., Weyand, I., and Kühn, H. (1986). *FEBS Lett.* **207**, 292–295.

Williams, D. S. (1991). *Bioessays* **13**, 171–178.

Witt, P. L., and Bownds, M. D. (1987). *Biochemistry* **26**, 1769–1776.

Wolbring, G., and J., C. N. (1991). *Eur. J. Biochem.* **201**, 601–606.

Wong, S., and Molday, R. S. (1986). *Biochemistry* **25**, 6294–6300.

Yamada, T., Takeuchi, Y., Komori, N., Kobayashi, H., Sakai, Y., Hotta, Y., and Matsumoto, H. (1990). *Science* **248**, 483–486.

Yamamoto, K., and Shichi, H. (1983). *Biophys. Struct. Mech.* **9**, 259–267.

Yokoyama, R., and Yokoyama, S. (1990a). *Proc. Natl. Acad. Sci. U.S.A.* **87**, 9315–9318.

Yokoyama, R., and Yokoyama, S. (1990b). *Vision Res.* **30**, 807–816.

Zhukovsky, E. A., and Oprian, D. D. (1989). *Science* **245**, 928–930.

Zuker, C. S., Cowman, A. F., and Rubin, G. M. (1985). *Cell* **40**, 851–858.

Zuker, C. S., Montell, C., Jones, K., Laverty, T., and Rubin, G. M. (1987). *J. Neurosci.* **7**, 1550–1557.

Subunit Structure and Transmembrane Signaling of the Erythropoietin Receptor

Mark O. Showers and Alan D. D'Andrea

Division of Pediatric Oncology, Dana Farber Cancer Institute, Children's Hospital, Harvard Medical School, Boston, Massachusetts 02115

I. Introduction

The most important direct stimulus to the development of committed erythroid cells is the glycoprotein hormone, erythropoietin (EPO). Although a large literature exists describing the *in vivo* efficacy of EPO in various amenic states, such as kidney failure (Erslev, 1987; Eschbach *et al.*, 1987, 1989), little is known about the actual cellular physiology of this hormone. This is due, in part, to our incomplete understanding of the structure of the EPO-receptor (EPO-R) complex.

In this review we describe the advances in our understanding of the EPO-R, derived from the cloning of the murine EPO-R cDNA (D'Andrea *et al.*, 1989). The review will address the following points. First, the murine EPO-R is a 507-amino acid Type I membrane-spanning protein, which is a member of a large cytokine receptor superfamily (Bazan, 1989; Cosman *et al.*, 1990; D'Andrea *et al.*, 1990a). Transfection of COS cells with the EPO-R cDNA confers EPO binding on these cells. Second, the EPO-R, although originally cloned from an EPO-unresponsive murine erythroleukemia (MEL) cell line, is a functional receptor. That is, transfection of the murine lymphocyte cell line, Ba/F3, and the murine myeliod cell lines, 32-D and FDCP1, with the murine or human EPO-R cDNA confers EPO dependence on these cells (Jones *et al.*, in press). Third, the murine EPO-R is activated not only by EPO binding but also by binding of the envelope protein, gp55, of the Friend spleen-focus forming virus (SFFV) (Li *et al.*, 1990; D'Andrea *et al.*, in press). This activation by gp55 accounts for the early phase of Friend virus erythroleukemia and is a novel mechanism of retroviral transformation. Fourth, the second messenger mechanisms involved in signal transduction by the EPO-R or by other receptors of the cytokine receptor superfamily are poorly understood (D'Andrea and Zon, 1990). The cytoplasmic domain of the EPO-R, although lacking tyrosine kinase activity, is required for signal transduction and probably interacts with some other critical intracellular protein. Recent data demonstrated that activation of the EPO-R by either EPO or by gp55 stimulates cellular tyrosine kinase activity (Quelle and Wojchowski, 1991).

II. Role of Erythropoietin in Normal Hematopoiesis

In vitro culture techniques have provided a model for developmental changes during erythropoiesis. Erythropoiesis begins with an undifferentiated stem cell that, within 7–10 days, gives rise to mature erythrocytes (Russell, 1979). The first step is "commitment" to erythroid differentiation, governed by unknown factors that occur gradually in the precursor population during its amplification. Although EPO does not influence commitment, it is essential for the succeeding two stages of differentiation (Spivak *et al.*, 1990). First, it is required in a relatively high concentration (10 pM) for amplification of the early committed but not yet differentiated erythroid precursors (burst-forming units-erythroid, BFU-E). Second, lower levels (1 pM) of EPO are required to trigger differentiated cells (colony-forming units-erythroid, CFU-E) into final erythroid maturation. During these final stages hemoglobin transcript and polypeptide accumulate and the cells lose their EPO responsiveness (Goldwasser, 1984).

III. Structure of the Murine and Human Erythropoietin Receptor

The structure of the murine EPO-R has previously been reviewed (D'Andrea and Zon, 1990; D'Andrea and Jones, in press). The receptor is a 507-amino acid Type I membrane-spanning protein (Fig. 1). Cleavage of the 24-amino acid hydrophobic leader sequence leaves the following: (1) 223 amino acids outside the cell, containing the EPO binding domain, (2) a 24-amino acid transmembrane anchor sequence, and (3) a 236-amino acid cytoplasmic domain. There are four conserved cysteine residues (C1–4) and a tryptophan-serine-X-tryptophan-serine motif in the ectodomain, which is conserved among members of the EPO-R superfamily. The full-length human EPO-R cDNA has also recently been cloned (Jones *et al.*, in press; Winkelman *et al.*, 1990). The predicted structure of the EPO-R polypeptide is a 508-amino acid polypeptide with 85% amino acid identity to the mouse primary sequence (Fig. 2). This high percentage of amino acid identity is not surprising because recombinant human EPO is known to bind to both murine and human erythroid cells (D'Andrea and Zon, 1990) and be-

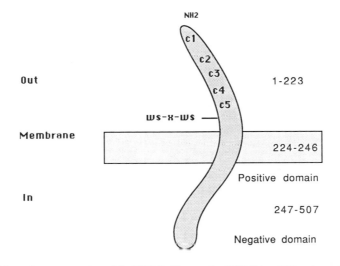

FIG. 1 Schematic representation of the EPO R. The murine EPO-R is a 507-amino acid polypeptide. There is a 24-amino acid hydrophobic, α-helical leader sequence that is cleaved, leaving a 223 amino acid ectodomain, a 24-amino acid transmembrane region, and a 236-amino acid cytoplasmic domain.

cause the human EPO-R confers EPO responsiveness on murine cell lines (Jones *et al.*, in press). Like the murine EPO-R cDNA, the human cDNA directs the synthesis of high- and low-affinity EPO-R.

A. Equilibrium Binding of Radiolabeled EPO to the EPO-R

Despite the cloning and sequencing of the murine and human EPO-R cDNAs, many questions remain unanswered about the structure of the EPO-R. First, the EPO-R is known to exist in both high-affinity (approximately 30 pM) and low-affinity (approximately 200 pM) states in some EPO-responsive cell lines (Sawyer *et al.*, 1987a,b), whereas it exists in only a low-affinity state (200 pM) in MEL cells (D'Andrea *et al.*, 1989a). Figure 3 shows the equilibrium-binding analysis and Scatchard analysis for MEL cells and for COS cells expressing the recombinant murine EPO-R. There are several possible explanations for this discrepancy in the affinity of EPO-R for these two cell lines. It is possible that the COS cells express a second subunit of the EPO-R which, when coupled with the cloned subunit, forms a high-affinity site. Alternatively, the high-affinity receptor in MEL cells may be occupied by some other polypeptide (such as the Friend SFFV gp55) and thereby be blocked for EPO binding. It is interesting that several other members of the cytokine receptor superfamily, including the receptors for interleukin (IL)-2, IL-3, IL-6, and granulocyte-macrophage–colony-stimulating factor

```
    K RVP   R  P                S S               S  S        Q
MDHLGASLWPQVGSLCLLLAGAAWAPPPNLPDPKFESKAALLAARGPEELLCFTERLEDL
         10                    30                  50


        S MDFx            G SR S S      V  S
VCFWEEAASAGVGPGNYSFSYQLEDEPWKLCRLHQAPTARGAVRFWCSLPTADTSSFVPL
         70                    90                  110


    Q   E   S     I          A   L R E GS           GA   T
ELRVTAASGAPRYHRVIHINEVVLLDAPVGLVARLADESGHVVLRWLPPPETPMTSHIRY
         130                   150                 170


        R  GT    V                  G                 S       A
EVDVSAGNGAGSVQRVEILEGRTECVLSNLRGRTRYTFAVRARMAEPSFGGFWSAWSEPV
         190                   210                 230


    A                L SL            T Q
SLLTPSDLDPLILTLSLILVVILVLLTVLALLSHRRALKQKIWPGIPSPESEFEGLFTTH
         250                   270                 290


    L R          GSS P       H       PR AVT  GD   A
KGNFQLWLYQNDGCLWWSPCTPFTEDPPASLEVLSERCWGTMQAVEPGTDDEGPLLEPVG
         310                   330                 350


              T  C   N  S        PT   A   P D       R    T
SEHAQDTYLVLDKWLLPRNPPSEDLPGPGGSVDIVAMDEGSEASSCSSALASKPSPEGAS
         370                   390                 410


PS              C RA  P                               G    VH DS
AASFEYTILDPSSQLLRPWTLCPELPPTPPHLKYLYLVVSDSGISTDYSSGDSQGAQGGL
         430                   450                 470


        H      V DS   H G          509
SDGPYSNPYENSLIPAAEPLPPSYVACS*    509
         490
```

FIG. 2 Comparison of human and murine EPO-R amino acid sequence. The human EPO-R sequence is shown. Amino acid differences in the murine sequence are shown above the appropriate amino acid residues.

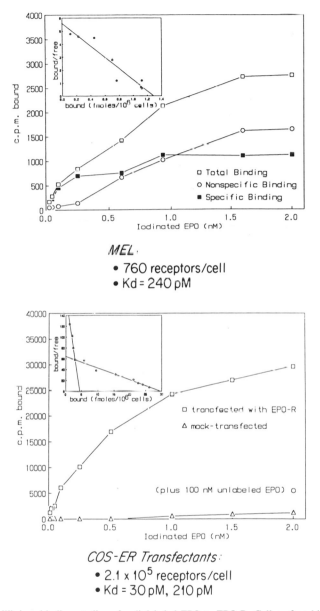

FIG. 3 Equilibrium binding studies of radiolabeled EPO to EPO-R. Cell surface binding studies with radioiodinated EPO reveal only low affinity receptors (240 pM) on murine erythroleukemia (MEL) cells but high and low affinity receptors on COS cells expressing the recombinant murine EPO-R.

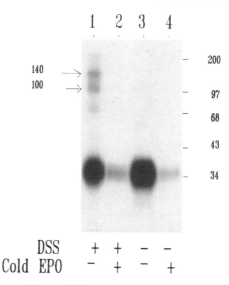

FIG. 4 Identification of two major cross-linked complexes of the EPO-R. COS cells expressing the recombinant murine EPO-R were incubated with radiolabeled EPO at a 1 nM concentration. Binding was performed in the absence (lanes 1 and 3) or presence (lanes 2 and 4) of excess cold EPO. Unbound EPO was washed and cross-linking was performed with disuccinimydyl suberate (DSS). Two major EPO-R cross-linked complexes of 100 kDa and 140 kDa were observed, analagous to those observed by Sawyer *et al.* (1987a).

(GM-CSF), exist in both high- and low-affinity states. For these receptors second subunits have been identified, which account for the formation of multiple affinities for the respective ligand (Smith, 1989; Gorman *et al.*, 1990; Hayashida *et al.*, 1990; Hibi *et al.*, 1990).

B. Cross-linking of Radiolabeled EPO to the EPO-R

A second unanswered question regarding EPO-R structure relates to the pattern of radiolabeled EPO cross-linked complexes observed (Fig. 4). When radiolabeled EPO is cross-linked to the EPO-R, two cross-linked complexes of approximately 140 kDa and 100 kDa are observed (Sawyer *et al.*, 1987; Mayeux *et al.*, 1987a,b; Todokora *et al.*, 1987, 1988). After subtraction of the molecular weight of EPO (34,000), one concludes that there are two receptor subunits (66 kDa and 106 kDa) (D'Andrea *et al.*, 1989a). No correlation exists between the presence of two cross-linked complexes and two receptor affinities. For instance, MEL cells demonstrate only the lower affinity receptor but yield both cross-linked complexes.

The identity of these two cross-linked complexes of the EPO-R remains a central unanswered question in the study of the EPO-R. The smaller complex (100 kDa) may be equal to EPO (34 kDa) cross-linked to the known receptor (66 kDa), but we cannot account for the larger complex (140 kDa). Several groups have shown that, in the absence of reducing agent, the cross-linked EPO and EPO-R migrates as a single 240-kDa band, suggesting that the two complexes exist as a disulfide-linked heterodimer (McCaffrey et al., 1989). Other groups have not observed such higher molecular weight species (Sawyer, 1989; Sawyer et al., 1988).

Sawyer (1989) has shown that proteolytic fragments of the 140 kDa and 100 kDa cross-linked complexes are shared, suggesting that the polypeptides are highly related. We have demonstrated that the smaller cross-linked complexes (100 kDa) are decreased in size, to a complex of 70 kDa, when a truncated (cytoplasmic tail minus) form of the EPO-R cDNA is used for transfection (A. D. D'Andrea, unpublished observation). This proves that the smaller complex contains the cloned EPO-R polypeptide. Also, both cross-linked complexes are immunoprecipitated by an antiserum against the amino terminus of the predicted EPO-R sequence, confirming that the EPO-R cross-linked complexes are bound together, even in detergent-solubilized extracts.

The structure of the EPO-R has been examined by ligand blot analysis (Atkins et al., in press). Solubilized cell membrane proteins from an EPO-dependent cell line were separated by sodium dodecyl sulfate (SDS)-polyacrylamide gel electrophoresis, transferred to nitrocellulose, and probed with [^{125}I]EPO. A single 61-kDa protein bound [^{125}I]EPO, and this binding was inhibited by antibodies known to inhibit EPO binding to EPO-R.

Although the identity of the two EPO-R cross-linked complexes is still unclear, certain conclusions can be made. The smaller (100 kDa) complex is composed of EPO (34 kDa) plus the cloned EPO-R (66 kDa). The larger cross-linked complex (140 kDa), although it is not immunoreactive itself with the anti–EPO-R antiserum, does co-immunoprecipitate with the smaller complex, suggesting that it is a closely bound second subunit. The larger (140 kDa) complex may be composed of a second EPO-R subunit analogous to the gp130 recently described (Taga et al., 1989; Hibi et al., 1990).

C. Cytokine Receptor Superfamily

Murine and human EPO-R were found to share regions of amino acid similarity with the ligand binding subunits of the receptors for the hematopoietic growth factors IL-2 (Hatakeyama et al., 1989), IL-3 (Itoh et al., 1990), IL-4 (Idzerda et al., 1990; Mosley et al., 1990), IL-6 (Yamasaki et al., 1988), granulocyte CSF

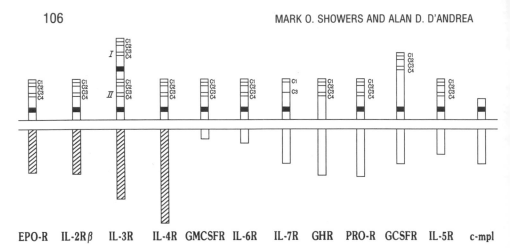

EPO-R   IL-2Rβ   IL-3R   IL-4R   GMCSFR   IL-6R   IL-7R   GHR   PRO-R   GCSFR   IL-5R   c-mpl

FIG. 5   The cytokine receptor superfamily. The homologous domains of the receptors for EPO, interleukin-2, interleukin-3, interleukin-4, interleukin-5, interleukin-6, interleukin-7, growth hormone, prolactin, granulocyte-CSF, and granulocyte-macrophage (GM)–CSF are shown. The structural basis for the superfamily is based primarily on conserved features of the ectodomain, including the WS-X-WS motif, shown with a black box, and the conserved cysteine residues. The external domain of the IL-3R is duplicated (regions I and II).

(Fukunaga *et al.*, 1990), GM-CSF (Gearing *et al.*, 1990), and the receptors for growth hormone and prolactin (Leung *et al.*, 1987). Also, a new truncated cytokine receptor (*v-mpl*) for an unidentified ligand was identified which can immortalize hematopoietic progenitors (Sonyri *et al.*, 1990). All cytokine receptors are Type I membrane-spanning proteins (Fig. 5) and are synthesized with a cleavable signal sequence. In the extracellular domain members of the receptor superfamily have four cysteine residues spaced approximately 9, 27, and 15 amino acids apart and also a tryptophan-serine-X-tryptophan-serine motif positioned outside the membrane. Other conserved amino acids, including several proline residues, have been observed among the extracytoplasmic domains of these receptors (Goodwin *et al.*, 1990). The sequence similarities probably reflect shared structural motifs among these receptors which may, in turn, extend to conserved structural features among the respective growth factors themselves (Bazan, 1989, 1990). For instance, the crystal structure of growth hormone reveals an antiparallel four-helix bundle core. Similarly, modeling of the tertiary structure of EPO and IL-6 reveals a growth hormone-like helix bundle fold. It is likely that many members of this family are multisubunit receptors. The additional subunits may account for the variable hormone affinities and responses observed. For instance, second units of the IL-6R (Hibi *et al.*, 1990) and the GM-CSFR (Hayashida *et al.*, 1990) have been identified, which alter the binding affinity of the receptor complex for its respective ligand.

Although the cytokine receptor superfamily does not share amino acid homology with members of the tyrosin receptor superfamily (Hanks *et al.*, 1988), these two receptor superfamilies do have some common characteristics. Both superfamilies contain receptors that possess a large glycosylated, extracytoplasmic ligand-binding domain, a single hydrophobic transmembrane region, and a variable cytoplasmic tail. Yet, whereas members of the tyrosine kinase superfamily have a cytoplasmic domain with a tyrosine kinase catalytic domain (Ullrich and Schlessinger, 1990), the cytokine receptor cytoplasmic domains apparently contain no such enzymatic activity. A subfamily of the cytokine receptors (EPO-R, IL-2R, IL-3R) does share some sequence similarity in the cytoplasmic region (D'Andrea et al., 1990a). Due to their configurations, however, members of both superfamilies are considered membrane-associated allosteric enzymes. Ligand binding to the extracytoplasmic region causes an activation that must be translated across the membrane barrier in both cases.

D. Chromosomal Localization of the EPO-R Gene

By somatic cell hybrid analysis (Budarf et al., 1990) and by in situ hybridization (Winkelmann et al., 1990; Hoatlin et al., 1990), the human EPO-R gene was found to map to human chromosome 19pter-q12. By interspecific backcross mapping the murine EPO-R locus was found to be tightly linked to the murine low density lipoprotein (Ldlr) locus on mouse chromosome 9. These regions of human chromosome 19 and mouse chromosome 9 have already been shown to be syntenic. There are no known disease states suggestive of an EPO-R mutation that map to this chromosome region in mouse or human. It will be of interest to determine whether genes encoding other members of the cytokine receptor superfamily are clustered in this region.

E. Advantages of Studying the EPO-R over Other Members of the Cytokine Receptor Superfamily

Several advances in our understanding of the cellular physiology of EPO and the cellular development of red blood cells make the EPO-R an ideal system to study the molecular basis of EPO signal transduction. First, it is well known that EPO does not act at a discrete stage of erythropoiesis but, instead, supports erythroid maturation over a period of 7–10 days (Russell, 1979). Interestingly, the sensitivity of erythroid cells to EPO varies over a 10-fold range during differentiation. Second, unlike several other growth factors receptors, the EPO-R is confined to erythroid cells, megakaryocytes, and mast cells (L. Zon, personal communication). Third, excellent in vitro models of erythroid differentiation exist, such as the MEL system (Marks and Rifkin, 1978) and the friend virus anemia strain (FVA) erythroblast system (Sawyer et al., 1987a), allowing study of the developmental fate of the EPO-R during red blood cell development. Likewise, the assembly of the red cell cytoskeleton during erythroid differentiation is well studied and could account for the selective loss of the EPO-R during cell development. Finally, there are several ways of activating the EPO-R. The EPO-R becomes active after binding EPO or after binding the Friend SFFV gp55 (Li *et al.*, 1990; D'Andrea *et al.*, in press). This latter mechanism of EPO-R stimulation by gp55 is unique among members of the cytokine receptor superfamily. Alternatively, the EPO-R can be activated by a single amino acid point mutation in the extracytoplasmic region (Yoshimura *et al.*, 1990). Alternative mechanisms of receptor activation will provide a convenient crosscheck of EPO-signaling mechanisms.

F. Transcriptional Control of the EPO-R

During erythroid development, expression of high-affinity EPO-R increases upon progression from late BFU-E to CFU-E. With further maturation there is a decrease in high-affinity EPO-R. Eventually, by the time the cells have differentiated to the point of reticulocytes, no EPO-R is present.

To understand this complex control of EPO-R expression, the normal EPO gene has been cloned from mouse (Youssouffian *et al.*, 1990) and human (Wong *et al.*, in press). The promoter of the murine EPO receptor gene has been studied in detail (Youssoufian *et al.*, 1990). The proximal promoter is notable for the presence of an SP-1 site and a GATA element, the binding site for the major erythroid transcription factor GF-1 (Ery-1, NF-E1). The GATA element has been previously identified in the promoters and enhancers of several erythroid-specific genes from human, mouse, and chicken. The promoter of the EPO-R gene confers erythroid-specific expression. EPO-R promoter–reporter fusions expressing growth hormone are inactive in nonerythroid cells.

It has been shown that mutations of the GATA element in the EPO-R promoter prevent GF-1 binding and remove promoter activity (L. Zon, personal communication). The presence of the GATA element in the EPO-R promoter suggests that the EPO-R is turned on later in differentiation, after the stem cell has committed to the erythroid lineage and has expressed the GF-1 transcription factor. This observation is consistent with the notion that EPO is a viability agent for erythroid cells but does not itself influence the commitment of stem cells in the bone marrow to the erythroid lineage.

G. Developmental Regulation of the EPO-R during Erythropoiesis

Several studies have examined the change in EPO-R cell surface expression during erythroid differentiation (Broudy *et al.*, 1990; Landshultz *et al.*, 1989). As a general rule, there appears to be an initial increase in EPO-R expression during erythroid differentiation (BFU-E to CFU-E transition) followed by a decline in EPO-R expression and complete loss by the retriculocyte stage. Although EPO-R expression is controlled, in part, by transcriptional control mechanisms, other cell-biologic factors may control cell surface expression during erythropoiesis, such as proteolytic degradation of the EPO-R or secretion of the EPO-R. A truncated EPO-R transcript encoding a secreted EPO-R has been identified (Kuramochi et al., 1990).

One system that allows detailed study of EPO-R developmental regulation is that of hamster yolk sac erythroid cells grown in vitro (Boussios et al., 1989). For these cells affinity for EPO decreases during gestation, but the number of cell surface receptors reaches a maximum at gestational day 10. Another system, that of EPO-unresponsive MEL cells, has been employed by Tojo et al. (1988). This study shows that surface EPO receptors increase in number after dimethyl sulfoxide (DMSO) induction, although the affinity of the EPO-R does not appear to change.

Another report, employing cell cytometry, examines sorted cell populations from human bone marrow. Cells that express increasing numbers of cell surface EPO-R also express decreasing CD34 expression. CD34 is an early marker of primitive hematopoietic cells, and this observation supports the model of increasing EPO-R surface expression during erythropoiesis. Regardless of the actual mechanisms that control EPO-R affinity and cell surface expression, it is clear that EPO responsiveness is maintained throughout most of erythroid development.

H. Interaction of the EPO-R with the gp55 Glycoprotein of the Friend Virus Spleen Focus-Forming Unit

One insight into EPO-R structure has been derived from the observation that the murine and human EPO-R bind to and are activated by the membrane glycoprotein, gp55, of the Friend SFFV (Li et al., 1990, 1991; D'Andrea et al., 1992). SFFV is a defective murine C-type retrovirus that causes a multistage erythroleukemia in mice and erythroplastosis in bone marrow cultures (Ruscetti et al., 1990). The SFFV env gene encodes a membrane glycoprotein, gp55 (also called gp52, Chung et al., 1989), which is located on the cell surface and within rough endoplasmic reticulum (Dresler et al., 1978; Gliniak et al., 1989). Gp55 is essential for the induction of erythroleukemia in vivo and the EPO-independent eyrthroblast proliferation in vitro. By co-transfecting both EPO-R and the gp55 into an IL-3–dependent lymphocyte cell line, it was shown that the physical interaction of these two proteins gives rise to constitutive, EPO-independent cell growth. When these cells are transfected with gp55 alone, they remain IL-3–dependent. That is, the gp55 specifically activates the EPO-R and not the IL-3 receptor that is endogenous to these cells (Fig. 6 and 7).

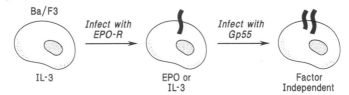

FIG. 6   Schematic model of the interaction between EPO-R and the gp55 from Friend spleen focus-forming virus in transfected cells. The IL-3–dependent pro-B lymphocyte line, Ba/F3, after transfection (or infection) with the EPO-R cDNA, expressed cell surface EPO-R and will grow either in EPO or IL-3. Subsequently, expression of gp55 will transform the cells to factor independence. Also (not shown) if the parent cell line Ba/F3 is transfected with the gp55 cDNA directly, the cells remain IL-3–dependent.

The gp55 may mimic EPO itself, although there is no amino acid homology shared between these two proteins. Alternatively, gp55 may have sequences derived from (and therefore mimic) a hypothetical second subunit of the EPO-R (Li et al., 1990). The interaction between the EPO-R and gp55 is also observed in MEL cells and may account for the absence of high-affinity binding sites on MEL cells and for the EPO-independent growth of these cells. It is not clear whether the gp55 stimulation of EPO-R alone accounts for the leukemic phenotype of MEL cells or whether a second event, such as proviral integration of SFFV (Moreau-Gachelin et al., 1988), contributes to the phenotype. Recent evidence suggests that the interaction between the EPO-R and the gp55 occurs within the endoplasmic reticulum (Yoshimura et al., 1990). That is, the gp55 binds to the high-mannose, endoglycosidase H-sensitive form of the EPO-R. No evidence exists at present for the formation of EPO-R–gp55 complexes at the cell surface. The interaction between gp55 and EPO-R is a novel mechanism of retroviral transformation leading to growth factor independence. It is possible that analogous retroviruses transform other cell types in a similar manner by using other receptors of the cytokine superfamily, such as the IL-2R, IL-3R, or IL-4R.

IV. Functional Domains of the Erythropoietin Receptor Polypeptide

Figure 2 shows that the EPO-R is composed of an extracytoplasmic region, a transmembrane region, and a cytoplasmic region. In terms of the known functions of the EPO-R, the receptor must contain the following: (1) a domain required for EPO binding (in the extracytoplasmic region), (2) a domain required for gp55 binding, and (3) a domain in the cytoplasmic region required for signal transduction.

A. EPO-Binding Domain

Little information exists regarding the specific amino acid sequences of the extracytoplasmic region required for EPO binding. Modeling studies (Bazan, 1990a,b) have suggested that the EPO-R, like the other members of the cytokine receptor superfamily, has a binding segment composed of two discrete folding domains that share significant sequence and structural resemblance. Each folding domain is composed of seven β-strands in conserved regions of the chain. The EPO-R is thereby predicted to contain a V-shaped crevice with the

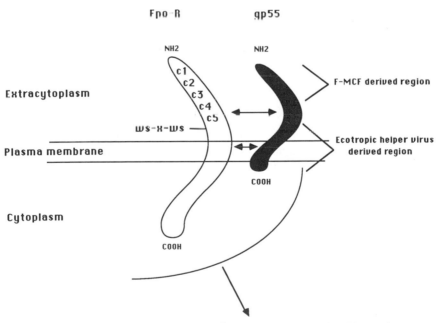

FIG. 7   Schematic representation of the physical interaction between the EPO-R and the gp55 of the Friend spleen focus-forming virus (SFFV). The EPO-R is 507 amino acids and the gp55 is 409 amino acids. Both are Type I membrane-spanning proteins. The gp55 has only a 2-amino acid cytoplasmic domain and, as predicted from this model, the cytoplasmic domain of the EPO-R is not required for gp5 binding. The physical interaction between these molecules occurs either in the transmembrane region or in the ectodomains, perhaps via disulfide linkages.

WS-X-W-S serving as the base of the crevice. Consistent with this model, we have recently shown that an antiserum against the amino terminus of the EPO-R does not block EPO binding (A. D. D'Andrea, unpublished observation). Future studies of site-directed mutagenesis or formation of chimeric receptors with the IL-3R should further delineate the EPO-binding region.

More is known regarding the region of the ligand, EPO, responsible for the EPO-R recognition than about the receptor itself. The general consensus is that whereas antibodies against the amino terminus of EPO do not block receptor binding (D'Andrea et al., 1990b), antibodies against amino acids 99–129 of EPO do neutralize the hormone activity (Sytkowski and Donahue, 1987).

Less is known about the region of the extracytoplasmic domain of the EPO-R that is required for EPO binding. We have generated a secreted form of the EPO-R, with a stop codon at amino acid 223. This truncated EPO-R is sufficiently secreted from stably transfected cells and binds radiolabeled EPO in the media. This result suggests that the transmembrane region of the EPO-R is not required for the proper folding of EPO-R.

B. Gp55-Binding Domain

The gp55 binding domain is another important domain of the EPO-R in the region of interaction with gp55 glycoprotein of the Friend SFFV unit (Fig. 7). As stated, although either EPO-binding or gp55-binding activates the EPO-R to signal cell growth, there is no amino acid similarity between EPO and gp55. One important question is whether gp55 actually competes with EPO binding or whether it binds to another discrete region of the EPO-R. Because the primary site of interaction of EPO with EPO-R is at the cell surface while the primary site of interaction between the EPO-R and gp55 is inside the cell, in the endoplasmic reticulum, it has been difficult to do EPO competition studies *in vitro*. To answer this question of binding competition adequately, a cell-free system examining EPO binding and gp55-binding will be required. Until such a system is developed, site-directed mutagenesis of the EPO-R has provided some clues to gp55 interaction. For instance, gp55 fails to bind to the secreted form of the EPO-R described herein, suggesting that the transmembrane anchor of the EPO-R is required for gp55 binding (A. D. D'Andrea, unpublished observation). Consistent with this finding are the results of Chung et al. (1989), which show by genetic arguments that the transmembrane region of the gp55 from the polycythemia strain of the Friend SFFV is required to promote EPO-independent growth. These results suggest that the membrane anchor is important in gp55 association; however, we cannot rule out the possibility that the amino terminus of the EPO-R and gp55 also interact.

The cytoplasmic region of the EPO-R is composed of a nonoverlapping positive and negative growth regulatory domain.

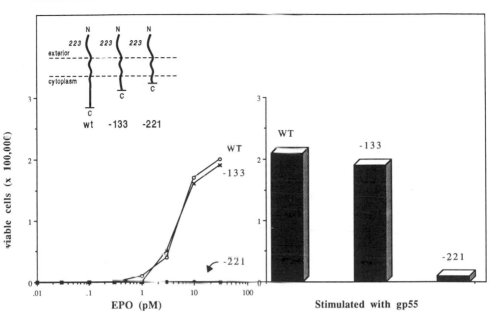

FIG. 8   The proximal cytoplasmic domain of the EPO-R is required for EPO-dependent and gp55-dependent cell growth. The wild-type, −133 and −221 truncated mutants, shown in the upper left-hand corner, were introduced into Ba/F3 cells. Only the wild-type and −133 mutant EPO-R conferred EPO-dependent growth, wherreas the −221 mutant, although capable of binding EPO, could not confer EPO-dependent growth. Likewise (right-hand panel), gp55 fails to activate the −221 mutant of EPO-R.

Another approach to understanding signal transducin by the EPO-R has been through structural and functional analysis of the cytoplasmic region of the EPO-R. Using a model system, the expression of wild-type and mutant EPO-R cDNAs in a IL-3–dependent, pro B lymphocyte cell line, Ba/F3, we have identified two critical domains of the EPO-R cytoplasmic region. First, membrane-proximal positive transducting domain of ≤103 amino acids in a region highly similar to the IL-2R beta chain (p75) (D'Andrea *et al.*, 1989b) is required for EPO-mediated signal transduction. This positive regulatory domain is demonstrated by the experiment in Figure 8. In this experiment we have compared the wild-type EPO-R with two truncated forms of the EPO-R expressed in Ba/F3 cells. This severely truncated EPO-R, although capable of binding EPO, fails to provide EPO-dependent growth or gp55-dependent growth on the Ba/F3 cells.

We have also demonstrated a carboxyl-terminated negative regulatory domain of 40 amino acids by the experiment described in Figure 9. In this experiment the truncated mutants −40 and −91, lacking the 40 or 91 amino acids at the carboxy terminus of the EPO-R, confer hypersensitive EPO growth, with maximal EPO responsiveness in the 1 p*M* range. Although there was a 10-fold increase in EPO responsiveness for these subclones, there were no differences in EPO-R number and affinity, there were no differences in EPO-R carbohydrate processing, and there no differences in the rates of endocytosis of radiolabeled EPO (Table I). Again, as shown on the right panel of Figure 9, the gp55 activates only the wild-type, -40 and -91 mutants but not the -221 mutant.

Based on these experiments we conclude that there exists a positive and a negtaive regulatory domain in the EPO-R cytoplasmic region (Fig. 10). The positive regulating domain is analogous to a critical region of signal transduction in the cytoplasmic region of the IL-2R beta chain (p75) (Hatakeyama, 1989b) and may be a domain for interaction with a critical cytoplasmic tyrosine kinase polypeptide. The negative regulatory domain is serine-rich and is probably a site of critical phosphorylation. It is analogous to the negative regulatory domain found at the carboxy terminus of the EGF-R (Honegger *et al.*, 1988) and the CSF-1R (Roussel *et al.*, 1987).

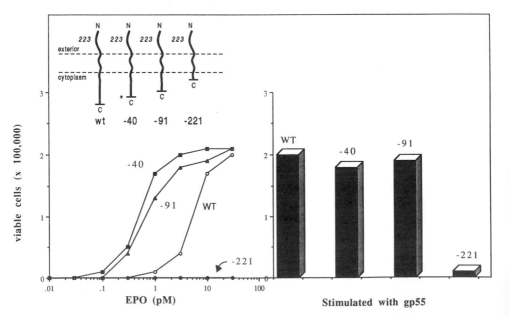

FIG. 9 The distal cytoplasmic domain of the EPO-R determines EPO sensitivity. The wild-type and the -40, -91, and -221 truncated mutants were transfected into Ba/F3 cells. The -40 and -91 mutants conferred hypersensitive EPO responsiveness with maximal growth in only 1 p*M* of EPO. The wild-type, -40, and -90 mutants were all activated by gp55 binding, but the -221 mutant was not.

TABLE I

Effect of Truncation of the Carboxy Terminal 40 Amino Acids on EPO-R Processing

	Wild-type EPO-R	−40 EPO-R Mutant
Ba/F3 surface binding	1080 receptors/cell $K_d = 309$ pM	1680 receptors/cell $K_d = 291$ pM
Processed primarily to endoglycosidase H-resistant forms	+	+
Endocytosis of radiolabeled EPO	+	+

C. Point Mutation of the Extracytoplasmic Region of the EPO-R

A constitutive mutant of the EPO-R has recently been reported (Yoshimura *et al.*, 1991) which confers EPO-independent growth on Ba/F3 cells. The point mutation (Arg^{221} → Cys^{221}) creates a new cysteine residue in the extracytoplasmic region of the EPO-R, a site for potential intrachain or interchain disulfide bonding, which may be important in the activation events.

Interestingly, the constitutive EPO-R is inefficiently processed to a cell surface, complex carbohydrate forms, analogous to the inefficient carbohydrate processing of normal EPO-R in the presence of gp55. This result suggests that EPO-R signaling may be occurring from an intracellular compartment.

The mechanisms by which the point mutation activates this EPO-R are not clear. It is likely that the mutation induces a conformational change in the receptor that results in activation. Whether or not this mutation affects receptor aggregation or association with other signaling polypeptides remains to be tested.

V. Transmembrane Signaling by the Erythropoietin Receptor

Although several studies have investigated the cellular action of the EPO ligand–receptor complex, no consistent mechanism of signaling has been obvious (D'Andrea and Zon, 1990). A major difficulty encountered in many of the published studies has been the small number of surface receptors for EPO and the unavailability of a purely EPO-dependent cell line. A concensus of the published studies suggests that although activation of the EPO-R does not immediately activate adenylate cyclase (Bonanou-Tzedaki *et al.*, 1986) or guanylate cyclase, cAMP and cGMP probably play a modulatory role later in the signaling pathway

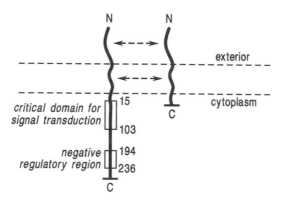

FIG. 10   Schematic model of EPO-R bound to gp55, showing nonoverlapping functional domains of the cytoplasmic region. The cytoplasmic region of the EPO-R, although required for EPO-induced signaling, is not required for gp55 binding.

(Bonanou-Tzedaki *et al.*, 1987). This conclusion was also reached by Kuramochi *et al.* (1990), who showed that cAMP-elevating agents, such as forskolin, stimulated the effects of EPO on erythroid differentiation. Also, although extracellular calcium is an absolute requirement for EPO signaling, the actual time course of calcium flux has been controversial and may also be a late event in the signaling pathway. Phospholipase C and phosphoinositol mechanisms do not appear to mediate the noted calcium flux.

One of the earliest events in EPO signaling is the rapid dephosphorylation of protein substrates. Choi *et al.* (1987) have described a protein of 43 kDa, which is rapidly dephosphorylated (on serine residues) in erythroid cell lines in response to EPO.

Other groups have concentrated on the later events in signal transduction. One approach to understanding the physiologic role of EPO is to identify EPO-responsive genes. If EPO is a viability agent, then it will induce genes, the protein products of which provide growth and maintenance of the cell. If EPO is a proliferative agent, then it will induce genes involved in DNA replication and cell division. This gene family may be similar to the set of competence genes induced by PDGF binding to the platelet-derived growth factor receptor (PDGF-R). If EPO is a differentiation agent, one would expect EPO to induce red blood cell-specific genes, such as glycophorin A and hemoglobin. Several groups have shown that binding of EPO to the EPO-R rapidly induces increased steady-state levels of *c-myc* and *c-myb* transcripts in erythroid cell (Chern *et al.*, 1991). This early EPO response appears in both EPO-responsive Rauscher cells and in Friend MEL cells.

At present, the best candidate mechanism for early EPO-mediated signal tranduction is via activation of protein kinase activity. It is known that other members of the cytokine receptor superfamily, particularly members with conserved cytoplasmic regions, such as the IL-2R beta chain and the IL-3R, have demonstrated tyrosine kinase activation. Binding of IL-2 to the IL-2R activates tyrosine kinase activity (Smith, 1989; Mills et al., 1990) and the p75 subunit is itself tyrosine phosphorylated. The binding of IL-3 to the IL-3R also activates tyrosine kinase activity rapidly, and a 140-kDa polypeptide is phosphorylated. Because the EPO-R cDNA confers EPO dependence upon IL-3–dependent cells, some common features of EPO and IL-3–mediated signal transduction must exist.

Specific substrates for tyrosine kinase activity have been identified. For instance, c-raf, a cytoplasmic serine kinase, which is itself tyrosine phosphorylated (Li et al., 1991), is activated in response to IL-3 or GM-CSF binding to its respective receptor (Carroll et al., 1990) or in response to IL-2 binding (Turner et al., 1991). Other data suggest that the EPO-R activates kinase activity as well. For instance, kinase inhibitors block the EPO-induced c-myc response with a rank order of potency suggestive of a role for protein kinase C (Spangler and Sytkowski, 1990). Another study shows that EPO induces protein kinase C activation in the nuclei of erythroid progenitor cells (Mason Garcia et al., 1990).

EPO has been shown to rapidly induce tyrosine kinase activity in early IL-3–dependent cells (Quelle and Wojchowski, 1991). Using a cell line with a proliferative response to both IL-3 and EPO, this work demonstrated that there is some overlap among the specific substrates that are phosophorylated in response to these growth factors. The six major substrates were tyrosine phosphorylated in response to low (100–500 pM) EPO concentrations. Maximum levels of EPO-dependent phosphorylation were achieved within only 3–7 minutes. We have obtained similar results with EPO stimulation of the EPO-dependent Ba/F3-EPO-R cell lines (D. Linnekin et al., manuscript in preparation). Interestingly, EPO-induced serine phosphorylation of a 68-kDa substrate was also observed.

Given that stimulation of the EPO-R activates tyrosine kinase activity, the next critical step in understanding signaling will be the identification of a specific tyrosine kinase associated with the EPO-R. Although the EPO-R itself does not have a tyrosine kinase catalytic domain, it may bind to a cytoplasmic or membrane-associated tyrosine kinase in a manner similar to the association of CD4 with lck (Veillette et al., 1988; Rudd et al., 1988). We have identified a critical domain of the cytoplasmic region of the EPO-R which is required for signaling and may associate with such a tyrosine kinase (D'Andrea et al., 1992).

VI. Conclusion

Our initial goal in cloning the EPO-R was to understand the role of EPO in the physiologic control of erythropoiesis. Study of the cloned EPO-R has provided a molecular explanation for a few of the physiologic observations about EPO that have been made over the last several decades.

First, *in vitro* studies have shown clearly that EPO does not have a role in commitment of the pluripotent stem cell to the erythroid lineage. Accordingly, molecular studies have shown that transcriptional activation of the EPO-R gene occurs only after commitment and only after expression of the erythroid transcription factor, GATA-1. Second, the FVA erythroblast system demonstrates that EPO is a viability agent for erythroblasts. Accordingly, the EPO-R, expressed in Ba/F3 cells, confers EPO dependence on these cells, whereas the cells do not differentiate in response to EPO. How a single EPO-R polypeptide can mediate both proliferative and differentiative effects in proerythroblasts remains a central unanswered question. Finally, *in vitro* culture has shown that erythroid cells have altered EPO sensitivity, depending on their stage of differentiation. Accordingly, the recent identification of a serine-rich negative regulatory domain of the EPO-R could provide a clue to the cellular control of EPO sensitivity during erythroid differentiation. The next major steps in the study of EPO physiology should derive from study of the second messenger-signaling mechanisms activated by the EPO-R.

References

Adachi, A., Sakai, K., Kitamura, N., Nakanishi, S., Niwa, O., Matryama, M., and Ishimoto, A. (1984). *J. Virol.* **50**, 813–821.

Atkins, H., Brandy, V. C., Papyannopoulou, T. (1991). *Blood* **77**, 2577–2582.

Bargman, C., Hung, M., and Weinberg, R. A. (1986). *Cell* **45**, 649–657.

Bazan, J. F. (1989). *Biochem. Biophys. Res. Commun.* **164**, 788–793.

Bazan, J. F. (1990a). *Proc. Natl. Acad. Sci. U.S.A.* **87**, 6934–6938.

Bazan, J. F. (1990b). *Immunol. Today* **11**, 350–354.

Bonanou-Tzedaki, S. A., Setchenska, M. S., and Arnstein, H. R. (1986). *Eur. J. Biochem.* **155**, 363–370.

Bonanou-Tzedaki, S. A., Sohi, M. K., and Arnstein, H. R. (1987). *Exp. Cell Res.* **170**, 276–289.

Borman, B. J., Knowles, W. J., and Marchesi, V. T. (1989). *J. Biol. Chem.* **264**, 4033–4037.

Bosselman, R., Van Straaten, F., Van Beveren, C., Verma, I., and Vogt, M. (1982). *J. Virol.* **44**, 19–31.

Boussios, T., Bertles, J. R., and Goldwasser, E. (1989). *J. Biol. Chem.* **264**, 16017–16021.

Boutin, J. M., Jolicoeur, C., Okamura, H., Gagnon, J., Edery, M., Shirota, M., Banville, D., Dusanter-Fourt, I., Djiane, J., and Kelly, P. A. (1988). *Cell* **53**, 69–77.

Broudy, V. C., Nakamoto, B., Lin, N., and Papyannopoulou, T. (1990). *Blood* **75**, 1622.

Budarf, M., Huebner, K., Emanuel, B., Croce, C. M., Copeland, N. G., Jenkins, N. A., D'Andrea, A. D. (1990). *Genomics* **8**, 575–578.

Carroll, M. P., Clark-Lewis, I., Rapp, U. T., and May, W. S. (1990). *J. Biol. Chem.* **265**, 19815–19817.

Chern, Y., Spangler, R., Choi, H. S., and Sytkowski, A. J. (1991). *J. Biol. Chem.* **266**, 2009–2012.

Choi, H. S., Wojchowski, D. M., and Sytokowski, A. J. (1987). *J. Biol. Chem.* **262**, 2933–2936.

Chung, S. W., Wolff, L., and Ruscetti, S. K. (1989). *Proc. Natl. Acad. Sci. U.S.A.* **86**, 7957–7960.

Cosman, D., Lyman, S. D., Idzerda, R. L., Beckman, M. P., Park, L. S., Goodwin, R. G., and March, C. J. (1990). *Trends Biosci.* **15**, 265–270.

D'Andrea, A. D., and Jones, S. S. (1991). *Semin. Hematol.* **28**, 152–157.

D'Andrea, A. D., and Zon, L. I. (1990). *J. Clin. Invest.* **86**, 681–690.

D'Andrea, A. D., Lodish, H. F., and Wong, G. G. (1989a). *Cell* **57**, 277–285.

D'Andrea, A. D., Fasman, G. D., and Lodish, H. F. (1989b). *Cell* **58**, 1023–1024.

D'Andrea, A. D., Fasman, G. D., and Lodish, H. F. (1990a). *Curr. Opin. Cell Biol.* **2**, 648–651.

D'Andrea, A. D., Szklut, P. J., Lodish, H. F., and Alderman, E. F. (1990b). *Blood* **75**, 874–880

D'Andrea, A. D., Yoshimura, A., Youssoufian, H., Zon, L. I., Koo, J. W., and Lodish, H. F. (1992). *Mol. Cell Biol* **12**, 2949–2957.

Dresler, S., Ruta, M., Murray, M. J., and Kabat, D. (1978). *J. Virol.* **30**, 564–575.

Erslev, A. (1987). *N. Engl. J. Med.* **316**, 101–103.

Eschbach, J. W., Egric, J. C., Downing, M. R., Browne, J. K., and Adamson, J. W. (1987). *N. Engl. J. Med.* **316**, 73–78.

Eschbach, J. W., Kelly, M. R., Haley, N. R., Abels, R. I., and Adamson, J. W. (1989). *N. Engl. J. Med.* **321**, 158–163.

Fukunaga, R., Ishizaka-Ikeda, E., Seto, Y., and Nagata, S. (1990). *Cell* **61**, 341–350.

Gearing, D. P., King, J. A., Gough, N. M., and Nicola, N. A. (1989). *EMBO J.* **8**, 3667–3676.

Gliniak, B. C., and Kabat, D. (1989). *J. Virol.* **63**, 3561–3568.

Goldwasser, E. (1984). *Blood Cells* **10**, 147–162.

Goodwin, R. G., Friend, D., Ziegler, S. F., Jerzy, R., Falk, B. A., Gimpel, S., Cosman, D., Dower, D. K., and Park, L. S. (1990). *Cell* **60**, 941–951.

Gorman, D. M., Itoh, N., Kitamura, T., Schreurs, J., Yonehara, S., Yahara, I., Arai, K., and Miyajima, A. (1990). *Proc. Natl. Acad. Sci. U.S.A.* **87**, 5459–5463.

Hanks, S. K., Quinn, A. M., and Hunter, T. (1988). *Science* **241**, 42–52.

Hatakeyama, M., Tsudo, M., Minamoto, S., Kono, T., Joi, I., Miyata, T., Miyasaka, M., and Taniguchi, T. (1989a). *Science* **244**, 551–556.

Hatakeyama, M., Mori, H., Doi, T., and Taniguchi, T. (1989b). *Cell* **59**, 837–845.

Hayashida, K., Kitamura, T., Gorman, D. M., Arai, K., Yokota, T., and Miyajima, A. (1990). *Proc. Natl. Acad. Sci. U.S.A.* **87**, 9655–9659.

Hibi, M., Murakami, M., Saito, M., Hirano, T., Taga, T., and Kishimoto, T. (1990). *Cell* **63**, 1149–1157.

Hoatlin, M. E., Kozak, S. L., Lilly, F., Chakraborti, A., Kozak, C. A., and Kabat, D. (1990). *Proc. Natl. Acad. Sci. U.S.A.* **87**, 9985–9989.

Honegger, A., Dull, T. J., Szapary, D., Komoriya, A., Kris, R., Ullrich, A., and Schlessinger, J. (1988). *EMBO J.* **7**, 3053–3060.

Idzerda, R. L., March, C. J., Mosley, B., Lyman, S. D., Vanden Bos, T., Gimpel, S. D., Din, W. S., Grabstein, K. H., Widner, M. B., Park, L. S., Cosman, D., and Beckman, M. P. (1990). *J. Exp. Med.* **171**, 861–873.

Itoh, N., Yonehara, S., Schreurs, J., Gorman, D. M., Maruyamna, K., Ishii, A., Yahara, I., Arai, K. I., and Miyajima, A. (1990). *Science* **247**, 324–327.

Jones, S. S., D'Andrea, A. D., Haines, A. L., and Wong, C. C. (1990). *Blood* **76**, 31–35.

Koch, W., Zimmermann, W., Oliff, A., and Friedrich, R. (1984). *J. Virol.* **45**, 828–840.

Krantz, S. B., and Goldwasser, E. (1984). *Proc. Natl. Acad. Sci. U.S.A.* **81**, 7574–7578.

Kuramochi, S., Ikawa, Y., and Todokoro, K. (1990). *J. Mol. Biol.* **216**, 567–575.

Landschultz, K. T., Noyes, A. N., Rogers, O., and Boyer, S. H. (1989). *Blood* **73**, 1476–1486.

Leung, D. W., Spencer, S. A., Cachianes, G., Hammonds, R. G., Collins, C., Henzel, W. J., Barnard, W. J., Waters, M. J., and Wood, W. I. (1987). *Nature* **330**, 537–543.

Li, J. P., Bestwick, R. K., Spiro, C., and Kabat, D. (1978). *J. Virol.* **61**, 2782–2792.

Li, J. P., and Baltimore, D. (in press). *J. Virol.*

Li, J. P., D'Andrea, A. D., Lodish, H. F., and Baltimore, D. (1990). *Nature* **343**, 762–764.

Li, P., Wood, K., Mamon, H., Haser, W., and Roberts, T. (1991). *Cell* **64**, 479–482.

Manolios, N., Bonifacino, J. S., and Klausner, R. D. (1990). *Science* **249**, 274–277.

Marks, P. A., and Rifkind, R. A. (1978). *Annu. Rev. Biochem.* **47**, 419–448.

Mason-Garcia, M., Weill, C. L., and Beckman, B. S. (1990). *Biochem. Biophys. Res. Commun.* **168**, 490–497.

Mayeux, P., Billat, C., and Jacquot, R. (1987a). *FEBS Lett.* **211**, 229–233.

Mayeux, P., Billat, C., and Jacquot, R. (1987b). *J. Biol. Chem.* **262**, 13985–13990.

McCaffery, P. J., Fraser, J. K., Lin, F. K., and Berridge, M. V. (1989). *J. Biol. Chem.* **164**, 10507–10512.

McGrath, M., and Weissman, I. (1978). *In* "Cold Spring Harbor Symposium: Normal and Neoplastic Hematopoietic Cell Differentiation" p. 557. Cold Spring Harbor Laboratory, New York.

McGrath, M., and Weissman, I. (1979). *Cell* **17**, 65–75.

Mills, G. B., May, C., McGill, M., Fungi, M., Baker, M., Sutherland, R., and Green, W. C. (199). *J. Biol. Chem.* **265**, 3561–3567.

Moreau-Gachelin, F., Tovitian, A., and Tambourin, P. (1988). *Nature* **331**, 277–280.

Mosley, B., Beckmann, M. P., March, C. J., Idzerda, R. L., Gimpel, S. D., VandenBox, T., Friend, D., Alpert, A., Anderson, D., Jackson, J., Wignall, J. M., Smith, C., Gallis, B., Sims, J. E., Urdal, D., Widmer, M. B., Cosman, D., and Park, L. S. (1989). *Cell* **59**, 335–348.

Quelle, F. W., and Wojchowski, D. M. (1991). *J. Biol. Chem.* **266**, 609–614.

Roussel, M. F., Dull, T. J., Rettenmier, C. W., Ralph, P., Ullrich, A., and Sherr, C. J. (1987). *Nature* **325**, 549–552.

Rudd, C. E., Trevillyn, J. M., Dasgupta, J. D., Wong, L. L., and Schlossman, S. F. (1988). *Proc. Natl. Acad. Sci. U.S.A.* **85**, 5190–5194.

Ruscetti, S. K., Janesch, N. J., Chakraborti, A., Sawyer, S. T., and Hankins, W. D. (1990). *J. Virol.* **43**, 1057–1062.

Russell, E. (1979). *Adv. Genet.* **20**, 373–459.

Sasaki, R., Yamagawa, S., Hitoma, K., and Chiba, H. (1987). *Eur. J. Biochem.* **168**, 43–48.

Sawyer, S. T. (1989). *J. Biol. Chem.* **264**, 13343–13347.

Sawyer, S. T., Krantz, S. B. and Luna, J. (1987a). *Proc. Natl. Acad. Sci. U.S.A.* **84**, 3690–3694.

Sawyer, S. T., Krantz, S. B., and Goldwasser, E. (1987b). *J. Biol. Chem.* **262**, 5554–5562.

Sawyer, S. T., Hosoi, T., and Krantz, S. B. (1988). *Blood* **72**, 133a.

Smith, K. A. (1989). *Annu. Rev. Cell Biol.* **5**, 397–425.

Sonyri, M., Gibon, I., Penciolelli, J. F., Heard, J. M., Tambourin, P., and Wendling, F. (1990). *Cell* **63**, 1137–1147.

Spangler, R., and Sytkowski, A. J. (1990). *Blood* **76**, 471a.

Spivak, J. L., Pham, T., Isaacs, M., and Hankins, W. D. (1991). *Blood* **77**, 1228–1233.

Steele, R. E. (1989). *TIBS* **14**, 201–202.

Sytkowski, A. J., and Donahue, K. A. (1987). *J. Biol. Chem.* **262**, 1161–1168.

Taga, T., Hibi, M., Hirata, Y., Yamasaki, K., Yasudawa, K., Matsuda, T., Hirano, T., and Kishimioto, T. (1989). *Cell* **58**, 573–581.

Todokoro, K., Kanazawa, S., Amnuma, H., and Ikawa, Y. (1987). *Exp. Hematol.* **15**, 833–837.

Todokoro, K., Kanazawa, S., Amnuma, H., and Ikawa, Y. (1988). *Biochim. Biophys. Acta* **943**, 326–330.

Tojo, A., Fukamachi, H., Saito, T., Kasuga, M., Urabe, A., and Takaku, F. (1988). *Cancer Res.* **48**, 1812–1822.

Turner, B., Rapp, U. App, H., Greene, M., Dobashi, K., and Reed, J. (1991). *Proc. Natl. Acad. Sci. U.S.A.* **88**, 1227–1231.

Ullrich, A., and Schlessinger, J. (1990). *Cell* **61**, 203–212.

Veillette, A., Bookman, M. A., Horak, E. M., and Bolen, J. B. (1988). *Cell* **55**, 301–308.

Winkelmann, J. C., Penny, L. A., Deaven, L. L., Forget, B.G., and Jenkins, R. B. (1990). *Blood* **76**, 24–30.

Wong, G. G., Jones, S. S., and D'Andrea, A. D. (in press). *In* "Erythropoietin: Molecular, Cellular, and Clinical Biology" (A. J. Ersler and J. W. Adamson, eds.).

Yamasaki, K., Taga, T., Hirata, Y., Yawata, H., Kawanishi, Y., Seed, B., Taniguchi, T., Hirano, T., and Kishimoto, T. (1988). *Science* **241**, 825–828.

Yoshimura, A., D'Andrea, A. D., and Lodish, H. F. (1990). *Proc. Natl. Acad. Sci. U.S.A.* **87**, 4139–4143.

Yoshimura, A., and Longmore, G., and Lodish, H. F. (1991). *Nature* **348**, 647–649.

Youssoufian, H., Zon, L. I., Orkin, S. H., D'Andrea, A. D., and Lodish, H. F. (1990). *Mol. Cell. Biol.* **10**, 3675–3682.

Cytokine Receptors: A New Superfamily of Receptors

Jolanda Schreurs,* Daniel M. Gorman,† and Atsushi Miyajima†
*Department of Protein Chemistry, Chiron Corporation, Emeryville, California 94608
†Department of Molecular Biology, DNAX Research Institute of Molecular and Cellular Biology, Palo Alto, California 94304

I. Introduction

The interleukins (ILs) are a class of soluble polypeptide growth factors that were originally defined as products of activated macrophages, T or B cells, and modulators of the function and proliferation of lymphoid cells. They are now more broadly categorized as lymphokines–cytokines, which can act on a wide variety of cells (Arai et al., 1990). A multitude of factors, as many as 45, are elaborated by activated cells of the immune system and additional factors are being described at a rapid rate, including the recently discovered molecules: the c-kit ligand (Sl factor) (Zsebro et al., 1990; Hung et al., 1990; Anderson et al., 1990), IL-10 (Fiorentino et al., 1989; Moore et al., 1990), and IL-11 (Paul et al., 1990). Functionally, these various molecules serve as stimulators or inhibitors of growth, differentiation, and cellular function (Arai et al., 1990; Nicola, 1989).

Hematopoiesis is a dynamic process that produces all the blood cells from multipotential stem cells. As the development of various lineages of cells is regulated by a number of cytokines, two types of hematopoiesis may be distinguished based on cytokine production: constitutive versus inducible. Normal constitutive hematopoiesis in the bone marrow may be regulated by soluble cytokines produced by stromal cells, as well as by direct contact with membrane or extracellular matrix-associated proteins on the stromal cells. However, hematopoiesis associated with the inflammatory reaction is mainly regulated by lymphocytes and is inducible by antigen (Miyajima et al., 1988). Cytokines are produced from multiple types of cells and in turn target a variety of cells that manifest cell-type specific as well as pleiotropic activities. A complicated cellular network among various kinds of blood cells is formed through the various cytokines. As a consequence, aberrations in the elaboration of cytokines, the expression of their receptors, and/or the signal transduction pathways also subserve the neoplastic process (Pierce, 1989).

An understanding of the normal and neoplastic hematopoietic process demands the biochemical elucidation of the structure and composition of the polypeptides that comprise the functional, high-affinity ligand-binding receptor complexes, as well as those enzymes that catalyze the intracellular signals transduced by the receptors. The cloning of the genes encoding the ligand-binding and signal-transducing components of the lymphokine–cytokine receptors has led to the discovery of a new superfamily of growth factor receptors (Gearing *et al.*, 1989; Itoh *et al.*, 1990; Idzerda *et al.*, 1990; Bazan, 1989). Unlike the growth factor receptors with intrinsic tyrosine kinase activity exemplified by the platelet-derived growth factor (PDGF), insulin, epidermal growth factor (EGF), and macrophage colony-stimulating factor (MCSF) receptors, the cytokine receptors do not encode for a tyrosine kinase. Yet, many of the cytokine receptors have been shown to act through a tyrosine phosphorylation process, indicating that some portion of the signal transduction process is mediated by an independent, receptor-associated tyrosine kinase. Substantial evidence also exists for the association and activation of receptors with serine–threonine kinase(s), as well as with the guanosine triphosphate (GTP) binding protein, p21ras, suggesting that the cytokine receptors utilize more than one signal transduction mechanism. The forthcoming challenges will be to identify and delineate the various roles and interrelationships among the signaling systems and to assess whether their actions occur in parallel or in series.

The intent of this chapter is to provide a summary of the following: (1) the characteristics that define the structure and composition of the receptors and their ability to bind ligand and transduce signals, (2) known signal transduction pathways, and (3) the evidence for oncogenic processes dependent on aberrant cytokine–receptor interactions. For the purposes of this review, we will restrict ourselves to the discussion of well-characterized receptors for the classically described cytokines, the interleukins IL-2–7, as well as the growth factors granulocyte colony-stimulating factor (G-CSF), granulocyte-macrophage–colony stimulating factor (GM-CSF), and erythropoietin (EPO). Because two other chapters in this volume focus on the receptors for G-CSF and EPO, discussion of these two receptors will be limited to comparative properties. Based on structural and pharmacological properties, we define two subdivisions of the cytokine receptors based on the subunit composition required to produce high-affinity ligand binding. The functional, high-affinity binding of the receptors for IL-2, IL-3, GM-CSF, IL-5, and IL-6 receptors clearly requires the cooperation of several independent subunits. In contrast, other receptors, such as that for IL-4 and G-CSF, appear to require the expression of only a single protein to define a high-affinity binding site, although evidence points to the existence of homodimeric states.

TABLE I

Members of the Cytokine Receptor Family

Receptors	MW[a]	Human		Mouse	
		AA[b]	K_d^c	AA[b]	K_d^c
IL-2R					
α	55	272	10 nM	268	10 nM
β	75	551	>1 μM	539	ND
γ	64	369	ND	?	?
α/β			200 pM		?
βγ	—	—	600 pM	?	?
αβγ	—	—	70 pM	?	?
IL-3R					
α	70	387	100 nM	396	40 nM
β	120	897	ND	878	20 nM
α/β	—	—	100 pM	—	200 pM
IL-4R	130	825	80 pM	810	165 pM
IL-5R					
α	60	420	1 nM	468	6 nM
β	120	897	ND	896	ND
α/β	—	—	ND	—	100 pM
IL-6R					
α	80	468	5 nM	460	?
β	130	918	ND	?	?
α/β	—	—	100 pM	—	?
IL-7R	75	459	180 pM	459	150 pM
			10 nM		10 nM
GM-CSFR					
α	80	400	5 nM	387	50 nM
β	120	897	ND	896	ND
α/β	—	—	100 pM	—	30 pM
G-CSFR	140	836		837	290 pM
EpoR	100	508	?	507	30 pM
					210 pM

ND, not detectable.

[a]Approximate molecular weight of mature proteins (kDa).

[b]Total amino acid residues, including the signal peptide.

[c]Approximate dissociation constants when expressed in COS7 cells.

II. Family of Lymphokine–Cytokine Receptors

A. Expression Cloning

Since 1988 the genes encoding receptors for a large number of lymphokines have been identified and cloned primarily through expression cloning techniques

Sequence alignment figure. Structural regions labeled A, B, D, E (top) and E′, F′, G′ (bottom). Gap lengths shown in parentheses.

Top block (regions A, B, D, E)

Sequence		A (ss)		B		D (ss)		E
hIL2Rβ	33	FTCFYNS	(–)	RANISC VW	(25)	CELLPVSQ	(1)	SWACNLILGAPD
βc I	32	LRCYNDY	(–)	TSHITCRW	(27)	CDLSDDMP	(10)	PRRCVPRRCVIP
βc II	247	LECFFDG	(–)	AAVLSCSW	(26)	CSPVLREG	(6)	RHHCQIPVPDPA
hIL4R	31	PTCVSDY	(–)	MSISTCEW	(27)	CIPENNGG	(1)	GCVCHLLMDDVV
mIL5Rα	128	LTCTTHT	(11)	QVSLRCTW	(24)	CQEYSRDA	(3)	NTACWFPRTFIN
hIL6Rα	118	LSCFRKS	(1)	LSNVVCEW	(30)	CQYSQESQ	(–)	KFSCQLAVPEGD
hIL6Rβ	131	LSCIVNE	(–)	GKKMRCEW	(25)	CKAKRDTP	(–)	–TSCTVDYSTVY
hIL7R	39	FSCYSQL	(5)	QHSLTCAF	(22)	CLNFRKLQ	(15)	SNICVKVGEKSL
hGMCSFRα	123	FSCFIYN	(–)	ADLMNCTW	(26)	CPYYIQDS	(2)	HVGCHLDNLSGL
hGCSFR	128	LSCLMNL	(1)	TSSLICQW	(32)	CVPKDG–	(–)	QSHCCIPRKHLL
hEPOR	49	LLCFTER	(–)	LEDLVCFW	(36)	CRLHQAPT	(5)	RWFCSLPTADTS
Consensus		L TsC F		S L C W		C		C LP L

Bottom block (regions E′, F′, G′)

Sequence		E′		F′		G′		
hIL2Rβ	103	LTP	–	DTQYEFQVRVKPL	(2)	EFTTWSPWSQPLAFRT	(9)	TM
βc I	102	LMP	–	SSTYVARVRTRLA	(4)	LSGRPSKWSPEVCWDS	(201)	TM
βc II	87	LEP	–	STRYWARVRVRTS	(3)	YNGIWSEWSEARSWDT	(2)	TM
hIL4R	96	LKS	–	GISYRARVRAWAQ	(1)	YNTTWSEWSPSTKWHN	(9)	TM
mIL5Rα	93	IDD	–	VSTYSIQVRAAVS	(4)	MPGRWGEWSQPIYVGK	(9)	TM
hIL6Rα	98	AWS	–	GLRHVVGLRAQEE	(1)	GQGEWSEWSPEAMG–T	(45)	TM
hIL6Rβ	98	LKP	–	FTEYVFRIRCMKE	(1)	GKGYWSDWSEEASGIT	(298)	TM
hIL7R	78	LQP	–	AAMYEIKVRSIPD	(2)	FKGFWSEWSPSYYFRT	(11)	TM
hGMCSFRα	97	PSS	–	EPRAKHSVKIRAA	(2)	RILNWSSWSEAIEFGS	(3)	TM
hGCSFR	102	LLP	–	ATAYTLQIRCIRW	(1)	LPGHWSDWSPSLELRT	(298)	TM
hEPOR	94	LRG	–	RTRYTFAVRARMA	(3)	FGGFWSAWSEPVSLLT	(6)	TM
Consensus		L P		TRY RQVRAR		G WSEWSPP F T		

(Table I). Expression cloning has largely supplanted traditional biochemical purification because the amounts of receptor are relatively low (100–20,000 receptors/cell). The evolution of various expression cloning techniques has resulted in a process whereby the gene for a receptor can be isolated in a relatively short period of time. In particular, the transient expression system using COS7 cells has been successfully applied for receptor gene cloning. Seed and Aruffo (1987) developed a method to enrich for receptor genes from total cDNA libraries. COS7 cells are transfected with a cDNA library made in an SV40-based expression vector and cells expressing the receptor are enriched by panning, using the antibody against the receptor. Plasmid DNAs from the enriched COS7 cells are recovered and used for the next enrichment. By repeating the cycle several times, the receptor cDNA can be highly enriched. Biotinylated ligands (Harada *et al.*, 1990) or fluorescenated ligands (Yamasaki *et al.*, 1988) can also be used as probes for enrichment. Another method also utilizes transient expression in COS7 cells. A cDNA library is divided into small pools and pools of cDNA are transfected into COS7 cells. The receptor expression is detected using radioactive ligand binding either by autoradiography (Sims *et al.*, 1988; Gearing *et al.*, 1989) or by direct counting of cell bound radioactivity (D'Andrea *et al.*, 1989). Positive pools are further divided into smaller pools and screening is continued until a single clone is obtained. Using these expression cloning techniques, a number of cytokine receptor genes have been isolated as described next.

B. Definition of the Cytokine Receptor Superfamily

The first indication for the existence of a new superfamily of receptors resulted from an exploration of the structural and evolutionary relationships of the sequences for the growth hormone, prolactin, IL-6, EPO (Showers and D'Andrea, this volume), and IL-2 receptors (Bazan, 1989). Linked pattern-matching techniques led to a consensus sequence profile for the extracellular domains of the aforementioned receptors. The designation "cytokine receptors" or "hematopoietic receptors" was subsequently applied (Gearing *et al.*, 1989; Itoh *et al.*, 1990; Idzerda *et al.*, 1990) upon the elucidation of the related sequences for the IL-3, IL-4, GM-CSF, G-CSF, IL-5, and IL-7 receptors.

FIG. 1 Conserved amino acid sequence of cytokine receptors. Human cytokine receptors except for IL-5 receptor are compared. βc is the common β-subunit of the IL-3 and GM-CSF receptors. The conserved cysteine residues and the WS × WS sequences are boxed. Possible disulfide bridges are shown as bars with SS. A, B, D, E, E', F', and G' indicate the blocks shown in Figure 3. Additional members of the cytokine receptor family, including the IL-2Rγ chain, show this same pattern of conserved amino acids. Reproduced with permission from the *Annual Review of Immunology* (Vol. 10), 1992 by Annual Reviews, Inc.

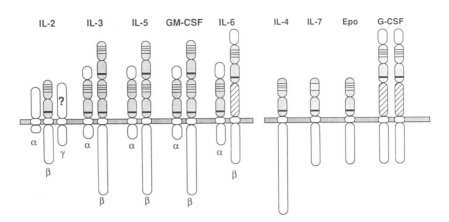

FIG. 2 Structure of cytokine receptors. Shaded area shows the region conserved in various cytokine receptors. Four conserved cysteine residues are indicated by bars. The WS × WS sequences are shown by thick bars. Hatched area in the IL-6 receptor and the G-CSF receptor indicates the region homologous to contactin. The recently identified (Takeshita *et al.*, 1992) γ-subunit of the IL-2 receptor is also shown. Reproduced with permission from the Annual Review of Immunology vol. 10 by Annual Reviews, Inc.

A structural interrelationship exists between the extracellular domains of these receptors (Figs. 1 and 2) as described next. Briefly, a general feature of the cytokine receptors is that a stretch of about 200 amino acid residues contain four spatially conserved cysteines in the N-terminal side and an additional strongly conserved sequence, WS × WS (where × defines any amino acid) in the C-terminal side. The extracellular domains are predicted to form antiparallel β-sandwiches with similarity to Type III fibronectin.

C. IL-2 Receptor

IL-2, a product of activated T cells (Smith, 1989), is a 15.5-kDa protein required for the proliferation and differentiation of cells in the T and B lymphoid lineage. Antigenic or mitogenic T cell activation stimulates the release of IL-2, which then interacts with a high-affinity receptor (K_d = 20–50 pM) to stimulate cell proliferation. This high-affinity receptor is found primarily on antigenically or mitogenically activated T cells, and not on resting T cells, leading to a positive feed-forward loop and rapid expansion of the T cell population (Smith, 1989). A second, abundant low-affinity IL-2 binding protein is found (10–50,000 receptors/cell; K_d = 10 nM) in various cells, including T, B, and some myeloid cells. Moreover, on certain cells, such as large granular lymphocytes, a third intermediate affinity class of the IL-2 receptor ($K_d = 1$ nM) is present. A composite model of the IL-2 receptor is that the low- and intermediate-affinity receptors interact to form the high-affinity receptor. The structural reconstruction of this complex multisubunit receptor has been facilitated by the molecular cloning of the genes for three subunits, p55 (α-subunit), p75 (β-subunit), and p64 (γ-subunit Takeshita et al., 1992).

The α-subunit, p55, was identified with the use of a monoclonal antibody, anti-TAC (T Activated Cells), which blocked high-affinity IL-2 binding and blocked IL-2–dependent proliferation (Waldmann, 1989). Cloning of the cDNA for the TAC molecule revealed that the α-subunit is a unique protein with a small cytoplasmic domain (Leonard et al., 1984; Nikaido et al., 1984; Cosman et al., 1984). Transfection studies showed that IL-2Rα bound IL-2 with low affinity, did not internalize, and was not able to transduce biological signals (Greene et al., 1985; Sabe et al., 1984).

A second binding site referred to as p75 or the IL-2Rβ chain was identified through cross-linking studies (Sharon et al., 1986, Tsudo et al., 1986, Dukovich et al., 1987; Teshigawara et al., 1987; Robb et al., 1987). Analysis of the binding characteristics of IL-2Rβ was facilitated by the discovery that certain cell lines could independently express either IL-2Rβ (large granular lymphocytes and a gibbon T cell line, MLA-144) or IL-2Rα (a human T cell line, MT-1) (Bich-Thuy et al., 1987; Robb et al., 1987; Teshigawara et al., 1987). In these cells lines the expression of only IL-2Rα manifests a binding site with low affinity for IL-2 ($K_d - 10$ nM), whereas expression of IL-2Rβ results in the appearance of a binding site with intermediate affinity ($K_d = 1$ nM). Co-expression of both subunits was hypothesized to cause the formation of the high-affinity binding site ($K_d = 10–50$ pM) (Tsudo et al., 1987; Teshigawara et al., 1987; Dukovich et al., 1987), resulting from the combined fast on-rate of IL-2Rα and slow off-rate of IL-2Rβ (Wang and Smith, 1987). This hypothesis was unequivocally verified by cloning and expression of the gene for IL-2Rβ (Hatakeyama et al., 1989a). The gene for human IL-2Rβ encodes a 525-amino acid protein, with a 214-amino acid extracellular domain, a single transmembrane-spanning region, and a large 286-amino acid intracellular domain. The cytoplasmic domain of the IL-2Rβ is particularly rich in prolines (42 of 286) and serines (30 of 286), similar to the IL-3, IL-4, G-CSF, and EPO receptor intracellular domains. Co-expression of the cloned α- and β-subunits in COS7 cells conferred a high-affinity binding site for IL-2, whereas transfection of only the cloned IL-2Rβ cDNA in nonlymphoid lines, such as COS7 or L929 fibroblasts, resulted in cell surface expression as detected by monoclonal antibodies, yet did not result in the capacity to bind IL-2 (Hatakeyama et al., 1989; Minamoto et al., 1990). Subsequent analysis of the binding characteristics of the fibroblast-expressed and detergent-solubilized recombinant IL-2Rβ did show that it binds IL-2, albeit with a dramatically diminished affinity ($K_d = 1$ μM) (Tsudo et al., 1990). In striking contrast, expression of IL-2Rβ in T lymphoid cells, such as Jurkat cells, resulted in an intermediate-binding affinity for IL-2 ($K_d = 1$ nM), and the ability to internalize IL-2

(Hatakeyama *et al.*, 1989a; Tsudo *et al.*, 1989). In conclusion, lymphoid cells appear to express some element, absent in fibroblastic cells, which modulates the structure of IL-2Rβ, leading to an enhanced recognition of IL-2.

It is also apparent that this third element, the γ-subunit, can modulate the high-affinity binding state of the α–β complex in addition to its ability to modulate the intermediate affinity state of IL-2Rβ. Thus, after co-expression of the α- and β-subunits in cells ordinarily nonresponsive to IL-2, the dissociation constant was consistently higher (K_d = 215 pM) than that found for the native receptor in T cells, such as HUT-102 cells, (K_d = 10–50 pM) (Minamoto *et al.*, 1990; Ringheim *et al.*, 1990). A marked difference in the rate of dissociation characterizes the two different forms: dissociation of IL-2 from α–β as expressed in CHO cells is rapid ($\tau_{1/2}$ = 19.5 minutes) compared with that from the native receptor in HUT-1 cells ($\tau_{1/2}$ = 170 minutes) (Ringheim *et al.*, 1990). These studies indicate that additional molecules interact with α and β to fine-tune the affinity of the binding site. The reconstituted α–β complex in fibroblasts is incapable of mediating signal transduction (Minamoto *et al.*, 1990). The mutational analyses of murine IL-2 by Zurawski and Zurawski (1989) and Zurawski *et al.* (1990) provided further insight into the pharmacological nature of the γ-subunit, demonstrating that this third receptor component is important for both ligand binding and signal transduction: IL-2 mutants at position Gln[141] are defective in binding to the native high-affinity IL-2R but bind normally to recombinant IL-2R α and α–β. Studies of the recently cloned molecule, p64 (Takeshita *et al.*, 1992), corroborate its role as the γ-subunit of the IL-2R.

D. IL-3 Receptor

Interleukin-3 is a primary product of activated T cells and mast cells (Schrader, 1986; Metcalf, 1988; Arai *et al.*, 1990) and is thought to function as a mediator of the inflammatory response, rather than as a factor in the constitutive hematopoietic process in bone marrow (Miyajima *et al.*, 1988). IL-3 or multi-CSF stimulates colony formation of multiple different myeloid lineages as well as primitive stem cell proliferation (Schrader, 1986; Metcalf, 1988; Arai *et al.*, 1990). It is now clear that factors in addition to IL-3, including the c-kit ligand and IL-6, act on primitive stem cells. Singly, none of these factors gives optimal stimulation; rather combinations result in synergistic effects on the stem cells defined as Thyl^{lo} Scal$^+$ Lin$^-$ (Heimfeld *et al.*, 1991). In addition to its effects on early stem cells, IL-3 modulates the growth and functions of a variety of progenitor and differentiated cells, including pre-B cells, eosinophils, mast cells, megakaryocytes, and granulocytes–macrophages. Interestingly, in a small number of B-lineage acute lymphocytic leukemias (ALL), a translocation [t(5;14)(q31;q32)] joins the IgH and the IL-3 genes and results in a constitutive and serologically detectable increase in IL-3 levels (Meeker *et al.*, 1990). This distinct subtype of acute leukemia is also characterized by acute eosinophilia. Because both pre-B cells and eosinophils are responsive to IL-3, it is conjectured that an autocrine–paracrine process may govern both the B cell ALL and the eosinophilia.

The functions of mouse and human IL-3 are mediated by a low abundance (100–5000 receptors/cell), high-affinity (K_d = 100–300 pM) receptor (Palaszynski and Ihle, 1984; Park et al., 1986; Nicola and Metcalf, 1986; Kuwaki et al., 1989). In addition, IL-3 interacts with a low-affinity (K_d = 10 nM) receptor, which is best characterized by its rapid rate of dissociation (Schreurs et al., 1990). Original descriptions of the high-affinity mouse IL-3 receptor indicated that it was a molecule of 60–75 kDa (Park et al., 1986; Nicola and Peterson, 1986; May and Ihle, 1986; Sorensen et al., 1986). However, subsequent studies showed that the ligand binding moiety of the IL-3 receptor was of substantially higher mass, 110–150 kDa (Schreurs et al., 1989, 1990; Isfort et al., 1988a), and upon cross linking formed three cross-linked complexes of net molecular mass of 140, 120, and 70 kDa. Moreover, dissociation experiments showed that the low-affinity binding site was also a molecule of 120–140 kDa (Schreurs et al., 1990), with a cross-linking pattern identical to the high-affinity binding molecule previously characterized (Miyajima et al., 1989), leading to the postulate that interconversion could occur between the low- and high-affinity states of the receptor. The human IL-3 receptor on a human IL-3–dependent cell line, TF1, is similar to the mouse receptor in that a low abundance, high-affinity site (K_d = 170 pM) is cross-linked to give two bands of net mass of 135 and 70 kDa (Kuwaki et al., 1989). Description of a second low-affinity site on human acute myeloblastic leukemia (AML) cells (Park et al., 1989) suggests that the structural organization of the mouse and human receptors is similar.

An agonistic monoclonal antibody, F9, stimulated the growth of IL-3–dependent cells and caused the tyrosine phosphorylation of an identical repertoire of substrates as did IL-3, and furthermore was competitive with IL-3 for binding to the receptor. In cross-linking studies the antibody competed with [^{125}I]IL-3 binding to the three complexes described herein, lending support to the model that the functional receptor consisted of multiple proteins (Schreurs et al., 1989). The controversy over the mass of the native IL-3–binding protein was resolved by the subsequent cloning of the gene for an IL-3–binding protein, AIC2A, which upon cross-linking with IL-3 gave a similar pattern as that found on IL-3–dependent cells (Itoh et al., 1990).

The gene for the IL-3 receptor was identified via expression cloning (Itoh et al., 1990) using a monoclonal antibody, anti-Aic2 (Yonehara et al., 1990), made against an IL-3–dependent mast cell line, IC2. Expression of the Aic-2 antigen was found to correlate with the known distribution of the high-affinity IL-3 receptor; moreover, the anti-Aic2 antibody partially blocked IL-3 binding. A cDNA encoding the Aic2 antigen (AIC2A) was isolated by expression in COS7 cells and was shown to encode an IL-3–binding activity with the characteristics of a low-affinity receptor: $K_d = 10$ nM and a rapid off-rate for IL-3 of about 2–3 minutes (at 37°C) (Itoh *et al.*, 1990). AIC2A does not have any characteristics associated with a high-affinity, functional receptor, such as the ability to stimulate a tyrosine kinase activity or to internalize. The complete nucleotide sequence revealed an open reading frame encoding a mature protein of 856 amino acids (M_r = 94.7 kDa), which is subsequently glycosylated (13 kDa) (Murthy *et al.*, 1990) to give a fully processed receptor of about 110–120 kDa (J. Schreurs *et al.*, 1991). The extracellular region of 417 amino acids contains two relatively homologous segments of about 200 amino acids, both of which have a significant similarity to the extracellular 200-amino acid domains of various cytokine receptors (Itoh *et al.*, 1990). The intracellular domain of 413 amino acids does not have any consensus sequence for kinases but contains an unusually high proportion of prolines and serines (17% and 12%, respectively), which are also found in the intracellular domains of the IL-4, IL-2, G-CSF, and EPO receptors.

The high-affinity IL-3 receptor, purified by biotin-IL-3 affinity chromatography, utilizes the AIC2A protein as a binding element (Schreurs *et al.*, 1991). Because the functional IL-3 receptor binds IL-3 with high affinity and induces protein tyrosine phosphorylation, the high-affinity functional receptor must be a multisubunit complex that includes the AIC2A protein. This hypothesis was confirmed using the human IL-3 receptor as a model.

The expression cloning of the IL-3-binding protein cDNA, AIC2A, led to the simultaneous identification of a second cDNA, AIC2B, which was recognized by anti-Aic2 (Gorman *et al.*, 1990). AIC2B has 91% amino acid sequence identity with AIC2A, and these two cDNAs were derived from two distinct genes. Although the mRNA for AIC2B is more abundant than AIC2A, no specific binding for IL-3 or other lymphokines has been demonstrable. Interestingly, however, AIC2B has proven to be a second component of the multisubunit GM-CSF and IL-5 receptors.

In contrast to mouse, only one type of cDNA clone (KH97) with 56% and 55% identity, respectively, to AIC2A and AIC2B, was isolated from a human cDNA library (Hayashida *et al.*, 1991). The protein has an extracellular domain of 440 amino acids, a single transmembrane domain, and an intracellular domain of 411 amino acids. Structurally, the KH97 protein, like AIC2A and AIC2B, has two repeats of the 200 amino acid residues that have the characteristic motif of the cytokine receptor family in the extracellular domain. Interestingly, the KH97 protein proved not to have any IL-3-binding capacity but instead functions as the second subunit (β) of the human IL-3 receptor. Kitamura *et al.* (submitted) isolated a cDNA that confers high-affinity human IL-3 binding when co-expressed with the KH97 cDNA in COS7 cells. The cDNA encodes a mature protein of about 70 kDa (α-subunit), which has the common extracellular motif of the cytokine receptor family and has a small cytoplasmic domain. The α-subunit in itself weakly binds IL-3 ($K_d = 100$ nM) but upon co-expression with the β-subunit confers high-affinity IL-3 binding ($K_d = 100$ pM). Thus, it is now clear that the high-affinity IL-3 receptor is composed of two subunits, α and β. Interestingly, the mouse IL-3-binding protein, AIC2A, homologous to the β-subunit of the human IL-3 receptor, retains the capacity to bind IL-3 with low affinity ($K_d = 10$ nM). Recently it was shown that the other component of the mouse IL-3 receptor is homologous to the human α-subunit and the mouse α-subunit forms high affinity IL-3 receptors with either AIC2A or AIC2B (Hara and Miyajima, 1992).

E. GM-CSF Receptor

GM-CSF, originally identified as a factor that stimulates colony formation of granulocytes and macrophages, is a pleiotropic factor and has many overlapping biological activities with IL-3. GM-CSF regulates the proliferation, differentiation, and function of macrophages, neutrophils, and eosinophils as well as some cells of nonhematopoietic origin, such as endothelial cells, fibroblasts, and osteoblasts (Metcalf, 1988). In addition, GM-CSF may also be important as a growth factor in early hematopoietic events. GM-CSF appears to act as an autocrine growth factor for some cases of AML and may be responsible for the overwhelming cellular proliferation characteristic of AML.

The GM-CSF receptor exists in both high- and low-affinity forms (Walker and Burgess, 1985). Initial descriptions showed that the activities of this growth factor are transduced by a low abundance (100–5000 sites/cell), high affinity ($K_d = 10$–50 pM) receptor in both the mouse and human (Walker and Burgess, 1985; Chiba *et al.*, 1990). A second low-affinity ($K_d = 1$–5 nM) binding site exists on most, but not all, cells that express the high-affinity binding site. On human placenta only a low-affinity site was detected, leading to the hypothesis that more than one subunit might ultimately form the functional high-affinity receptor (Chiba *et al.*, 1990). Multiple reports have shown bidirectional cross-competition of binding between human IL-3 and human GM-CSF at both 4°C and 37°C (Park *et al.*, 1989; Lopez *et al.*, 1989; Budel *et al.*, 1990; Onetto-Pothier *et al.*, 1990), leading to the suggestion that the ligand-binding domain of the receptors recognizes both ligands. However, the data show that the ability to cross-compete is dependent on the number of high-affinity receptors, strongly suggesting that the "cross-competition" is not a true competitive interaction between the ligands to the receptors.

Molecular cloning studies have unequivocally shown that the high-affinity receptor for GM-CSF is composed of at least two subunits (Hayashida *et al.*, 1991). The gene for the low-affinity, human GM-CSF receptor (α-subunit) ($K_d = 5$ nM) (Gearing *et al.*, 1989) was shown to encode for a protein of 378 amino acids ($M_r = 43.7$ kDa) with 11 potential glycosylation sites; the mature native receptor is expressed as an 80-kDa molecule. The receptor has a 300-amino acid extracellular domain with a typical sequence of the cytokine receptor family, a single transmembrane domain, and a short intracellular domain of 54 amino acids. The chromosomal location of the GM-CSFRα maps to the pseudoautosomal region of the X and Y chromosomes (Gough *et al.*, 1990), of particular interest as a percentage of M2 acute myeloid leukemias, which are unresponsive to GM-CSF, have a loss in the X or Y chromosome.

The high-affinity human GM-CSF receptor was reconstituted by co-transfection of the gene for GM-CSFR α-subunit and the gene for a 120-kDa protein (β-subunit) encoded by the KH97 cDNA, isolated by hybridization with a probe derived from the cDNA for the mouse IL-3-binding protein, AIC2A (Hayashida *et al.*, 1990). Although co-expression of the 120-kDa protein with the low-affinity GM-CSF receptor α-subunit led to the formation of a high-affinity binding complex (α–β) ($K_d = 120$ pM); by itself the 120-kDa protein showed no detectable binding to GM-CSF. Kinetic parameters also showed that the dissociation rate of the ligand from the α–β complex was much slower ($\tau_{1/2} = 290$ minutes) than that from the 80-kDa binding protein alone ($\tau_{1/2} = 2$ minutes).

As described herein, human IL-3 and human GM-CSF cross-compete for a high-affinity binding site. The finding that KH97 is a common β-subunit (β_c) of both the IL-3 and GM-CSF receptors provides a possible molecular explanation for the cross-competition: the α-subunit of IL-3 receptor and the α-subunit of the GM-CSF receptor compete for β_c in the formation of high-affinity receptors. This hypothesis was proven by reconstitution of the high-affinity IL-3 and GM-CSF receptors in fibroblasts and demonstration of the cross-competition between IL-3 and GM-CSF (Kitamura *et al.*, 1991a).

Expression of the protein for the low-affinity human GM-CSFRα in a mouse GM-CSF-responsive hematopoietic cell line, FDC-P1, did not lead to a reconstitution of the high-affinity receptor (Metcalf *et al.*, 1990). However, the human ligand, at high concentrations ($K_d = 10$ nM), was able to stimulate the proliferation of FDC-P1 cells expressing the human GM-CSFRα as well as promoting a slow internalization of the receptor. Subsequent work by Kitamura *et al.* (1991b) demonstrated that β_c was critical for signal transduction: the human GM-CSFRα expressed in an IL-2–dependent mouse T cell line, CTLL, does not transmit a growth signal, but the high-affinity GM-CSF receptor α–β transmits a growth signal. In addition, co-transfection of the cDNAs for the human GM-CSFRα and the mouse AIC2B in CTLL led to the reconstitution of a functional receptor capable of stimulating cell proliferation, although high-affinity binding was not completely restored (Kitamura *et al.*, 1991b). These results indicate that the mouse AIC2B protein is a second subunit in the GM-CSF receptor complex, with signal-transducing properties that have been conserved evolutionarily between mouse and human. The inability to completely restore high-affinity binding may be a consequence of species selectivity because human and mouse GM-CSF are not functionally interchangeable.

F. IL-5 Receptor

A number of apparently unrelated factors, described as T cell-replacing factor, B cell growth factor II, and eosinophil activating factor, were all shown to be the functions of a single cytokine, encoded by the gene for IL-5 (Takatsu *et al.*, 1988; Yokota *et al.*, 1988). IL-5 is a principle factor required for the growth and differentiation of eosinophils, a specialized granulocyte found at low levels in healthy individuals and at high levels in individuals with parasitic infections and allergic diseases. IL-5 is also important for the induction of the growth and differentiation of activated B cells and is active on Ly-1$^+$ (CD5) B cells, a separate lineage of B cells primarily concentrated in the peritoneal cavity (Takatsu *et al.*, 1988).

Both high- (K_d = 100 pM) and low-affinity (K_d = 10 nM) binding sites for IL-5 have been found on a variety of mouse B cell lines (Mita *et al.*, 1988, 1989) and high-affinity receptors are present on eosinophilic subclones of HL-60 cells. The high-affinity receptor is rare, about 100–1000 receptors/cell, whereas the low-affinity receptor is present at about 5000–10,000 receptors/cell. The expression of the high-affinity receptor can be up-regulated in B cells by LPS stimulation and in HL-60 cells by butyrate treatment. That more than a single polypeptide is involved in the binding of IL-5 is suggested by a variety of evidence (Rolink *et al.*, 1989; Mita *et al.*, 1989): (1) cross-linking of IL-5 results in the identification of two molecules of about 60 and 130 kDa and (2) antibodies to the IL-5 receptor, which inhibit IL-5 biological activity and ligand binding, immunoprecipitate molecules with distinctly different molecular weights of 45–60 kDa and 110, 130, or 140 kDa (Rolink *et al.*, 1989; Hitoshi *et al.*, 1990).

The expression cloning of the murine IL-5 receptor (p60) convincingly demonstrates that the ligand-binding moiety, a glycoprotein of 415 amino acids, is inadequate of itself to reconstitute high-affinity binding (Takaki *et al.*, 1990). Structurally, the IL-5 receptor shares common structural features with other members of the cytokine receptor family. The extracellular domain is 322 amino acids, with a calculated mass of 45 kDa, and contains four potential glycosylation sites, which result in its posttranslational modification and a final mass, as observed in native cells of 60 kDa (Yamaguchi *et al.*, 1990). The recombinant receptor expressed in COS7 cells has a low affinity for IL-5 (K_d = 2–10 nM), whereas expression in the myeloid cell line, FDC-P1, gave both high- (K_d = 30 pM) and low-affinity (K_d = 6 nM) ligand binding. FDC-P1 cells transfected with the p60 IL-5 receptor cDNA showed two cross-linked bands: one for the p60 IL-5 receptor and the other for a 120-kDa protein. Moreover, IL-5–stimulated proliferation of those FDC-P1 transfectants, indicating that the reconstituted receptor is functional (Takaki *et al.*, 1990). The antibody R52.120, developed by Rolink *et al.* (1989), inhibited IL-5 binding as well as IL-3 binding. Interestingly, this antibody immunoprecipitated a doublet of 120-kDa proteins, which were shown to be AIC2A and AIC2B (Takaki *et al.*, 1991). Reconstitution experiments using the cloned p60 IL-5 receptor and the AIC2 cDNAs unequivocally demonstrated that AIC2B, but not AIC2A, is a second component of the IL-5 receptor (Takaki et al., 1991). Thus, AIC2B is a component in both the mouse IL-5 and GM-CSF receptors. This is in contrast to the situation in humans because only one type of AIC2 homolog has been isolated. However, IL-5 binding is inhibited by IL-3 in human basophils (Lopez et al., 1990), suggesting that the β-subunit (β_c, or KH97) of the IL-3 and GM-CSF receptors may also be involved in the high-affinity IL-5 receptor. This was demonstrated by the cloning of the human IL-5 receptor α-subunit and reconstitution of the high affinity IL-5 receptor with the common β-subunit (KH97) (Tavenier et al., 1991).

G. IL-6 Receptor

Like many of the interleukins, IL-6 was initially thought to be a number of discrete, unrelated molecules, derived from apparently unrelated sources and acting on functionally different cell types. The historically and biologically descriptive names for IL-6 include the following: interferon β/26K protein, B cell stimulatory factor 2 (BSF-2), plasmacytoma growth factor, hepatocyte-stimulating factor, and cytotoxic T cell differentiation factor. The variety of names underscores the wide diversity of functions for IL-6. As reviewed (Kishimoto and Hirano, 1988; Van Snick, 1990), IL-6 is produced by activated fibroblasts, endothelial cells, keratinocytes, monocytes–macrophages, T cells, mast cells, and folliculostellate cells of the anterior pituitary, where it acts as a pituitary secretagogue. Activation signals include viral infections, LPS, and/or the release of cytokines, such as IL-1, TNF, and IL-3. In the absence of these activation signals, there is little constitutive production of IL-6. A variety of tumors, including multiple myelomas, produce IL-6, which in turn acts in an autocrine fashion. An excess production of IL-6 may also be involved in the pathology of Castleman disease and in some cases of Kaposi's sarcoma.

Two classes of binding sites exist on both human and mouse cells: a high-affinity receptor (K_d = 10–50 pM) and a low-affinity receptor (K_d = 1–5 nM) (Yamasaki et al., 1988; Coulie et al., 1989). Cross-linking of IL-6 demonstrates that two molecular entities of 80 and 130 kDa interact with IL-6. Both subunits have been recently molecularly cloned. The recombinant p80 IL-6R (or the α-subunit) binds IL-6 with low affinity. The p80 IL-6R cDNA encodes for 468 amino acids, with a single transmembrane domain, a-340 amino acid extracellular domain, and a 120-amino acid intracellular moiety; upon glycosylation it is expressed as an 80-kDa cell surface protein. The 90-amino acid N-terminal region of the extracellular domain contains the characteristic motif of the immunoglobulin superfamily (Williams and Barclay, 1988). The remainder of the extracellular domain has homology to the cytokine–hematopoietin receptor superfamily. Thus, the extracellular domain of the IL-6 receptor is a composite receptor, but deletion analysis indicated that the immunoglobulin motif is not essential for IL-6 binding.

The 80-kDa receptor interacts with a second membrane glycoprotein of 130 kDa (gp130, or the β-subunit) (Taga *et al.*, 1989). This interaction is both ligand- and temperature-dependent, occurring only at 37°C and not at 4°C and, furthermore, is independent of the presence of the transmembrane or cytoplasmic portions of the 80-kDa receptor, as the soluble p80 IL-6 receptor is capable of modulating the signal transduction properties of gp130. It is of interest, therefore, to note that significant quantities of a soluble IL-6 receptor, with the same amino terminus as the recombinant protein, are found in normal human urine (Novick *et al.*, 1990).

The gene for gp130 was cloned using monoclonal antibodies obtained after immunization of mice with purified protein (Hibi *et al.*, 1990). This antibody blocked the formation of high-affinity sites; low-affinity binding was unaltered. Expression of gp130 alone gave no detectable binding to either IL-6 or any other cytokine, whereas co-expression of 80-kDa IL-6 R and gp130 in the Jurkat T cell line resulted in an increase in high-affinity binding (Hibi *et al.*, 1990), indicating that gp130 complements the low-affinity binding of the 80-kDa subunit. The gp130 cDNA encodes for a 918-amino acid protein, with a 597-amino acid etracellular region, a single transmembrane-spanning region, and a 277-amino acid cytoplasmic domain. The extracellular domain of gp130 contains the cytokine receptor motif and a sequence similar to contactin, which possesses fibronectin Type III modules. The overall structure of gp130 is similar to the G-CSF receptor. Transfection of gp130 cDNA in mouse IL-3–dependent BaF3 cells conferred a proliferative response to IL-6 in the presence of soluble IL-6 receptor (80 kDa subunit), indicating that gp130 is important for signal transduction (Hibi *et al.*, 1990).

H. IL-4 Receptor

Interleukin-4 is secreted by activated T cells and mast cells and functions as a growth and differentiation factor for T and B cells, thymocytes, mast cells, and macrophages (Yokota *et al.*, 1988). IL-4 induces class switching of immunoglobulins, expression of genes for such as CD23 and MHC class II, and acts as a differentiating factor for activated B cells and macrophages. However, it is important to differentiate the biochemical mode of action of IL-4 from growth factors, such as IL-2, IL-3, and GM-CSF. That is, IL-4 synergizes with these factors, decreasing the cellular concentration requirement for IL-3 and GM-CSF (Koyasu *et al.*, 1989); decreasing the amount of time cells spend in GI; and augmenting the maximal levels of cell proliferation above that induced by factors, such as IL-3 alone (Rennick *et al.*, 1987; Koyasu *et al.*, 1989). Biochemical studies of the IL-4 receptor support the concept that there are alternate signal transduction properties induced by the IL-4 receptor.

The IL-4 receptor exists as a low abundance, high-affinity (K_d = 20–80 pM) receptor on both hematopoietic and nonhematopoietic cells (Lowenthal *et al.*, 1988). The equilibrium dissociation constant is defined by rapid associa-tion ($\tau_{1/2}$ = 2 minutes) and slow dissociation ($\tau_{1/2} \geq 4$ hours) kinetics. Reports of a low-affinity form of the IL-4 receptor (Foxwell *et al.*, 1989) have as yet not been confirmed. The receptor was originally defined as a 60–75-kDa mol-ecule (Ohara and Paul, 1987; Park *et al.*, 1987); however, subsequent experi-ments showed it to be a 140-kDa molecule, which is rapidly degraded to the smaller 60–75-kDa forms. Cross-linking of radioiodinated IL-4 demonstrates several additional bands, which do not appear to be degradation products (Galizzi *et al.*, 1990a, unpublished work) and may therefore be receptor-asso-ciating molecules.

The molecular cloning of the gene for the mouse IL-4 receptor was achieved by (1) expression cloning using a biotinylated IL-4 probe (Harada *et al.*, 1990) and (2) amplification of the IL-4 receptor expression on the CTLL-19.4 cell line, followed by preparation of a subtracted cDNA probe, and screening of a cDNA library (Mosley *et al.*, 1989). Subsequently, the sequence for the human IL-4 receptor was also delineated and shown to have about 50% identity with the mouse receptor (Idzerda *et al.*, 1990; Galizzi *et al.*, 1990b). The nucleotide sequence showed that the mouse IL-4 receptor gene encodes a protein of 810 amino acids with a signal peptide of 25 amino acids. The receptor has a single transmembrane domain, with a 210-amino acid extracellular domain and a 553-amino acid cytoplasmic domain, resulting in a calculated mass of 85 kDa. Ra-dioligand cross-linking showed a molecule of net M_r of 120–140 kDa. The extracellular domain has the WS $\times$ WS box and seven cysteine residues, four of which represent the conserved cysteines found in the cytokine receptor fam-ily. The intracellular domain of the IL-4 receptor also has the serine and pro-line-rich character found in other cytokine receptors.

Interestingly, in contrast to most of the other ligand-binding subunits of the cytokine receptors, the cDNA for the IL-4 receptor encodes a protein with high affinity (K_d = 194 pM vs. native receptor K_d = 108 pM) for its ligand, IL-4. An alternatively spliced cDNA was isolated which contained a small insertion of 5′ to the transmembrane sequence, leading to the expression of a soluble IL-4 re-ceptor that binds IL-4 with high affinity and inhibits biological functions of IL-4 (Mosley *et al.*, 1989), indicating that the high-affinity receptor does not require an additional protein.

I. IL-7 Receptor

Interleukin-7 was originally identified by its ability to stimulate pre-B cell pro-liferation in Whitlock-Witte cultures (Namen *et al.*, 1988a). The cDNAs for both mouse (Namen *et al.*, 1988b) and human IL-7 (Goodwin *et al.*, 1989) were recently cloned and shown to encode for a 25-kDa molecule. The availability of recombinant IL-7 led to the demonstration that IL-7 had a broad spectrum of actions, including the stimulation of peripheral T cells as well as resting fetal and adult thymocytes (Conlon *et al.*, 1989; Chazen *et al.*, 1989); radiolabeling studies increased the known variety of receptor-positive cells to include cells of the myeloid lineage (Park *et al.*, 1990).

IL-7 binds to pre-B cells, some T cell lines, and macrophages, but not to mature B cells. Two discrete binding sites were found, and cross-linking studies showed a major protein of 75–79 kDa and a minor protein of 160 kDa (Park *et al.*, 1990). Using direct expression cloning, Goodwin *et al.* (1990) isolated the genes for both the human and mouse IL-7 receptors. The cloned cDNA encodes a protein of 459 amino acids and the extracellular domain contains the motif of the cytokine receptor family. The N-terminal domain containing two cysteine residues is related to the CD5 cell surface antigen (Bazan, 1990). Transfection of the cloned gene into COS7 cells led to a complex binding pattern similar to the natural IL-7 receptor: two discrete binding classes were identified (K_{d1} = 220 pM and K_{d2} = 24 nM). The cytoplasmic portion of the recombinant receptor was not necessary for the expression of a functional binding moiety, and a soluble receptor lacking the transmembrane moiety was also able to compete for ligand binding to cells. Interestingly, the soluble form of the IL-7 receptor appears frequently in at least three different cDNA libraries; however, because only a single mRNA transcript could be seen by Northern analysis (the 94 nucleotide difference between full-length and soluble receptor, precludes their separation) (Goodwin *et al.*, 1990), it is not clear whether the soluble receptor occurs naturally as an alternative splice variant. Cross-linking of the transfected receptor with radioligand results in the appearance of two bands of about 68 kDa and 153 kDa (Goodwin *et al.*, 1990). Noncovalent dimerization of the IL-7 receptor may confer high-affinity binding; alternatively, a second component may be endogenously present in COS7 cells.

III. Molecular Properties of Cytokine Receptors

A. Extracellular Structure

Comparisons between the members of the cytokine receptor superfamily have utilized conventional alignment methods as well as more sensitive sequence and structural pattern-matching methods to map consensus features (Bazan, 1989, 1990). Linked pattern-matching generates an alignment pattern for the extracellular domains of the aforementioned receptors, leading to a consensus sequence profile. Fourteen blocks of conserved amino acid residues were found within the approximately 210-amino acid residues of the extracellular domains of those cytokine receptors (Fig. 3) (Bazan, 1990). Each of the blocks defines a core sequence, an amphiphilic pattern of alternating hydrophobic–hydrophilic residues,

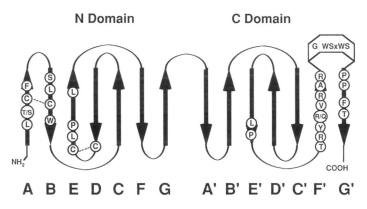

FIG. 3 Secondary structure of the extracellular domain of the cytokine receptors. The cytokine receptor consensus motif of about 200 amino acid residues is divided into 14 blocks (A to G'). Two seven-block stretches (N domain, A to G; C domain, A' to G') are shown. The location of the most conserved amino acid residues is shown.

consistent with the formation of β-strands. The blocks are separated by variable-length linker regions (Fig. 3). Moreover, an internal alignment of the sequences by a pattern-matching algorithm showed that there was an internal duplication of a seven-block stretch. Cysteine residues in blocks A, B, D, and E are well conserved in all the members of this family (Figs. 1 and 3). The elucidation of the disulfide-bonding pattern for the growth hormone receptor (Fuh *et al.*, 1990) provides additional information concerning the spatial relationships between the β-strands and the constraints governing the folding of the native protein. Disulfide bridges occur between adjacent cysteines (C1–C2) and (C3–C4), and (C5–C6); the latter pair are unique to the growth hormone receptor and are not found in the other cytokine receptors (Fig. 3). Furthermore, a unique WS × WS motif (× defines any amino acid) close to the transmembrane region was found in all the receptors (Figs. 1–3). Interestingly, the nucleotide sequence in this WS × WS region showed an almost complete conservation in the codon usage pattern.

Little analysis of the physical structure of these receptors has as yet been undertaken. However, some supportive evidence for the possibility that the extracellular region of the cytokine receptors is composed of at least two domains comes from an analysis of the exon–intron boundaries of the IL-2 receptor β, AIC2A, AIC2B, IL-4 receptor, EPO receptor, and growth hormone receptor. Exon–intron boundaries separate structural domains within proteins (Baron *et al.*, 1991). Without any exception so far, the two tandem seven-block stretches are separated by an intron sequence (Godowski *et al.*, 1989; Shibuya *et al.*, 1990; Youssoufian *et al.*, 1990; Gorman *et al.*, 1992).

The family of cytokine receptors is structurally related to another group of proteins, which include tissue factor and the interferon-α/β and -γ receptors (referred to as class II, cytokine receptors). Sequence modeling studies (Bazan,

1990) of the cytokine receptor family (class I) showed that the tandem 100-amino acid motif was also present within the class II group. The block alignment shows that there is a significant conservation of amino acid residues between the two classes of receptor. For instance in block A there exists a conserved proline; in block B there is a conserved hydrophobic site, aliphatic site, and tryptophan residue. Although cysteines 3 and 4 of the class I receptors are retained in the class II family, cysteines 1 and 2 have been evolutionarily lost. In addition, for the class II receptors there are two additional cysteines in the second repeat motif, blocks F and G. Interestingly, the exon–intron boundaries of tissue factor (class II) are positioned identically to those for the growth hormone receptor (class I). The fibronectin type III (FBN-III) structure, a 90-amino acid module, is also distantly related to the class I and II receptors, as determined by diagnostic sequence pattern-matching (Patthy, 1990). Modeling of secondary structure leads to the prediction that FBN-III is composed of seven conserved β-strands, analagous to that previously predicted for the cytokine receptors. Experimental data to date, primarily circular dichroism, are consistent with a β-stranded structure for the FBN-III structure.

B. Cytoplasmic Sequence

Whereas there is nothing remarkably different about the cytokine receptor TM sequences, it is of interest that an unusually high proportion of cysteines (four each) was found in the sequences for the G-CSFR and the IL-4R and both require a single molecular species to bind their respective ligand with high affinity. In contrast, the receptors for IL-2, IL-3, GM-CSF, and IL-6 have between 0 and 2 cysteines.

In contrast to the extracellular domains of the various members of the cytokine receptor family, the intracellular domains diverge more widely. On a gross level it is apparent that a significant difference exists even in the length of this region, such that some of the receptors have a short cytoplasmic tail of 54-amino acid residues, whereas others have as many as 530 amino acids. Overall, it appears to be true that those receptors with short cytoplasmic tails, such as the α-subunits of the IL-3, GM-CSF and IL-5 receptors, do not have signal-transducing properties by themselves, but there are some conserved sequences (FPPVPaPK) just below the transmembrane domains of the two receptors (Takaki et al., 1990; Kitamura et al., 1991).

Mutagenesis experiments have been used to delineate some of the functional features of the cytoplasmic domain. The region adjacent to the transmembrane domain of the IL-2 and EPO receptors shares significant homology; truncations of this region of the IL-2 receptor abrogate IL-2–mediated signal transduction (Hatakeyama et al., 1989b). This same region in the EPO receptor is also important for signal transduction (D'Andrea et al., 1991). Moreover, the COOH-terminal 40-amino acid residues of the cytoplasmic domain of the EPO receptor appears to exert a negative control on EPO-dependent signaling (D'Andrea *et al.*, 1991). Although the IL-4 receptor has a large cytoplasmic domain of 530 amino acids, the proliferative signal of IL-4 can be mediated by a short stretch of about 100 amino acids in the N-terminal cytoplasmic domain (N. Harada *et al.*, in press). However, because the readout of these deletion analyses is "proliferation," it is not clear yet whether other functions may be mediated by the same or different regions of the intracellular domains.

The serine- and proline-rich character of the IL-2, IL-3, IL-4, G-CSF, and EPO receptors is also of interest. The negative regulatory region of the EPO receptor contains a serine-rich region. Interestingly, there are a significant number of paired Pro-Ser and Pro-Thr residues, (11 in total for AIC2A; 9 for the IL-4 receptor; and 6 for the G-CSF receptor). Pro-Ser/Thr couplets are recognition sequences for a recently described class of Ser-Thr kinases (Maller, 1990), suggesting that this common modality between the cytokine receptors may be important for signal transduction, receptor cross-talk, or down-regulation.

C. Multimeric Receptor Complex

Many of the receptors in the cytokine receptor family have a multisubunit structure (Fig. 2) and have complex binding profiles characterized by a low abundance high-affinity receptor ($K_d = 20$–200 pM) and a high abundance, low-affinity receptor ($K_d = 1$–20 nM). A characteristic feature of the multimeric receptors is that the high-affinity state is dominated by an extremely slow rate of dissociation and a rapid rate of association. The typical low-affinity ligand binding domain (α-subunit) has a rapid rate of dissociation, whereas the rate of association is the same as that found in the high-affinity state. Reconstitution with the second chain (β-subunit) of the complex, typically by itself a nonligand binding or ultra low-affinity moiety ($K_d \sim 10^{-6}M$), leads to the appearance of the characteristic slow rate of dissociation. Because the β-subunit does not bind ligand, it is apparent that simple ligand binding for detection will not suffice in the traditional expression cloning methods. Rather, detection via binding requires reconstitution of the high-affinity receptor with the α-subunit as demonstrated for the human GM-CSF receptor (Hayashida *et al.*, 1991), human IL-6 receptor (Hibi *et al.*, 1990), human IL-3 receptor (Kitamura *et al.*, 1991a), and IL-5 receptor (Takaki *et al.*, 1991; Tavernier *et al.*, 1991).

This structural theme of the high-affinity cytokine receptors (Fig. 2) differs from that of the tyrosine kinase receptor family, as exemplified by the EGF and PDGF receptors in which high-affinity binding appears to be dependent on the noncovalent interaction of two homologous subunits (Bishayee *et al.*, 1989; Hammacher *et al.*, 1989; Heldin *et al.*, 1989). Because the β-subunits for the IL-2, IL-3, IL-5, IL-6, and GM-CSF receptors in and of themselves have no or very low intrinsic ability to bind ligand, it is intriguing to ask whether interaction of the subunits passively creates a new binding pocket with higher affinity or rather is the result of a dynamic topological change in the β-subunit binding pocket. Interestingly, all the receptor components, except for the IL-2Rα, have one or two units of the sequence motifs of the cytokine receptor family in their extracellular domains (Fig. 2). Therefore, there may be a common mechanism for the association of the α- and β-subunits as well as for the ligand–receptor interaction. The overall structural conservation among the cytokines is that of the four α-helical bundle. Shanafelt *et al.* (1991) demonstrated that the N-terminal α-helix is responsible for the interaction of GM-CSF and IL-5 to the β-subunit of the receptors, AIC2B. GM-CSF has strict species specificity for receptor binding. A chimeric GM-CSF having the N-terminal α-helix from mouse GM-CSF and the rest of the sequence from human GM-CSF binds the human GM-CSF receptor α-subunit with a normal low affinity. Interestingly, however, the chimeric GM-CSF binds with high affinity to the receptor composed of human GM-CSF receptor α-subunit and mouse AIC2B protein, that is, the β-subunit of both the IL-5 and GM-CSF receptors, indicating that the first N-terminal helix of GM-CSF is responsible for the interaction with the β-subunit. This model was further supported by the finding that the N-terminal helix of mouse IL-5 is capable of replacing the mouse GM-CSF sequence of the chimeric GM-CSF: the hybrid molecule with the N-terminal helix from mouse IL-5 and the rest of the sequence from human GM-CSF binds with high affinity to the receptor composed of human GM-CSF receptor α-subunit and mouse AIC2B. Thus, it appears likely that the N-terminal helix of IL-5 and GM-CSF is responsible for the interaction with the β-subunit, consistent with the identification of Asp^{34} found in the first α-helix of mouse IL-2 as the only residue interacting directly with the IL-2 receptor β-subunit by exhaustive mutagenesis (Zurawski *et al.*, 1990).

For the IL-2 receptor β chain, it would appear that a new functional ligand-binding surface structure is created as a consequence of the interaction with the putative IL-2Rγ-subunit, implying further cooperative interactions. Although high-affinity receptors for the IL-4 and G-CSF seem to be composed of a single molecular species, the possibility exists that high-affinity receptors are formed by dimerization (Fukunaga *et al.*, 1990a, b). At this time it is not clear which class the EPO receptor belongs to, although binding data suggest a complex set of interactions. Moreover, the constitutive activation of the EPO receptor by the viral protein, gp55 (Li *et al.*, 1990), suggests that a cellular homolog for gp55 may act as a second chain for the complex receptor.

D. Receptor-Associated Molecules

The multimolecular nature of the cytokine receptor family is important both for high-affinity ligand binding and signal transduction. Although the receptor elements important for high-affinity binding of ligand have been identified, none of these components has any known ablity to transduce signals, such as phosphorylation or nucleotide binding. Therefore, it appears likely that additional components are required for functional receptors. Next, we describe the current status of receptor-associating molecules, using the IL-2, IL-3 and GM-CSF receptors as examples.

1. IL-2 Receptor

Exhaustive mutagenesis predicts the presence of a γ-subunit, which is important for signal transduction (Zurawski *et al.*, 1990). Saragovi and Malek (1990) showed that affinity-isolated IL-2 receptors co-purified with a molecule of 22 kDa, which is disulfide linked as a heterodimer in complex with IL-2Rβ. In addition, molecules of 40–45 kDa and 90–105 kDa also co-purified with the receptor (the latter molecule may be identical to ICAM-1). Another candidate for the γ-chain was a 70-kDa molecule defined by a set of monoclonal antibodies that enhance the affinity and processing of the IL-2 receptor complex; yet the distribution of the antigen does not correspond to that of the high-affinity IL-2 receptor (Nakamura *et al.*, 1989). Sugamura *et al.* (1990) presented evidence for the existence of a novel 64-kDa, putative γ-chain. A monoclonal anti-IL-2Rβ antibody (TU11) co-immunoprecipitated p55 (α) and p75 (β) as well as a 64-kDa molecule that is detectable in IL-2–dependent cells but not in COS transfectants of α and β (Sugamura *et al.*, 1990). Takeshita *et al.* (1992) demonstrated that p64 represents the third element of the high-affinity receptor complex. The other molecules may be responsible for events associated with signal transduction.

A 95-kDa molecule that associates with the IL-2 receptor has been shown to be the same as ICAM-1 (CD54) (Burton *et al.*, 1990). Antibodies (OKT27 and OKT27b) to this 95-kDa molecule co-precipitated IL-2Rα (p55), and p95 expression was elevated in parallel with IL-2Rα upon T cell activation (Szollosi *et al.*, 1987; Edidin *et al.*, 1988). Fluorescence energy transfer and photobleaching experiments strongly indicated that p95 and p55 were intimately associated (Szollosi *et al.*, 1987; Edidin *et al.*, 1988). The identity of p95 as ICAM-1 was made by purification, microsequencing, and comparison with known ICAM-1 antibodies. The reason for ICAM-1 association with the IL-2 receptor is as yet unclear; anti-ICAM-1 antibodies do not affect IL-2–mediated functions. It is postulated that ICAM-1, through its interaction with LFA-1, may juxtapose the IL-2R opposite to the site of IL-2 release, presumed to be a local and directional process (Poo *et al.*, 1988). Two IL-2 receptor-associating molecules of 97 and/or 57 kDA appear to be the receptor-associated tyrosine kinase; both molecules bind ATP and mediate kinase functions (Michiel *et al.*, 1990).

2. IL-3 and GM-CSF Receptors

A variety of monoclonal antibodies have been derived which are able to block IL-3 and/or GM-CSF–dependent proliferation, independent of any ability to bind to the known ligand-binding domains of these receptors, suggesting that

these antibodies recognize a receptor-associated structure(s). One set of monoclonal antibodies (2F2 and 4G8), made against a mouse mast cell line, exclusively blocks IL-3–dependent proliferation of 37°C, yet only one, 4G8,
blocks IL-3 binding (Morel *et al.*, 1991), indicating that the method by which
proliferation is blocked does not result from direct competition at the binding
site. The 4G8–2F2 antibodies recognize a 130-kDa molecule, which is distinct from the recombinant IL-3 receptor (AIC2A) and is found expressed on
mast cells at levels approximately 10-fold greater than AIC2A. Another set of
monoclonal antibodies (anti-MaG-1, TGI-1, TGI-5, and TGI-6) to a 66-kDa
molecule blocks the proliferation of AML-193 cells in response to both GM-
CSF and IL-3, yet do not affect G-CSF or IL-2 stimulation of a different cell
line (Taniyama *et al.*, 1989). The nature of the antigen is as yet unknown;
however, this work accentuates the functional overlap between the IL-3 and
GM-CSF receptors as discussed herein. A human hematopoietic progenitor-
specific monoclonal antibody, HIM1, inhibits binding of human IL-3 to KG1
cells (Emanuel *et al.*, 1990), yet also sees an antigen on PMNs, which do not
have detectable binding of IL-3. The molecular characteristics of this antigen
are not known.

E. Receptor Cross-talk

The presence of multiple cytokine receptors on a single cell suggests that cell
growth and differentiation are determined by an integration of signaling systems. The ability of various cytokines to interact, either synergizing or interfering, is a consequence of interactions that occur at several steps. Here we
describe these interactions at the receptor level.

Evidence for cross-talk at the receptor level exists for the IL-3, GM-CSF, G-
CSF, and M-CSF receptors. Walker *et al.* (1985) presented data for hierarchical
cross-desensitization in bone marrow cell populations, in which IL-3 is capable
of down-regulating the receptors for GM-CSF, G-CSF, and M-CSF, but the con-
verse is not true. Similarly, GM-CSF could down-regulate G-CSF and M-CSF
receptors, but the inverse was again not possible. Only the high-affinity GM-
CSF receptor was down-regulated. Because this hierarchical down-regulation
process reflects the hematopoietic development process, it appears that receptor
cross-talk may have important implications for hematopoietic development and
differentiation steps. Loss–reduction of the differentiation-specific G-CSF and
M-CSF receptors should slow the development of the differentiated cells of the
granulocyte and/or monocyte–macrophage pathways. Gliniak and Rohrschneider (1990) demonstrated that IL-3 and GM-CSF down-regulate the expression
of the M-CSF receptor by dominantly destabilizing the M-CSF receptor mRNA.
Reports of bidirectional competition (Park *et al.*, 1989; Lopez *et al.*, 1989;
Onetto-Pothier *et al.*, 1990; Gesner *et al.*, 1989; Budel *et al.*, 1990) between the
IL-3 and GM-CSF receptors led to a hypothesis that a common binding element

for these two ligands existed. The molecular cloning of the receptor components and reconstitution of the high-affinity receptors has unequivocally demonstrated that the β-subunits of the IL-3, GM-CSF, and, IL-5 receptors are identical. This finding provides a molecular basis of the cross-competition: the primary ligand-binding component (α-subunit) of the IL-3, IL-5, and GM-CSF receptors compete for the common β-subunit on the same cell.

IL-1 is known to synergize with IL-3 to stimulate hematopoeitic cells (Moore, 1989). One possible mechanism for the synergy is receptor up-regulation. For example, the TF-1 cell line proliferates in response to IL-3, IL-5, and GM-CSF. IL-1 by itself is incapable of maintaining cells, but it augments the proliferative potential of the TF-1 cells in response to these cytokines. Interestingly, IL-1 up-regulated the high-affinity receptors for IL-3, IL-5, and GM-CSF by enhancing the expression of the larger components (120 kDa) of these three receptors, leading to a hypothesis that these three receptors share a common β-subunit (Kitamura *et al.*, 1991c).

The signal transduction pathways of the GM-CSF and IL-3 receptors are similar; both ligands induce the tyrosine phosphorylation of similar sets of proteins (Isfort and Ihle, 1990). Moreover, pre-incubation with GM-CSF leads to a loss in the ability of IL-3 to stimulate tyrosine phosphorylation (J. Schreurs, unpublished). Interestingly, IL-3 and IL-2 (Farrar and Ferris, 1989) as well as IL-3 and IL-5 (Murata *et al.*, 1990) share a strong overlap in the panel of substrates that are tyrosine phosphorylated. Thus, an alternate mechanism for receptor cross-talk between the IL-3 and GM-CSF receptors is desensitization, proteolysis, or internalization of the receptors and, consequently, a loss of high-affinity ligand-binding receptor component; phosphorylation may be a signal controlling this process.

IV. Signal Transduction

Although cytokines have a wide variety of biological activities that depend on the cell types, only one or two types of ligand-binding proteins have been identified for each cytokine (Table 1, Fig. 2). Therefore, the pleiotropic function of the cytokines must be mediated by different signal transduction pathways or the expression of genes for cell-specific functions in different cells. In addition, the ability of various cytokines to interact, either synergizing or interfering, is a consequence of overlaps in the various signal transduction pathways.

A. Structural Alteration of Cytokine Receptors

Transfer of signal through the membrane requires some structural or conformational changes in the receptors themselves. To date, a few basic mechanisms explain the signal transduction properties of the best characterized families of receptors, including the tyrosine kinase receptors, the G protein-linked receptors, and the ion channel receptors. These mechanisms include the following: (1) the ligand-regulated oligomerization–dimerization of the tyrosine kinase receptors, which is then permissive of intersubunit phosphorylation and activation; (2) ligand-mediated conformational changes, such as that found for the nicotinic cholinergic receptor to permit ion flow; and (3) ligand-mediated activation of second messengers through the G proteins.

Evidence for a structural alteration in the cytokine receptors requisite for signal transduction has come from studies of viral genes. The Friend spleen focusforming virus (SFFV) containing the gene product, gp55, interacts with a cytokine receptor, the EPO receptor, and leads to growth factor independence; gp55, in and of itself, is inactive and it is important that both gp55 and the EPO receptor be co-expressed to achieve growth factor independence (Li et al., 1990). It appears likely that gp55 physically interacts with the EPO receptor and induces a structural alteration that allows it to induce signals without EPO. A single point mutation of the extracellular domain of the EPO receptor constitutively activates the EPO receptor (Yoshimura et al., 1990), also indicating that a structural alteration of the receptor is important for signaling.

The myeloproliferative leukemia virus, an acute leukemogenic murine replication-defective retrovirus, contains an oncogene with sequence homology to the cytokine receptor family (Souyri et al., 1990). This virus has the capacity to immortalize both erythroid and myeloid progenitor cells as well as committed cells, making them autonomous of growth factors. It appears that the v-mpl oncogene is an env fusion protein that contains a truncated form of an as yet unidentified hematopoietic growth factor receptor. The env fusion protein has an open reading frame of 284 amino acids, encoding a single transmembrane-spanning region bounded by a 109-amino acid extracellular domain and a 119-amino acid cytoplasmic domain. It shares in common with the cytokine receptors the extracellular WS × WS box with the surrounding amino acids as well as having high levels of prolines and serines in the cytoplasmic domain. It appears likely that v-mpl has a constitutively active conformation of the unidentified cytokine receptor or that it interacts with some cytokine receptor and activates it independent of the presence of ligand.

B. Tyrosine Kinases

Tyrosine kinase function is strongly implicated in the mechanism of action of a variety of the cytokines discussed herein, including IL-2, IL-3, IL-4, IL-5, GM-CSF, and EPO (Morla et al., 1988; Koyasu et al., 1987; Ferris et al., 1988; Isfort et al., 1988b; Gomez-Cambronero; Cambronero et al., 1989; Sharon et al., 1989; Mills et al., 1989; Asao et al., 1990; Cannistra et al., 1990; Quelle and Wojochowski, 1991; Kanakura et al., 1991; Murata et al., 1990). The importance of tyrosine kinase activity for ligand-induced cell proliferation is highlighted by a variety of observations, including: (1) specificity, factors that do not stimulate cell proliferation do not cause tyrosine phosphorylation (Kanakura et al., 1991); (2) inhibition of phosphatases augments factor-induced phosphorylation and cell proliferation (Kanakura et al., 1991; Tojo et al., 1987); and (3) tyrosine kinase oncogenes make cell lines factor-independent (Pierce et al., 1985; Watson et al., 1987; Kipreos and Wang, 1988).

As the cytokine receptors, the genes of which have been cloned have no apparent tyrosine kinase moiety (Hanks et al. 1988), it appears likely that there is an independent receptor-associating kinase, like but not the same as, the T cell-specific CD4- and CD8-associated p56lck (Turner et al., 1990). Tyrosine phosphorylation mediated by cytokine interaction with receptor is a rapid event occurring within 1 minute of addition of ligand (37°C), suggesting a close kinetic link between receptor and kinase (Morla et al., 1988). Yet tyrosine phosphorylation also can occur at 4°C; a 2-hour incubation with IL-3 at 4°C results in levels of phosphorylation similar to that observed after 1 minute of stimulation at 37°C (J. Schreurs, unpublished).

Numerous substrates are phosphorylated after ligand addition, as monitored both by Western blotting and immunoprecipitation using antiphosphotyrosine antibodies. Interestingly, both one- and two-dimensional gel electrophoresis shows the considerable overlap between the phosphotyrosine substrates for the IL-3, IL-2, GM-CSF, EPO, and IL-5 receptor-associated tyrosine kinase(s). It is interesting that the constitutive phosphorylation of 150 and 55 kDa proteins is correlated with factor independence for cells, including Rauscher erythroleukemia cells, IC2, FDCP1, and DA1 cells (Isfort et al., 1988; Koyasu et al., 1987; Quelle and Wojochowski, 1991). In contrast to the aforementioned receptors, the pattern of tyrosine phosphorylation stimulated by IL-4 is substantially different (Isfort and Ihle, 1990; Morla et al., 1988). Other differences in the signal transduction properties for IL-4 also exist with reference to stimulation of p21ras, suggesting that the difference in biochemical signal transduction pathways may be responsible for the role of IL-4 as a synergy–inhibitory factor.

There has been progress toward the identification of cytokine receptor-associated tyrosine kinases. Ligand affinity chromatography has been used to purify an IL-2 receptor-associated tyrosine kinase activity (Michiel et al., 1990) and tyrosine phosphorylated substrates of 55 and 105–115 kDa (Shackelford and Trowbridge, 1991). Tyrosine phosphorylation was stimulated by ligand upon reconstitution in an in vitro system, resulting in the phosphorylation of both the IL-2 receptor and two additional molecules of 97 and 57 kDa. The 97 and 57 kDa molecules were both able to recognize a tyrosine kinase substrate as well as to bind ATP. Moreover, the 97-kDa kinase was found associated with the IL-2 receptor in normal human T lymphocytes. This is interesting in light of the fact that the major tyrosine phosphorylated substrate for IL-3 and GM-CSF is also a 92-kDa molecule (Isfort and Ihle, 1990).

A few of the substrates for the receptor-associated tyrosine kinases have also been identified. Ligand stimulation leads to the tyrosine phosphorylation of the IL-2Rβ chain (Sharon *et al.*, 1989; Mills *et al.*, 1989; Asao *et al.*, 1990; Sugamura *et al.*, 1990; Shackelford and Trowbridge, 1991) and the IL-3 receptor (Isfort *et al.*, 1988a; Sorensen *et al.*, 1989; J. Schreurs *et al.*, 1991). Another IL-2 receptor-associated substrate is PI-3 kinase, which is phosphorylated in response to IL-2 (Augustine *et al.*, 1990).

The functional consequence of the tyrosine phosphorylation of the receptors themselves is not yet understood but may be important for their activation, desensitization, internalization, or proteolysis. An examination of the role of tyrosine phosphorylation in other classical receptor systems provides some insight. For a number of tyrosine kinase receptors, such as the PDGF and insulin receptors, a tyrosine phosphorylation-dependent process is responsible for a change in receptor conformation, as defined by an enhanced accessibility of certain antibodies to the native, nondenatured receptor (Keating *et al.*, 1988; Pang *et al.*, 1985) and leading in the case of the PDGF receptor, to the exposure of a binding site for raf-1 (Morrison *et al.*, 1989). Tyrosine phosphorylation is also important for insulin and EGF receptor down-regulation and proteolysis (Russell *et al.*, 1987; Felder *et al.*, 1990).

C. Serine–Threonine Kinases

Ligand stimulation of the IL-2, IL-3, GM-CSF, and EPO receptors results in serine–threonine phosphorylation in addition to Tyr phosphorylation. Thus, IL-3 stimulates the serine–threonine phosphorylation of a variety of substrates of 150 and 70 kDa (Evans *et al.*, 1986; Klemm and Elias, 1988; Koyasu *et al.*, 1987; Choi *et al.*, 1987). A major tyrosine phosphoprotein of 150 kDa, thought to be analogous to the recently purified IL-3 receptor (Isfort *et al.*, 1988a; Sorenson *et al.*, 1989; Schreurs *et al.*, 1991), is also serine phosphorylated upon treatment of cells with IL-3 (Koyasu *et al.*, 1987). GM-CSF stimulates the serine phosphorylation of a 68-kDa protein (Sorenson *et al.*, 1989). Moreover, the cAMP-dependent protein kinase and protein kinase C (PKC) inhibitors, H7 and H8, enhance the effects of GM-CSF on neutrophil function, possibly through their inhibition of receptor internalization. However, the inhibitors do not inhibit GM-CSF stimulated priming, that is, GM-CSF does not stimulate neutrophil function through PKC, but may be negatively down-regulated by PKC (Khwaja *et al.*, 1990).

The IL-2 receptor is also known to stimulate a serine–threonine kinase, leading to the phosphorylation of 63, 67, 85, and 136 kDa molecules (Mire *et al.*, 1985; Gaulton and Eardley, 1986). It had been shown that activators of PKC stimulated the phosphorylation of the same subset of proteins as did IL-2, and as PKC translocation resulted from IL-2 treatment (Farrar *et al.*, 1985; Evans and Farrar, 1987), it was suggested that IL-2 stimulated the PKC pathway.

However, a T lymphocyte clone, which apparently lacks PKC, as determined by several experimental measures, can proliferate in response to IL-2 (Mills et al., 1988).

Advances have been made toward the identification of the IL-3R–associated serine–threonine kinase. Carroll et al. (1990) have shown in receptor-purified preparations that IL-3 stimulates the serine–threonine kinase, raf-1. A potentially different serine–threonine kinase, which is associated with the IL-3 receptor, is co-purified by ligand affinity chromatography; experiments in vitro show that there is specific serine–threonine phosphorylation of several substrates of 40, 100, and 120 kDa (Schreurs et al., 1991). Furthermore, the co-association– activation of the serine–threonine kinase is temperature-dependent (requires IL-3 treatment at 37°C) and occurs subsequent to the tyrosine phosphorylation of the IL-3 receptor, an event that is not temperature-dependent. This is particularly in- triguing, in light of previous observations that the serine–threonine phosphoryla- tion of the M-CSF receptor is associated in a temperature-dependent fashion with a serine–threonine kinase (Baccarini et al., 1990).

D. Phosphatidylinositol Turnover

Both IL-2 and IL-3 have been shown to translocate and activate PKC (Whetton et al., 1988; Farrar and Anderson, 1985; Farrar et al., 1985) in IL-3–dependent cells. PKC is the major receptor for phorbol esters, which also cause the translo- cation of PKC and enhance the survival and proliferation of IL-3–dependent cells, suggesting that PKC activation may be important in IL-3–dependent sig- nal transduction. Other targets of phorbol esters do exist, however. Interestingly, the mechanism whereby IL-2 and IL-3 stimulate PKC translocation is indepen- dent of the major inositol phosphate hydrolysis pathway (Whetton et al., 1988; Kozumbo et al., 1987; Mills et al., 1986), leading to speculation that another system to generate PKC-stimulating second messenger must exist, potentially phosphotidylcholine.

E. GTP-Binding Proteins

Guanine nucleotide-binding proteins have been implicated in the signal trans- duction events associated with most receptors, including the traditional multi- transmembrane receptors, the insulin receptor (Moises and Heidenreich, 1990), the PDGF receptor, as well as receptors in the cytokine receptor family. The na- ture of the interacting GTP-binding protein differs among these families. Thus, the serpentine receptors interact with a trimeric GTP-binding protein containing an α-, β-, and γ-subunit, in which the differing α-subunits, such as G_s, G_i, G_q, and G_o, specify the final effector protein that is modified by agonist stimulation of the receptor. Interestingly, evidence for the association of receptors, such as the insulin and IGF-II receptors, with a 41-kDa GTP-binding molecule (Moises and Heidenreich, 1990) and the ability of IGF-II and a peptide fragment of the IGF-II receptor to activate G_i and thereby inhibit adenylate cyclase (Okamoto et al., 1990), illustrates that the trimeric G proteins may also interact with receptors of a nonserpentine nature. A second group of GTP-binding proteins is exemplified by the small 20–22-kDa proteins, such as rho, rap, and ras (Barbacid, 1987). Stimulation by PDGF and EGF causes an enhanced formation of p21*ras*-GTP (Satoh et al., 1990a, b), suggesting an important role for p21ras in signal transduction for normal proliferation and malignant transformation.

Evidence for the association of the cytokine receptors with GTP-binding proteins is recent. Gomez-Cambronero et al. (1989) demonstrated that GM-CSF–stimulated neutrophils had higher GTPase activities, the levels of which were significantly reduced by pertussis toxin treatment. McColl et al. (1989) showed that only pertussis toxin and not cholera toxin inhibited GM-CSF–dependent effects. A definitive study of IL-2, IL-3, IL-4, and GM-CSF effects in T, B, and myeloid cell lines showed that IL-2, IL-3, and GM-CSF enhanced the formation of active p21ras-GTP, whereas IL-4 did not, suggesting that the former three growth factors utilize p21ras as a signal transducer, whereas IL-4 does not utilize this same pathway (Satoh et al., 1991).

V. Conclusion

Molecular cloning of the genes for cytokine receptors has revealed that many cytokine receptors are composed of multiple subunits. Now high-affinity receptors have been reconstituted with molecularly cloned receptor components. However, functional receptors capable of transmitting signals require more components. Obviously, the forthcoming challenge will be to identify these components and to understand how cytokines generate intracellular signals. Because cytokines have numerous functions, a single receptor may be linked to multiple signal transduction pathways. It will be an interesting subject to study how the expression of the receptors and the linkage between receptors to signal transduction pathways is regulated during the developmental process of hematopoietic cells.

References

Anderson, D. M., Lyman, S. D., Baird, A., Wignall, J. M., Eisenman, J., Rauch, C., March, C. J., Boswell, H. S., Gimpel, S. D., Cosman, D., and Williams, D. E. (1990). Cell 63, 235–243.
Arai, K., Lee, F., Miyajima, A., Miyatake, S., Arai, N., and Yokota, T. (1990). Annu. Rev. Biochem. 59, 783–836.

Asao, H., Takeshita, T., Nakamura, M., Nagata, K., and Sugawara, K. (1990). *J. Exp. Med.* **171**, 637–644

Augustine, J. A., Sutor, S. L., Leibson, P. J., and Abraham, R. T. (1990). *Lymphokine Res.* **9**, 549.

Baccarini, M., Sabatini, D. M., App, H., Rapp, U. R., and Stanley, E. R. (1990). *EMBO J.* **9**, 3649–3657.

Barbacid, M. (1987). *Annu. Rev. Biochem.* **56**, 779–827.

Baron, Norman, D. G., and Campbell, I. D. (1991). *Trends Biochem. Sci.* **16**, 13–17.

Bazan, J. F. (1989). *Biochem. Biophys. Res. Commun.* **164**, 788–795.

Bazan, J. F. (1990). *Proc. Natl. Acad. Sci. U.S.A.* **87**, 6934–6938.

Bich-Thuy, L. T., Dukovich, M., Peffer, N. J., Fauci, A. S., Kehrl, J. H., and Greene, W. C. (1987). *J. Immunol.* **139**, 1550–1556.

Bishayee, S., Majumbar, S., Khire, J., and Das, M. (1989). *J. Biol. Chem.* **264**, 11699–11705.

Budel, L. M., Elbaz, O., Hoogerbrugge, H., Delwel, R., Mahmoud, L. D., Lowenberg, B., and Touw, I. P. (1990). *Blood* **75**, 1439–1445.

Burton, J., Goldman, C. K., Rao, P., Moos, M., and Waldmann, T. A. (1990). *Proc. Natl. Acad. Sci. U.S.A.* **87**, 7329–7333.

Cannistra, S. A., Groshek, P., Garlick, R., Miller, J., and Griffin, J. D.(1990). *Proc. Natl. Acad. Sci. U.S.A.* **87**, 93–97.

Carroll, M. P., Clark-Lewis, I., Rapp, U. R., and May, W. S. (1990). *J. Biol. Chem.* **265**, 19812–19817.

Chazen, G. D., Pereira, G. M. B., LeGros, G., Gillis, S., and Shevach, E. M. (1989). *Proc. Natl. Acad. Sci. U.S.A.* **86**, 5923–5927.

Chiba, S., Shibuya, K., Tojo, A., Sasaki, N., Matsuki, S., Piao, Y. F., Miyagawa, K., Miyazono, K., and Takaku, F. (1990) *Cell Regul.* **1**, 327–335.

Choi, H. S., Wojochowski, D. M., and Sytkowski, A. J. (1987). *J. Biol. Chem.* **262**, 2933–2936.

Conlon, P. J., Morrissey, P. J., Nordan, R. P., Grabstein, K. H., Prickett, K. S., Reed. S. G., Goodwin, R., Cosman, D., Namen, A. E. (1989). *Blood* **74**, 1368–1373.

Cosman, D., Cerretti, D. D., Larsen, A., Park, L., March, C., Dowers, S., Gillis, S., and Urdal, D. (1984). *Nature* **312**, 768–771.

Coulie, P. G., Stevens, M., and Van Snick, J. (1989). *Eur. J. Immunol.* **19**, 2107–2114.

D'Andrea, A. D., Lodish, H. F., and Wong, G. G. (1989). *Cell* **57**, 277–285.

D'Andrea, A. D., Yoshimura, A., Youssoufian, H., Zon, L. I., Koo, J. W., and Lodish, H. F. (1991). *Mol. Cell. Biol.* **11**, 1980–1987.

Dukovich, M., Wano, Y., Thuy, J. B., Katz, P., Cullen, B. R., Kehrl, J. H., and Greene, W. C. (1987). *Nature* **327**, 518–522.

Edidin, M., Aszalos, A., Damjanovich, S., and Waldmann, T. A. (1988). *J. Immunol.* **141**, 1206–1210.

Emanuel, P. D., Peiper, S. C., Chen, Z., Sheng, D. C., and Zuckerman, K. S. (1990). *Proc. Natl. Acad. Sci. U.S.A.* **87**, 4449–4452.

Evans, S. W., and Farrar, W. L. (1987). *J. Cell Biochem.* **34**, 47–59.

Evans, S. W., Rennick, D., and Farrar, W. L. (1986). *Blood* **68**, 906–913.

Farrar, W. L., and Anderson, W. B. (1985). *Nature* **315**, 233–235.

Farrar, W. L., and Ferris, D. K. (1989). *J. Biol. Chem.* **264**, 12562–12567.

Farrar, W. L., Thomas, T. P., and Anderson, W. B. (1985). *Nature* **315**, 235–237.

Felder, S., Miller K., Moehren, G., Ullich, A., Schlessinger, J., and Hopkins, C. R. (1990). *Cell* **61**, 623–634.

Ferris, D. K., Willette-Brown, T., Martensen, T., and Farrar, W. L. (1988). *Biochem. Biophys. Res. Commun.* **154**, 991–996.

Fiorentino, D. F., Bond, M. W., and Mosmann, T. R. (1989). *J. Exp. Med.* **170**, 2081.

Foxwell, B. M. J., Woerly, G., and Ryffel, B. (1989). *Eur. J. Immunol.* **19**, 1637–1641.

Fuh, G., Mulkerrin, M. G., Bass, S., McFarland, N., Brochier, M., Bourell, J. H. Light, D. R., and Wells, J. A. (1990). *J. Biol. Chem.* **265**, 3111–3115.

Fukunaga, R., Ishizaka-Ikeda, E., Seto, Y., and Nagata, S. (1990a). *Cell* **61**, 341–350.

Fukunaga, R., Seto, Y., Mizushima, S., and Nagata, S. (1990b). *Proc. Natl. Acad. Sci. U.S.A.* **87**, 8702–8706.

Galizzi, J. P., Castle, B. E., Djossou, O., Harada, N., Cabrillat, H., Yahia, X., Barrett, R., Howard, M., and Banchereau, J. (1990a).*J. Biol. Chem.* **265**, 439–444.

Galizzi, J. P., Zuber, C. E., Harada, N., Gorman, D. M., Djossou, O., Kastelein, R., Banchereau, J., Howard, M., and Miyajima, A. (1990b). *Int. Immunol.* **2**, 669–675.

Gaulton, G. N., and Eardley, D. D. (1986). *J. Immunol.* **136**, 2470–2477.

Gearing, D. P., King, J. A., Gough, N. M., and Nicola, N. A. (1989). *EMBO J.* **8**, 3667–3676.

Gesner, T., Mufson, R. A., Turner, K. J., and Clark, S. C. (1989). *Blood* **74**, 2652–2656.

Gliniak, B. C., and Rohrschneider, L. R. (1990). *Cell* **63**, 1073–1083.

Godowski, P. L., Leung, D. W., Meacham, L. R., Galgani, J. P., Hellmiss, R., Keret, R., Rotwein, P. S., Parks, J. S., Laron, Z., and Wood, W. I. (1989). *Proc. Natl. Acad. Sci. U.S.A.* **86**, 8083–8087.

Gomez-Cambronero, J., Yamazaki, M., Metwally, F., Molski, T. F. P., Bonak, V. A., Huang, C. K., Becker, E. L., and Sha'afi, R. I. (1989). *Proc. Natl. Acad. Sci. U.S.A.* **86**, 3569–3573.

Goodwin, R. G., Lupton, S., Schmierer, A., Hjerrild, K. J., Jerzy, R., Clevenger, W., Gillis, S., Cosman, D. and Namen, A. E. (1989). *Proc. Natl. Acad. Sci. U.S.A.* **86**, 302–306.

Goodwin, R. G., Friend, D., Ziegler, S. F., Jerzy, R., Falk, B. A., Gimpel, S. D., Cosman, D., Dower, S. K., March, C. J., Namen, A. E., and Park, L. S. (1990). *Cell* **60**, 941–951.

Gorman, D. M., Itoh, N., Kitamura, T., Schreurs, J., Yonehara, S., Yahara, I., Arai, K. I., and Miyajima, A. (1990). *Proc. Natl. Acad. Sci. U.S.A.* **87**, 5459–5463.

Gorman, D. M., Itoh, N., Jenkins, N. A., Gilbert, D. J., Copeland, N. G., and Miyajima, A. (1992). *J. Biol. Chem.* **267** (in press).

Gough, N. M., Gearing, D. P., Nicola, N. A., Baker, E., Pritchard, M., Callen, D. F., Sutherland, G. R. (1990). *Nature* **345**, 734–736.

Greene, W. C., Robb, R. J., Svetlik, P. B., Rusk, G. M., Depper, J. M., and Leonard, W. J. (1985). *J. Exp. Med.* **162**, 363–368.

Hammacher, A., Mellstrom, K. Heldin, C. H., and Westermark, B. (1989). *EMBO J.* **8**, 2489–2495.

Hara, T. and Miyajima, A. (1992). *EMBO J.* **11**, 1875–1884.

Hanks, S. K., Quinn, A. M., and Hunter, T. (1988). *Science* **241**, 42–52.

Harada, N., Castle, B. E., Gorman, D. M., Itoh, N., Schreurs, J., Barrett, R. L., Howard, M., and Miyajima, A. (1990). *Proc. Natl. Acad. Sci. U.S.A.* **87**, 857–861.

Harada, N., Yang, G., Miyajima, A., and Howard, M. *J. Biol. Chem.* (in press).

Hatakeyama, M., Tsudo, M., Minamoto, S., Kono, T., Doi, T., Miyata, T., Miyasaka, M., and Taniguchi, T. (1989a). *Science* **244**, 551–556.

Hatakeyama, M., Mori, H., Doi, T., and Taniguchi, T. (1989b). *Cell* **59**, 837–845.

Hayashida, K., Kitamura, T., Gorman, D. M., Arai, K., Yokota, T., and Miyajima, A. (1991). *Proc. Natl. Acad. Sci. U.S.A.* **87**, 9655–9659.

Heimfeld, S., Hudak, S., Weissman, I., and Rennick, D. (1991). *Proc. Natl. Acad. Sci. U.S.A.* **88**, 9902–9906.

Heldin, C. H., Ernlund, A., Rorsman, C., and Ronnstrand, L. (1989). *J. Biol. Chem.* **264**, 8905–8912.

Hibi, M., Murakami, M., Saito, M., Hirano, T., Taga, T., and Kishimoto, T. (1990). *Cell* **63**, 1149–1157.

Hitoshi, Y., Yamaguchi, N., Mita, S., Sonada, E., Takaki, S., Tominaga, A., and Takatsu, K. (1990). *J. Immunol.* **144**, 4218–4225.

Hung, E., Nocka, K., Beier, D. R., Chu, T. Y., Buck, J., Lahm, H. W., Weiner, D., Leder, P., and Besmer, P. (1990). *Cell* **63**, 225–233.

Idzerda, R. L., March, C. J., Mosley, B., Lyman, S. D., Vanden Bos, T., Gimpel, S. D., Din, W. S., Grabstein, K. H., Widmer, M. B., Park, L. S., Cosman, D., and Beckman, M. P. (1990). *J. Exp Med.* **171**, 861–873.

Isfort, R. J., Stevens, D., May, W. S., and Ihle, J. N. (1988a). *Proc. Natl. Acad. Sci. U.S.A.* **85**, 7982–7986.

Isfort, R., Abraham, R., Huhn, R. D., Frackelton, A. R., and Ihle, J. N. (1988b). *J. Biol. Chem.* **263**, 19203–19209.

Isfort, R. J., and Ihle, J. N. (1990). *Growth Factors* **2**, 213–220.

Itoh, N., Yonehara, S., Schreurs, J., Gorman, D. M., Maruyama, K., Ishii, A., Yahara, I., Arai, K. I., and Miyajima, A. (1990). *Science* **247**, 324–327.

Kanakura, Y., Druker, B., DiCarlo, J., Cannistra, S. A., and Griffin, D. (1991). *J. Biol. Chem.* **266**, 490–495.

Keating, M. T., Escobedo, J., and Williams, L. T. (1988). *J. Biol. Chem.* **263**, 12805–12808.

Khwaja, A., Roberts, P. J., Jones, H. M., Yong, K., Jaswon, M. S., and Linch, D. C. (1990). *Blood* **76**, 996–1003.

Kipreos, E. T., and Wang, J. Y. J. (1988). *Oncogene Res.* **2**, 277–284.

Kishimoto, T., and Hirano, T. (1988). *Annu. Rev. Immunol.* **6**, 485–512.

Kitamura, T., Sato, N., Arai, K. and Miyajima, A. (1991a). *Cell* **66**, 1165–1174.

Kitamura, T., Hayashida, K., Sakamaki, K., Yokota, T., Arai, K., and Miyajima, A. (1991b). *Proc. Natl. Acad. Sci. U.S.A.* **88**, 5082–5086.

Kitamura, T., Takaku, F., and Miyajima, A. (1991c). *Int. Immunol.* **3**, 571–577.

Koyasu, S., Tojo, A., Miyajima, A., Akiyama, T., Kasuga, M., Urabe, A., Schreurs, J., Arai, K., Takau, F., and Yahara, I. (1987). *EMBO J.* **140**, 456–464.

Koyasu, S., Miyajima, A., Arai, K., Okajima, F., Ui, M., and Yahara, I. (1989). *Cell Struct. Funct.* **14**, 459–471.

Kozumbo, W. J., Harris, D. J., Gromkowski, S., Cerottini, J. C., and Cerutti, P. A. (1987). *J. Immunol.* **138**, 606–612.

Kuwaki, T., Kitamura, T., Tojo, A., Matsuki, S., Tamai, Y., Miyazono, K., and Takaku, F. (1989). *Biochem. Biophys. Res. Commun.* **161**, 16–22.

Leonard, W. J., Depper, J. M., Crabtree, G. R., Rudikoff, S., Pumphrey, J., Robb, R. J., Kronke, M., Svetlik, P. B., Peffer, N. J., Waldmann, T. A., and Greene, W. C. (1984). *Nature* **311**, 626–635.

Li, J. P., D'Andrea, A. D., Lodish, H. F., and Baltimore, D. (1990). *Nature* **343**, 762–764.

Lopez, A. F., Eglinton, J. M., Gillis, D., Park, L. S., and Vadas, M. S. (1989). *Proc. Natl. Acad. Sci. U.S.A.* **86**, 7022–7026.

Lowenthal, J. W., Castle, B. E., Christiansen, J., Schreurs, J., Rennick, D., Arai, N., Takebe, Y., and Howard, M. (1988). *J. Immunol.* **140**, 456–464.

Maller, J. L. (1990). *Biochemistry* **29**, 3157–3166.

May, S., and Ihle, J. N. (1986). *Biochem. Biophys. Res. Commun.* **135**, 870–879.

McColl, S. R., Kreis, C., DiPersio, J. F., Borgeat, P., and Naccache, P. H. (1989). *Blood* **73**, 588–591.

Meeker, T. C., Hardy, D., Willman, C., Hogan, T., and Abrams, J. (1990). *Blood* **76**, 285–289.

Metcalf, D. (1988). "The Molecular Control of Blood Cells." Cambridge Press. Harvard University Press, Cambridge, Massachusetts.

Metcalf, D., Nicola, N. A., Gearing, D. P., and Gough, N. M. (1990). *Proc. Natl. Acad. Sci. U.S.A.* **87**, 4670–4674.

Michiel, D., Garcia, G., Evans, G., Linnekin, D., and Farrar, W. (1990). *Lymphokine Res.* **9**, 550.

Mills, G. B., Stewart, D. J., Mellors, A., and Gelfand, E. W. (1986). *J. Immunol* **136**, 3019–3024.

Mills, G. B., Girard, P., Grinstein, S., and Gelfand, E. W. (1988). *Cell* **55**, 91–100.

Mills, G. B., May, C., McGill, M., Fung, M., Baker, M., Sutherland, R., and Greene, W. C. (1989). *J. Exp. Med.* **265**, 3561–3567.

Minamoto, S., Mori, H., Hatakeyama, M., Kono, T., Doi, T., Ide, T., Uede, T., and Taniguchi, T. (1990). *J. Immunol.* **145**, 2177–2182.

Mire, A. R., Wickremasinghe, R. C., Michalevicz, R., and Hoffbrand, A. V. (1985). *Biochim. Biophys. Acta.* **847**, 159–163.

Mita, S., Tominaga, A., Hitoshi, K., Sakamoto, K., Honjo, T., Akagi, M., Tominaga, N., and Takatsu, K. (1988). *J. Exp. Med.* **168**, 863–878.

Mita, S., Tominaga, A., Hitoshi, Y., Sakamoto, K., Honjo, T., Akagi, M., Tominaga, A., and Takatsu, K. (1989). *Proc. Natl. Acad. Sci. U.S.A.* **86**, 2311–2315.

Miyajima, A., Miyatake, S., Schreurs, J., DeVries, J., Arai, N., Yokota, T., and Arai, K. (1988). *FASEB J.* **2**, 2462–2473.

Miyajima, A., Schreurs, J., Wang, H. M., Maruyama, K., Gorman, D., Koyasu, S., Yahara, I., Wang, J. Y. J., Ohta, Y., and Arai, K. (1989). *Adv. Immunopharmacol.* **4**, 87–93.

Moises, R. J., and Heidenreich, K. A. (1990). *J. Cell Physiol.* **44**, 538–545.

Moore, M. A. (1989). *Immunol. Res.* **8**, 165–175.

Moore, K. W., Vieira, P., Fiorentino, D. E., Troustine, M. L., Khan, T. A., and Mosmann, T. R. (1990). *Science* **248**, 1230–1234.

Morel, P., Schreurs, J., Townsend, K., Gross, M., Chiller, J. M., and Tweardy, D. J. (1991). *J. Immunol.* **146**, 2295–2304.

Morla, A., Schreurs, J., Miyajima, A., and Wang, J. Y. J. (1988). *Mol. Cell Biol.* **8**, 2214–2218.

Morrison, D. K., Kaplan, D. R., Escobedo, J. A., Rapp, U. R., Roberts, T. M., and Williams, L. T. (1989). *Cell* **58**, 649–657.

Mosley, B., Beckman, P., March, C. J., Idzerda, R. L., Gimmpel, S. D., Vanden Bos, T., Friend, D., Alpert, A., Anderson, D., Jackson, J., Wignall, J. M., Smith, C., Gallis, B., Sims, J. E., Urdal, D., Widmer, M. B., Cosman, D., and Park, L. S. (1989). *Cell* **59**, 335–348.

Murata, Y., Yamaguchi, N., Hitoshi, Y., Tominaga, A., and Takatsu, K. (1990). *Biochem. Biophys. Res. Commun.* **173**, 1102–1108.

Murthy, S. C., Mui, A. L. F., and Krystal, G. (1990). *Exp. Hematol.* **18**, 11–17.

Nakamura, Y., Inamoto, T., Sugie, K., Masutani, H., Shindo, T., Tagaya, Y., Yamauchi, A., Ozawa, K., and Yodoi, J. (1989). *Proc. Natl. Acad. Sci. U.S.A.* **86**, 1318–1322.

Namen, A. E., Schmierer, A. E., March, C. J., Overell, R. W., Park, L. S., Urdal, D. L., and Mochizuki, D. Y. (1988a). *J. Exp. Med.* **167**, 988–1002.

Namen, A. E., Lupton, S., Hjerrild, K., Wignall, J., Mochizuki, D. Y., Schmierer, A., Mosley, B., March, C. J., Urdal., D., Gillis, S., Cosman, D., and Goodwin, R. G. (1988b). *Nature* **333**, 571–573.

Nicola, N. A. (1989). *Annu. Rev. Biochem.* **58**, 45–78.

Nicola, N. A., and Metcalf, D. (1986). *J. Cell. Physiol.* **128**, 180–188.

Nicola, N. A., and Peterson, L. (1986). *J. Biol. Chem.* **261**, 12384–12389.

Nikaido, T., Shimizu, A., Ishica, N., Sabe, H., Teshigawara, M., Maeda, M., Uchiyama, T., Yodoi, J., and Honjo, T. (1984). *Nature* **311**, 631–635.

Novick, D., Engelmann, H., Wallach, D., Leitner, O., Revel, M., and Rubinstein, M. (1990). *J. Chromatogr.* **510**, 331–337.

Ohara, J., and Paul, W. E. (1987). *Nature* **325**, 537–540.

Okamoto, T., Katada, T., Murayama, Y., Ui, M., Ogata, E., and Nishimoto, I. (1990). *Cell* **62**, 709–717.

Onetto-Pothier, N., Aumont, N., Haman, A., Park, L., Clark, S. C., DeLean, A., and Hoang, T. (1990). *Leukemia* **4**, 329–336.

Palaszynski, E. W., and Ihle, J. N. (1984). *J. Immunol.* **132**, 1872–1878.

Pang, D. T., Sharma, B. R., Shafer, J. A., White, M. F., and Kahn, C. R. (1985). *J. Biol. Chem.* **260**, 7131–7136.

Park, L. S., Friend, D., Gillis, S., and Urdal, D. L. (1986). *J. Biol. Chem.* **261**, 205–210.

Park, L. S., Friend, D., Grubstein, K., and Urdal, D. L. (1987). *Proc. Natl. Acad. Sci. U.S.A.* **84**, 1669–1673.

Park, L. S., Friend, D., Price, V., Anderson, S. A., Singer, J., Prickett, K. S., and Urdal, D. L. (1989) *J. Biol. Chem.* **264**, 5420–5427.

Park, L. S., Friend, D., Schmierer, A., Dower, S. K., and Namen, A. E. (1990). *J. Exp. Med.* **171**, 1073–1089.

Patthy, L. (1990). *Cell* **61**, 13–14.

Paul, S. R., Bennett, F., Calvetti, J. A., Kelleher, K., Wood, C. R., O'Hara, R. M., Leary, A. C., Sibley, B., Clark, S. C., and Williams, D. A. (1990). *Proc. Natl. Acad. Sci. U.S.A.* **87**, 7512–7516.

Pierce, J. H. (1989). *Biochim. Biophys. Acta* **989**, 179–208.

Pierce, J. H., Di Fiore, P. P., Aaronson, S. A, Potter, M., Pumphrey, J., Scott, A., and Ihle, J. N. (1985). *Cell* **41**, 685–693.

Poo, W. J., Conrad, L., and Janeway, C. A. (1988). *Nature* **332**, 378–380.

Quelle, F. W., and Wojochowski, D. M. (1991). *J. Biol. Chem.* **266**, 609–614.

Rennick, D., Yang, G., Muller-Sieburg, C., Smith, C., Arai, N., Takabe, Y. and Gemmell, L. (1987). *Proc. Natl. Acad. Sci. U.S.A.* **84**, 6889–6893.

Ringheim, G. E., Friemark, B. D., and Robb, R. J. (1990). *FASEB J.* **4**, 2784.

Robb, R. J., Rusk, C. M., Yodoi, J., and Greene, W. C. (1987). *Proc. Natl. Acad. Sci. U.S.A.* **84**, 2002–2011.

Rolink, A. G., Melchers, F., and Palacios, R. (1989). *J. Exp. Med.* **169**, 1693–1701.

Russell, D. S., Gherzi, R., Johnson, E. L., Chou, C. K., and Rosen, O. M. (1987). *J. Biol. Chem.* **262**, 11833–11840.

Sabe, H., Kondo, S., Shimizu, A., Tagaya, Y., Yodoi, J., Kobayashi, N., Hatanaka, M., Matsunami, N., Maeda, M., Noma, T. (1984). *Mol. Biol. Med.* **2**, 379–396.

Saragovi, H., and Malek, T. R. (1990). *Proc. Natl. Acad. Sci. U.S.A.* **87**, 11–15.

Satoh, T., Endo, M., Nakafuku, M., Nakamura, S., and Kaziro, Y. (1990). *Proc. Natl. Acad. Sci. U.S.A.* **87**, 5993–5997.

Satoh, T., Endo, M., Nakafuku, M., Akiyama, T., Nakamura, S., and Kaziro, Y. (1990b). *Proc. Natl. Acad. Sci. U.S.A.* **87**, 7926–7929.

Satoh, T., Nakafuku, M., Miyajima, A., and Kaziro, Y. (1991). *Proc. Natl. Acad. Sci. U.S.A.* **88**, 3314–3318.

Schrader, J. W. (1986). *Annu. Rev. Immunol.* **4**, 205–230.

Schreurs, J., Sugawara, M., Arai, K., Ohta, Y., and Miyajima, A. (1989). *J. Immunol.* **142**, 819–825.

Schreurs, J., Arai, K., and Miyajima, A. (1990). *Growth Factors* **2**, 221–234.

Schreurs, J., Hung, P., May, W. S., Arai, K. and Miyajima, A. (1991). *Internat. Immunol.* **3**, 1231–1242.

Seed, B., and Aruffo, A. (1987). *Proc. Natl. Acad. Sci. U.S.A.* **84**, 3365–3369.

Shackelford, D. A., and Trowbridge, I. S. (1991). *Cell Regul.* **2**, 73–85.

Shanafelt, A. B., Miyajima, A., Kitamura, T., and Kastelein, R. A. (1991). *EMBO J.* **10**, 4105–4112.

Sharon, M., Klausner, R. D., Cullen, B. R., Chizzonite, R., and Leonard, W. J. (1986). *Science* **234**, 859–863.

Sharon Gnarra, J. R., and Leonard, W. J. (1989). *J. Immunol.* **143**, 2530–2533.

Shibuya, H., Yoneyama, M., Nakamura, Y., Harada, H., Hatakeyama, M., Minamoto, S., Kono, T., Doi, T., White, R., and Taniguchi, T. (1990). *Nucleic Acids Res.* **18**, 3697–3703.

Sims, J. E., March, C. J., Cosman, D., Widmer, M. B., MacDonald, H. R., McMahon, C. J., Grubin, C. E., Wignall, J. M., Jackson, J. L., Call, S. M., Friend, D., Alpert, A. R., Gillis, S., Urdal., D. L., and Dower, S. K. (1988). *Science* **241**, 585–589.

Smith, K. A. (1989). *Annu. Rev. Cell Biol.* **5**, 397–425.

Sorensen, P., Farber, N. M., and Krystal, G. (1986). *J. Biol. Chem.* **261**, 9094–9097.

Sorensen, P., Farber, N. M., and Krystal, G. (1989). *J. Biol. Chem.* **264**, 19253–19258.

Souyri, M., Vigon, I., Penciolelli, J. F., Heard, J. M., Tambourin, P., and Wendling, F. (1990). *Cell* **63**, 1137–1147.

Sugamura, K., Takeshita, T., Asao, H., Kumaki, S., Ohbo, K., Ohtani, K., and Nakamura, M. (1990). *Lymphokine Res.* **9**, 539–542.

Szollosi, J., Damjanovich, S., Goldmann, C. K., Fulwyler, M. J., Aszalos, A. A., Goldstein, G., Rao, P., Talle, M. A., and Waldmann, T. A. (1987). *Proc. Natl. Acad. Sci. U.S.A.* **84**, 7246–7250.

Taga, T., Hirata, Y., Yamasaki, K., Yasukawa, K., Matsuda, T., Hirano, T., and Kishimoto, T. (1989). *Cell* **58**, 573–581.

Takaki, S., Tominaga, A., Hitoshi, Y., Mita, S., Sonoda, E., Yamaguchi, N., and Takatsu, K. (1990). *EMBO J.* **9**, 4367–4374.

Takaki, S., Mita, S., Kitamura, T., Yonehara, S., Yameguchi, N., Tominaga, A., Miyajima, A., and Takatsu, K. (1991). *EMBO J.* **10**, 2833–2838.

Takatsu, K., Tominaga, A., Harada, N., Mita, S., Matsumoto, M., Takahashi, T., Kikuchi, Y., and Yamaguchi, N. (1988). *Immunol. Rev.* **102**, 107–135.

Takeshita, T., Asao, H., Ohtani, K., Ishii, N., Kumaki, S., Tanaka, N., Munakata, H., Nakamura, M., and Sugamura, K. (1992). *Science* **257**, 379–382.

Taniyama, T., Taki, S., Nagata, M., Yoshizawa, K., Hirayama, N., Hamuro, J., Uchiyama, T., Wong, G., and Rovera, G. (1989). *Growth Factors* **1**, 263–270.

Tavernier, J., Devos, R., Cornelis, S., Tuypens, T., Van der Heyden, J., Fiers, W., and Plaetinck, G. (1991). *Cell* **66**, 1175–1184.

Teshigawara, K., Wang, H. M., Kato, K., and Smith, K. A. (1987). *J. Exp. Med.* **165**, 223–238.

Tojo, A., Kasuga, M., Urabe, A., and Takaku, F. (1987). *Exp. Cell Res.* **171**, 16–23.

Tsudo, M., Kozak, R. W., Goldman, C. K., and Waldmann, T. A. (1986). *Proc. Natl. Acad Sci. U.S.A.* **83**, 9694–9698.

Tsudo, M., Kozak, R. W., Goldman, C. K., and Waldmann, T. A. (1987). *Proc. Natl. Acad. Sci. U.S.A.* **84**, 4215–4218.

Tsudo, M., Karasuyama, H., Kitamura, F., Nagasaka, Y., Tanaka, T., and Miyasaka, M. (1989). *J. Immunol.* **143**, 4039–4043.

Tsudo, M., Karasuyama, H., Kitamura, F., Tanaka, T., Kubo, S., Yamaura, Y., Tamatani, T., Hatakeyama, M., Taniguchi, T., and Miyasaka, M. (1990). *J. Immunol.* **145**, 599–606.

Turner, J. M., Brodsky, M. H., Irving, B. A., Levin, S. D., Permutter, R. M., and Littman, D. R. (1990). *Cell* **60**, 755–765.

Van Snick, J. (1990). *Annu. Rev. Immunol.* **8**, 253–278.

Waldmann, T. A. (1989). *Annu. Rev. Biochem.* **58**, 875–911.

Walker, F., and Burgess, A. W. (1985). *EMBO J.* **4**, 933–939.

Walker, F., Nicola, N., Metcalf, D., and Burgess, A. W. (1985). *Cell* **43**, 269–276.

Wang, H. M., and Smith, K. A. (1987). *J. Exp. Med.* **166**, 1055–1069.

Watson, J. D., Eszes, M., Overell, R., Conlon, P., Widmer, M., and Gillis, S. (1987). *J. Immunol.* **139**, 123–129.

Whetton, A. D., Monk, P. N., Consalvey, S. D., Huang, S. J., Dexter, T. M., and Downes, C. P. (1988). *Proc. Natl. Acad. Sci. U.S.A.* **85**, 3284–3288.

Williams, A. F., and Barclay, A. N. (1988). *Annu. Rev. Immunol.* **6**, 381–405.

Yamaguchi, N., Hitoshi, Y., Mita, S., Hosoya, Y., Murata, Y., Kikuchi, Y., Tominaga, A., and Takatsu, K. (1990). *Int. Immunol.* **2**, 181–187.

Yamasaki, K., Taga, T., Hirata, Y., Yawata, H., Kawanishi, Y., Seed, B., Taniguchi, T., Hirano, T., and Kishimoto, T. (1988). *Science* **241**, 825–828.

Yokota, T., Arai, N., DeVries, J., Spits, H., Banchereau, J., Zlotnik, A., Rennick, D., Howard, M., Takebe, Y., Miyatake, S., Lee, F., and Arai, K. (1988). *Immunol. Rev.* **102**, 137–187.

Yonehara, S., Ishi, A., Yonehara, M., Koyasu, S., Miyajima, A., Schreurs, J., Arai, K., and Yahara, I. (1990). *Int. Immunol.* **2**, 143–150.

Yoshimura, A., Longmore, G., and Lodish, H. F. (1990). *Nature* **348**, 647–649.

Youssoufian, H., Zon, L. I., Orkin, S., D'Andrea, A. D., and Lodish, H. F. (1990). *Mol. Cell. Biol.* **10**, 3675–3682.

Zsebro, K. M., Wypych, J., McNiece, I. K., Lu, H. S., Smith, K. A., Karkare, S. B., Birkett, N. C., Williams, L. R., Satyagal, V. H., Tung, W., Bosselman, R. A., Mendiatz, E. A., and Langley, K. E. (1990). *Cell* **63**, 195–201.

Zurawski, S. M., and Zurawski, G. (1989). *EMBO J* **8**, 2583–2590.

Zurawski, S. M., Imler, J. L., and Zurawski, G. (1990). *EMBO J.* **9**, 3899–3905.

Polymeric Immunoglobulin Receptor

Benjamin Aroeti, James Casanova, Curtis Okamoto, Michael Cardone, Anne Pollack, Kitty Tang, and Keith Mostov
Departments of Anatomy and Biochemistry, and Cardiovascular Research Institute, University of California, San Francisco, San Francisco, California 94143

I. Introduction

The most basic type of organization of cells into tissues is that of epithelia (Simons and Fuller, 1985). Epithelial cells line a cavity or cover a surface and as such they can form a selective barrier to the exchange of molecules between the lumen of an organ and an underlying tissue. For many decades physiologists have studied the movements of small molecules, such as water, ions, or sugars across epithelia. It has become increasingly clear that large molecules, such as proteins, can also cross an epithelial layer. One way that this can occur is by diffusion between cells, namely, by a paracellular route. However, in many types of epithelia the cells are closely attached to each other by tight junctions. These tight junctions form an effective seal between cells and usually preclude paracellular transport of macromolecules (Cereijido et al., 1989).

The other route of transport is across the cells themselves via vesicular carriers, in a process known as transcytosis (Mostov and Simister, 1985). The first step of this specialized pathway requires endocytosis. A ligand bound to its receptor is endocytosed via coated pits and delivered to the smooth endosomal compartment. In the endosome the internalized proteins can be sorted to a number of cellular destinations. They can be recycled to the plasma membrane from which they were internalized (e.g., the transferrin and low density lipoprotein [LDL] receptor) (Brown et al., 1983). The other route involves transport to lysosomes for degradation (for example, epidermal growth factor [EGF] receptor). Alternatively ligand–receptor complexes can be directly transported via vesicular carriers to the opposite pole of the cell, fuse with the plasma membrane, and then release their contents by exocytosis. In hepatocytes sorting into these three pathways have been shown to occur in an early endosomal compartment (Geuze et al., 1984).

157

In polarized epithelial cells transcytosis may occur in either direction across a cell, from apical to basolateral or basolateral to apical (Brändli *et al.*, 1990). In most epithelial cells the apical and basolateral surfaces maintain different protein and lipid compositions. Transcytosis of macromolecules is therefore considered to be one among several other mechanisms that are involved in establishing and maintaining of this compositional assymetry (Simons and Wandinger-Ness, 1990).

A wide variety of molecules have been shown to be transcytosed: insulin and serum albumin across endothelia (King and Johnson, 1985), EGF across kidney epithelia (Maratos-Flier *et al.*, 1987), EGF across intestinal epithelium, and transferrin across capillaries in the brain. In some cases transcytosis is a quantitatively major pathway. In hepatocytes, for example, newly synthesized apical (bile canalicular) membrane proteins are not delivered directly to the apical surface (Bartles *et al.*, 1987); rather both apical and basolateral proteins are transported to the basolateral (sinusoidal) surface. Apical proteins are then selectively retrieved and transcytosed to the apical plasma membrane. In these cells transcytosis is the only way for membrane proteins to reach the apical surface. In the intestinal cell line, CaCo2, a number of apical proteins utilize both the direct *trans*-Golgi network (TGN) to apical pathway and the indirect transcytotic pathway (Le Bivic *et al.*, 1990; Matter *et al.*, 1990) to reach the apical cell surface.

The best-studied examples of transcytosis is the transport of immunoglobulins that occurs in at least three situations in mammals: transport of IgA and IgM across various mucosa (Childers *et al.*, 1989), transport of IgG across the intestinal epithelium in newborn rats (Rodewald and Kraehenbuhl, 1984), and transport of IgG across the human placenta. The first process is discussed in detail next.

This review will focus on the cellular and molecular mechanisms involved in IgA transport across epithelial cells. We can only cover selected aspects of this problem. Therefore, the review will concentrate primarily on research done by our laboratory, which focused mainly on the identification of sorting signals associated with the polymeric immunoglobulin receptor.

II. Transcytosis of Polymeric Immunoglobulins

The major class of immunoglobulin found in a wide variety of muscosal secretions, such as gastrointestinal and respiratory secretions, is IgA (Brandtzaeg, 1981; Bienenstock, 1984; Ahnen *et al.*, 1985). IgA is produced by submucosal plasma cells that are often found in lymphoid aggregates, such as gut-associated lymphoid tissue (GALT) and bronchus-associated lymphoid tissue (BALT) (Bienenstock, 1984). After secretion from the plasma cells, IgA is taken up by an overlying epithelial cell, transported across the epithelium, and released into external secretions (Brandtzaeg, 1981). Here the IgA forms the first specific immunologic defense against infection. This transport system transports only polymeric immunoglobulins (Brandtzaeg, 1981). Dimers or higher oligomers of IgA are transported, as are pentamers of IgM, although transport of the latter is less efficient. All of these polymers contain the J (joining) chain.

This transport of polymeric immunoglobulins is now known to be a receptor-mediated event. In 1965 Tomasi *et al.* found that IgA isolated from human saliva contained an extra polypeptide that was a glycoprotein of 70 kDa. This protein, named secretory component (SC), was found to be synthesized by epithelial cells and added to IgA as it was transported across the cell. In 1974 Brandtzaeg examined the cellular location of SC by immunofluorescence and reported its presence on the basolateral surface of various epithelial cells. IgA from appropriate secretions could bind to the basolateral cell surface, and antisera to SC could block the binding of IgA. This led to the proposal that SC acted as a receptor mediating the uptake and transport of IgA across cells.

The hypothesis that SC was an IgA receptor presented an interesting paradox that formed the basis for many studies (Brandtzaeg, 1974; Mostov *et al.*, 1980). If SC were a receptor for IgA on the basolateral cell membrane, it would be expected to be an integral membrane protein that could only be solubilized with detergents, yet SC isolated from secretions was water-soluble and had no affinity for membranes. One proposed solution to this paradox was that SC was secreted at the basolateral surface and combined with IgA in the extracellular fluid or blood (Kuhn and Kraehenbuhl, 1979). The SC–IgA complex could then bind to an unidentified receptor on the basolateral cell surface and be transported to the luminal surface. A problem with this model was that no SC could be detected in blood.

An alternative hypothesis was that SC was part of a larger precursor, now known as the polymeric immunoglobulin receptor (pIg-R) (Mostov *et al.*, 1980). The first evidence in support of this model came from cell-free translation of rabbit liver and mammary mRNA (Mostov *et al.*, 1980). SC was found to be synthesized as a large precursor of about 90 kDa. In a cell-free membrane integration system, this precursor was shown to be a membrane-spanning protein, which had a cytoplasmic domain of about 10–15 kDa. Moreover, the precursor could specifically bind to IgA, suggesting that it was the true pIg-R.

We next studied the biosynthesis and processing of pIg-R in a human colon carcinoma cell line, HT29, which was known to secrete SC (Mostov and Blogel, 1982). Biosynthetic labeling of these cells revealed that the pIg-R was found initially as a single species of 90 kDa. Its carbohydrate side chains were subsequently modified to the complex type, which increased the apparent size of the protein to about 105 kDa. The pIg-R was then slowly cleaved to SC (70 kDa) and released from the cells. This cleavage is slower than that *in vivo* probably because the HT29 cells are not well differentiated or polarized. This type of pulse-chase analysis has been carried out by others using rabbit mammary

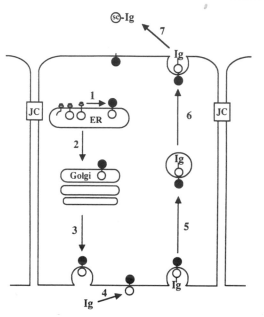

FIG. 1   The pathway of IgA transcytosis. An epithelial cell is shown, with the apical surface at the top and the basolateral surface at the bottom. Junctional complexes (JC) divide the two surfaces and join the cell to its neighbors.

cells in culture and intact rat liver, and the general observations have been confirmed (Solari and Kraehenbuhl, 1984; Sztul *et al.*, 1985a, b).

The current understanding of the general pathways taken by the pIg-R is summarized in Figure 1. An epithelial cell is shown with the apical surface at the top and the basolateral surface at the bottom. In step 1 the pIg-R is synthesized by membrane-bound polysomes of the rough endoplasmic reticulum (RER) as an integral membrane protein. A portion of the molecule extends into the lumen of the RER (open circle), a segment spans the membrane, and a portion is in the cytoplasm (filled circle). After transport through the Golgi apparatus (step 2), the receptor is targeted to the basolateral surface (step 3). Here the portion of the molecule formerly in the RER lumen is outside the cell, where it can bind IgA (step 4). The receptor–ligand complex is then endocytosed in coated vesicles (step 5) and transported by a variety of vesicles and tubules to the apical surface (step 6). At the apical surface (or perhaps during transport), the extracellular portion of the pIg-R is proteolytically cleaved from the transmembrane anchor. This cleaved fragment is the previously identified SC. It remains associated with the IgA in the extracellular secretions and has the additional function of stabilizing the IgA against denaturation or proteolysis in the harsh external environment.

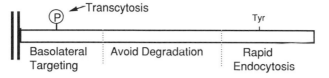

FIG. 2 Arrangement of sorting signals in the cytoplasmic domain of the pIg-R. The membrane is at the left and the carboxy terminus is at the right. The phosphate group is indicated by the P.

Our next step was to clone cDNA for pIg-R (Mostov *et al.*, 1984). We used rabbit liver mRNA, an abundant source of pIg-R mRNA. We obtained a full-length cDNA clone that encoded a protein of 755 amino acids (not including the 18 amino acid N-terminal signal sequence). The protein had a single membrane-spanning segment and a cytoplasmic C-terminal domain of 103 amino acids (Fig. 2). The extracellular ligand-binding portion, which is cleaved to generate SC, contains five homologous repeating domains of 10–110 residues each. These are members of the immunoglobulin superfamily and most clearly resemble immunoglobulin variable regions (Mostov *et al.*, 1984).

In the rabbit there appears to be only one gene for pIg-R (Deitcher and Mostov, 1986). However, in many rabbits there are four primary translation products, two of 70–75 kDa and two of 90–95 kDa. Part of the heterogeneity is due to the existence of multiple alleles of rabbit SC and most rabbits are heterozygous. However, even in homozygous rabbits, there are two primary translation products, one of 70–75 kDa and one of 90–95 kDa. The pIg-R mRNA can be alternately spliced to yield a pIg-R protein that lacks the second and third of the five immunoglobulin domains (Deitcher and Mostov, 1986). This alternately spliced form is the 70–75-kDa translation product, whereas the form with all five immunoglobulin domains is the 90–95-kDa product. Sequencing of a genomic clone revealed that the two immunoglobulin domains involved are encoded by a single large exon. This is a rare example of two immunoglobulin domains encoded on one exon.

III. Expression of Polymeric Immunoglobulin Receptor cDNA

The pathway of the pIg-R is unusual for a membrane protein in polarized cells because it is targeted first to the basolateral surface and then, following endocytosis, to the apical surface, rather than directly to its final destination. Postendocytotic sorting is complicated by the fact that internalized pIg-R enters endosomes that contain a variety of other receptors and ligands (Geuze *et al.*, 1984). The pIg-R is apparently sorted away from this admixture of receptors and ligands in tubular extensions of the endosomes, and is then packaged into carrier vesicles for transcytosis to the apical surface (Geuze *et al.*, 1984). These two features of the pIg-R pathway, sequential targeting to both surfaces and postendocytotic sorting into the transcytotic pathway, offer unique opportunities to study membrane trafficking. Our major strategy to study sorting of the pIg-R has been to express the cloned rabbit pIg-R cDNA in cultured cell lines and analyze the function of normal and mutant receptors. For expression we have used a retroviral vector system (Breitfeld *et al.*, 1989b).

First, the pIg-R cDNA was expressed in the fibroblast line, y2, which is a derivative of the NIH 3T3 mouse fibroblast line (Deitcher *et al.*, 1986). Although these cells have no obvious apical-basolateral polarity, the pIg-R appears to function normally. In a pulse-chase experiment the receptor was found to be first synthesized as a single species of about 90 kDa. In the subsequent chase it was converted to a heterogenous form of about 105 kDa, due to modification of the carbohydrates, and was eventually cleaved to SC that was released into the medium. Like authentic SC the SC released by these cells is heterogenous. This is apparently due both to heterogeneity in glycosylation and to heterogeneity in the exact site of cleavage from the transmembrane anchor.

In these fibroblasts when the pIg-R reaches the cell surface, it is not immediately cleaved to SC. This results in a pool of uncleaved receptor on the surface where it is capable of internalizing bound ligand. About 35% of the bound ligand can be internalized. It is then rapidly recycled back to the cell surface and then released into the medium without substantial degradation.

We next expressed the pIg-R in cultured epithelial cells (Mostov and Deitcher, 1986), specifically, the Madin-Darby canine kidney (MDCK) cell line, which has been widely used for studying cell polarity. When grown on porous filter supports, such as a Millipore filter, these cells form a well-polarized, electrically resistant epithelial monolayer (Simons and Fuller, 1985). Tight junctions separate the apical from the basolateral surfaces and, in effect, a simple epithelial tissue is reconstituted in culture. The monolayer is relatively impermeable, especially to macromolecules. Under these culture conditions one can experimentally access either the apical surface or, through the filter, the basolateral surface. In MDCK cells, as in fibroblasts, pIg-R is synthesized as a single species and then converted to a heterogenous form due to carbohydrate modification. Proteolytic cleavage also occurs in these cells, and the free SC is released almost exclusively into the apical medium. This is exactly as occurs *in vivo*: SC is released at the lumenal surface and not into the bloodstream. Moreover, if ^{125}I-labeled dIgA is added to the basolateral medium, it is taken up by the cells and transported into the apical medium. This transcytosis is saturated by a competing excess of unlabeled dIgA and does not occur in cells that do not express the receptor. Most importantly, the transport of dIgA is unidirectional, occurring only in the basolateral to apical direction. This process occurs with a half-time of about 30 minutes (Mostov and Deitcher, 1986).

IV. Regulation of Sorting of the Polymeric Immunoglobulin Receptor

We have carried out additional studies to determine why the pathway for transcytosis of pIg-R is unidirectional (Breitfeld *et al.*, 1989a). One possibility is that unidirectionality is conferred by the protease that cleaves the pIg-R to SC at the apical surface. To test this hypothesis, we took advantage of the observation that the cleavage of pIg-R to SC is inhibited by the microbial thiol protease inhibitor, leupeptin (Musil and Baenziger, 1987). We found that even though cleavage to SC was inhibited, transcytosis of ligand to the apical surface and its release into the apical medium were unaffected. In the absence of cleavage, ligand simply dissociates from the uncleaved pIg-R at the apical surface. More importantly, if ligand were added to the apical medium, apical to basolateral transcytosis was not observed. These results indicate that the unidirectionality of transport is conferred by signals other than proteolytic release of SC.

It appears that transcytosis is imperfect in terms of its efficiency. When a ligand molecule is endocytosed from the basolateral surface, it has three possible fates: transcytosis to the apical surface, recycling to the basolateral surface, or degradation. We have recently developed an assay that allows us to examine the fate of ligand endocytosed at the basolateral surface (Breitfeld *et al.*, 1989a). We found that about 55% of internalized ligand is transcytosed over a 2-hour time course, and about 20% recycles. Very little (3–5%) is degraded. The recycling of receptor to the basolateral surface provides a further opportunity to be reendocytosed and transcytosed. We also found that ligand could be endocytosed from the apical plasma membrane (Breitfeld *et al.*, 1989a) but that this apically internalized ligand mostly recycles back to the apical surface. Almost none is transcytosed to the basolateral domain. There is thus a clear difference between ligand endocytosed from the basolateral surface, which can go to either surface, and ligand endocytosed apically, which can only return to the apical domain. It appears that once the receptor reaches the apical plasma membrane, it is essentially "trapped"; it is then either cleaved to SC or endocytosed and recycled back to the apical surface. The molecular signals that account for these observations are not known.

It is interesting to note that treatment of the MDCK cells with the microtubule-depolymerizing drug nocodazole slows the rate of transcytosis by 60–70% but does not affect the overall accuracy of delivery (Breitfeld *et al.*, 1990a). This suggests that microtubules may facilitate the process of transcytosis but are not absolutely required for targeting of vesicles to the appropriate membrane. Importantly, transport of newly synthesized receptor from the Golgi to the basolateral membrane is completely unaffected by nocodazole treatment, suggesting that this process does not involve microtubules.

V. Sorting Signals in the Polymeric Immunoglobulin Receptor

A major goal of our research is to analyze the structural determinants inherent to molecules, such as the pIg-R, which direct them to the appropriate cellular locations. We assume that the pIg-R contains sorting signals that control its transport. The complexity of the pIg-R cellular itinerary suggests that it may contain multiple sorting signals. One obvious location for such signals is the 103-amino acid, C-terminal cytoplasmic domain. Being in the cytoplasm, this receptor tail would be accessible to interact with proteins in the cytoplasm that presumably constitute the cellular sorting machinery.

To address this issue, we first constructed a mutant pIg-R that lacked 101 of the 103 amino acids of the cytoplasmic domain (Mostov *et al.*, 1986) (Fig. 2). Oligonucleotide-directed *in vitro* mutagenesis was used to insert a stop codon two amino acids after the membrane-spanning segment. This truncated "tail-minus" receptor was expressed using the retroviral vector system in both non-polarized fibroblasts and in MDCK cells.

In fibroblasts the receptor is transported normally to the surface and is cleaved to SC. However, unlike the wild-type pIg-R, the tail-minus mutant is not endocytosed, an observation that is not completely surprising. In other systems, notably the LDL receptor, the cytoplasmic tail has been shown to be essential for endocytosis (Davis *et al.*, 1987). When expressed in MDCK cells, this tail-minus pIg-R does not appear at the basolateral surface; rather it is sent directly to the apical surface from the Golgi and is cleaved to SC. This result suggests that the cytoplasmic domain either contains a signal for basolateral targeting or that the cytoplasmic domain is necessary for basolateral targeting to occur. In a separate construction, we further truncated the receptor by deleting both the transmembrane and cytoplasmic domains, producing a soluble receptor (Mostov *et al.*, 1987) (Fig. 2). This anchor-minus receptor was secreted predominantly from the apical pole of MDCK cells. This suggests that the extracellular (or luminal) portion of the pIg-R may contain an intrinsic apical sorting signal.

We have made a large number of smaller mutations in the cytoplasmic domain of the pIg-R, which indicate that it contains at least four sorting signals (Breitfeld *et al.*, 1989c). The arrangement of these signals is depicted in Figure 2. Deletion of the carboxy-terminal 30 amino acids, which are encoded by a single exon, produces a receptor that follows the pathway of the wild-type receptor, except that the rate of endocytosis from the basolateral surface is decreased by about 60% (Breitfeld *et al.*, 1990b). Exactly the same phenotype is produced by mutation of a tyrosine residue in this segment to a serine. This result is consistent with previous observations in several other systems, which have shown that tyrosine residues are important for clathrin-mediated endocytosis (Davis *et al.*, 1987; Jing *et al.*, 1990) and suggests a similar role for tyrosine in the pIg-R. Mutation of a second, more membrane proximal, tyrosine reduces the endocytotic rate by only 5–10%. However, mutation of both tyrosines together virtually eliminates endocytosis, suggesting that both residues may play a role in this process.

Deletion of 37 residues from the middle section of the tail produces a pIg-R that is endocytosed normally but is then degraded (Breitfeld *et al.*, 1990b). It is unlikely that this receptor is simply malfolded, as it is normally delivered to the basolateral cell surface, binds ligand, and is endocytosed. Perhaps the receptor has a mechanism to avoid degradation, which has been disrupted by the mutation. Alternatively, a signal for degradation may have been artifically created.

Further mutational analysis indicates that only the 17 amino acids closest to the membrane are needed for basolateral targeting (Casanova *et al.*, 1991). A truncated receptor containing only these residues in the cytoplasmic domain is sorted basolaterally from the TGN, whereas deletion of these residues, leaving the remainder of the tail intact, produces a receptor that is targeted directly to the apical surface. Moreover, transplantation of this 17-amino acid signal to a heterologous, normally apical protein (placental alkaline phosphatase) is sufficient to redirect the chimeric protein to the basolateral surface.

Finally, the receptor has been shown to be phosphorylated on a serine residue in the cytoplasmic domain (Larkin *et al.*, 1986). Phosphorylation occurs at the basolateral surface and/or shortly after endocytosis. Mutation of this serine to an alanine, which cannot be phosphorylated, produces a receptor that is not efficiently transcytosed but rather recycles to the basolateral surface (Casanova *et al.*, 1990). In contrast, mutation of this serine to an aspartic acid, the negative charge of which may mimic that of the phosphate group, produces a receptor that is transcytosed more efficiently than wild-type. These results indicate that phosphorylation is the signal that directs the segregation of receptor into the transcytotic pathway.

The aspartic acid mutant also allows several indirect conclusions to be drawn. First, if the function of the pIg-R were simply to maximally transcytose IgA, why would the cell use phosphorylation rather than simply using aspartate at this site? The most likely explanation is that phosphorylation is used to regulate transcytosis, perhaps in response to external cues. Second, many other proteins are transcytosed in epithelial cells. These may not necessarily use a phosphorylation mechanism but may instead be analogous to the Asp mutant. Third, both the TGN and the basolateral endosome are organelles where apical-basolateral sorting takes place. However, the endosome relies on a negative charge from either phosphate or aspartate at a specific site in the pIg-R to send the molecule to the apical surface. The TGN, in contrast, ignores this charge, and sends even the Asp mutant to the basolateral surface. In other words, the negative charge does not inactivate the signal for TGN to basolateral targeting. Rather, it appears that targeting from the TGN and the endosome uses different mechanisms.

Having made substantial progress in defining these sorting signals, the next step is to analyze the cellular machinery that recognizes such signals and is ultimately responsible for sorting processes. A first step in this direction is the work of Sztul *et al.* (1991) who have isolated putative transcytotic vesicles from rat liver. These vesicles are enriched in a 108-kDa protein, which is apparently bound to the cytoplasmic face of the vesicular membrane. The possibility that this protein may be involved in either docking of the vesicle to its target membrane or in attachment to microtubules for transport is currently under investigation.

VI. Immunological Functions of Immunoglobulin A and Secretory Component

The immunological functions of IgA and SC have been thoroughly reviewed (Childers *et al.*, 1989) and will be mentioned here only briefly. The complex of IgA and SC is termed secretory IgA (sIgA). The major unique characteristic of sIgA, relative to other immunoglobulins, is its relative resistance to proteolysis and denaturation. This is largely the result of a protective, stabilizing effect of SC on IgA. This makes sIgA uniquely suited for action in a harsh mucosal environment where it is exposed to many proteases, extreme pH, and denaturants. A second unusual feature is that sIgA is at least tetravalent, so it is better at aggregating antigens.

Several mechanisms are known for how sIgA protects against bacteria, viruses, and other potentially harmful agents. Perhaps the most important is that sIgA prevents attachment of bacteria and viruses, causing immune exclusion; sIgA can also aggregate pathogens. At least in some cases sIgA can activate complement. sIgA has further been shown to be potentially active in both direct cell killing and antibody-dependent cytotoxicity. A different protective effect is that sIgA can inactivate various toxins and enzymes as well as prevent the systemic absorption of antigens. sIgA has an affinity for mucus, which can prevent it from being rapidly carried away and enable the sIgA to work synergistically with other nonspecific antimicrobial factors.

VII. Conclusion

As this review indicates, substantial progress has been made in understanding the cellular and molecular basis of the transcytosis of immunoglobulins. This knowledge is important for two reasons. First, transcytosis is one of several systems of protein traffic in epithelial cells. Analyzing the sorting signals and cellular machinery that decodes these signals will permit elucidation of the general principles that govern protein sorting. Second, transcytosis is important in the overall physiology of an organism. For example, polymeric immunoglobulins form an early specific immunologic defense against infection, and their transport is mediated by specific transcytotic events. Although much less is known about the regulation of their transcytosis, many other proteins are carried across epithelia by this mechanism to tissue sites where they are likely to carry out important functions. Analyzing transcytosis is thus an important area of connection between cell and molecular biology and the overall functioning of organs and organ systems.

Acknowledgments

Benjamin Aroeti was supported by a postdoctoral fellowship from the International Human Frontier Science Program Organization. Keith Mostov was supported by a Scaile Scholarship, Cancer Research Institute Investigator Award and NIH RO1 AI 25144.

References

Ahnen, D. J., Brown, W. R., and Kloppel, T. M. (1985). *Gastroenterology* **89**, 667–682.
Bartles, J. R., Ferraci, H. M., Stieger, B., and Hubbard, A. L. (1987). *J. Cell Biol.* **105**, 1241–1251.
Bienenstock, J. (1984). *Ann. Allergy* **53**, 535–539.
Brändli, A. W., Parton, R. G., and Simons, K. (1990). *J. Cell Biol.* **111**, 2902–2921.
Brandtzaeg, P. (1974). *J. Immunol.* **112**, 1553–1559.
Brandtzaeg, P. (1981). *Clin. Exp. Immunol.* **44**, 221–232.
Breitfeld, P. P., Harris, J. M., and Mostov, K. M. (1989a). *J. Cell Biol.* **109**, 475–486.
Breitfeld, P., Casanova, J. E., Harris, J. M., Simister, N. E., and Mostov, K. E. (1989b). *Methods Cell Biol.* **32**, 329–337.
Breitfeld, P. P., Casanova, J. E., Simister, N. E., Ross, S. A., McKinnon, W. C., and Mostov, K. E. (1989c). *Curr. Opin. Cell Biol.* **1**, 617–623.
Breitfeld, P. P., McKinnon, W. C., and Mostov, K. E. (1990a). *J. Cell Biol.* **111**, 2365–2373.
Breitfeld, P. P., Casanova, J. E., McKinnon, W. C., and Mostov, K. E. (1990b). *J. Biol. Chem.* **265**, 13750–13757.
Brown, M. S., Anderson, R. G. W., and Goldstein, J. L. (1983). *Cell* **32**, 663–667.
Casanova, J. E., Breitfeld, P. P., Ross, S. A., and Mostov, K. E. (1990). *Science* **248**, 742–745.
Casanova, J. E., Apodaca, G., Mostov, K. E. (1991). *Cell* **66**, 65–75.
Cereijido, M., Ponce, A., and Gonzalez-Marical, L. (1989). *J. Membrane Biol.* **110**, 1–9.
Childers, N. K., Bruce, M. G., and McGhee, J. R. (1989). *Annu. Rev. Microbiol.* **43**, 503–536.
Davis, C. G., Van Driel, I. R., Russell, D. W., Brown, M. S., and Goldstein, J. L. (1987). *J. Biol. Chem.* **262**, 4075–4082.
Deitcher, D. L., and Mostov, K. E. (1986). *Mol. Cell. Biol.* **6**, 2712–2715.
Deitcher, D. L., Neutra, M. R., and Mostov, K. E. (1986). *J. Cell Biol.* **102**, 911–919.
Geuze, H. J., et al. (1984). *Cell* **37**, 195–204.
Jing, S., Spencer, T., Miller, K., Hopkins, C., and Trowbridge, I. S. (1990). *J. Cell Biol.* **110**, 283–294.

King, G. L., and Johnson, S. M. (1985). *Science* **227**, 1583–1586.

Kuhn, L. C., and Kraehenbuhl, J. P. (1979). *J. Biol. Chem.* **254**, 11072–22081.

Larkin, J. M., Sztul, E. S., and Palade, G. E. (1986). *Proc. Natl. Acad. Sci. U.S.A.* **83**, 4759–4763.

LeBivic, A., Quaroni, A., Nichols, B., and Rodriguez-Boulan, E. (1990). *J. Cell Biol.* **111**, 1351–1361.

Maratos-Flier, E., Kao, B. Y., Verdin, E. M., and King, G. L. (1987). *J. Cell Biol.* **105**, 1595–1601.

Matter, K., Bucher, K., and Hauri, H. P. (1990). *EMBO J.* **9**, 3163–3170.

Mostov, K. E., and Blobel, G. (1982). *J. Biol. Chem.* **257**, 11816–11821.

Mostov, K. E., and Deitcher, D. L. (1986). *Cell* **46**, 613–621.

Mostov, K. E., and Simister, N. E. (1985). *Cell* **43**, 389–390.

Mostov, K. E., Kraehenbuhl, J. P., and Blobel, G. (1980). *Proc. Natl. Acad. Sci. U.S.A.* **77**, 7257–7261.

Mostov, K. E., Friedlander, M., and Blobel, G. (1984). *Nature* **308**, 37–43.

Mostov, K. E., De Bruyn Kops, A., and Deitcher, D. L. (1986). *Cell* **47**, 359–364.

Mostov, K., Breitfeld, P., and Harris, J. M. (1987). *J. Cell Biol.* **105**, 2031–2036.

Musil, L., and Baenziger, J. (1987). *J. Cell Biol.* **104**, 1725–1733.

Rodewald, R., and Kraehenbuhl, J. P. (1984). *J. Cell Biol.* **99**, 159S–164S.

Simons, K., and Fuller, S. D. (1985). *Annu. Rev. Cell Biol.* **1**, 243–288.

Simons, K., and Wandinger-Ness, A. (1990). *Cell* **62**, 207–210.

Solari, R., and Kraehenbuhl, J. P. (1984). *Cell* **36**, 61–71.

Sztul, E. S., Howell, K. E., and Palade, G. E. (1985a). *J. Cell Biol.* **100**, 1255–1262.

Sztul, E. S., Howell, K. E., and Palade, G. E. (1985b). *J. Cell Biol.* **100**, 1248–1254.

Sztul, E., Kaplin, A., Saucan, L., and Palade, G. (1991). *Cell* **64**, 81–89.

Tomasi, T. B., Jr., Tan, E. M., Solomon, A., and Prendergast, R. A. (1965). *J. Exp. Med.* **121**, 101–124.

Receptors for Nerve Growth Factor

Moses V. Chao, David S. Battleman, and Marta Benedetti
Department of Cell Biology and Anatomy, Cornell University Medical College, New York, New York 10021

I. Introduction

During development of the nervous system, a variety of genetic and epigenetic factors are known to determine the survival and maintenance of selective neuronal populations. Nerve growth factor (NGF) is a prototypic neurotrophic factor that is released by target cells, taken up in nerve terminals by specific receptors, internalized, and then transported to the cell body by retrograde transport. Abundant evidence has established this mechanism (Thoenen and Barde, 1980). NGF still serves as the primary example of a neurotrophic factor, the physiological role of which has been firmly established, and represents one of a family of similar growth factors, including brain-derived neurotrophic factor (BDNF) (Barde *et al.*, 1982; Leibrock *et al.*, 1989), neurotrophin-3 (NT-3), NT-4, and NT-5 (Hohn *et al.*, 1990; Maisonpierre *et al.*, 1990; Hallbook *et al.*, 1991; Berkemeier *et al.*, 1991).

Although the actions of NGF were originally thought to be restricted to the sympathetic and neural crest-derived sensory neurons, NGF also affects the cholinergic neuronal population of the basal forebrain. For example, cholinergic neurons in the septal-hippocampal pathway not only bind and internalize NGF but respond with the induction of choline acetyltransferase (Korsching, 1986; Thoenen *et al.*, 1987). NGF is believed to possess a far broader range of actions than previously believed, based upon its potential roles outside the nervous system (Levi-Montalcini *et al.*, 1990).

NGF interacts with two receptor glycoproteins in responsive cells, p75^{NGFR} and p140^{prototrk} (Bothwell, 1991). Both receptors are expressed on NGF-responsive cells, including sympathetic and neural crest-derived sensory neurons, cholinergic neurons of the basal forebrain, and also PC12 cells, a widely used cell culture model for studying the mechanism of action of NGF. This chapter will discuss the two receptors for NGF: (1) the low-affinity NGF receptor (p75^{NGFR}) and (2) the product of the proto-oncogene *trk* (p140^{prototrk}); also reviewed is how they were originally identified and the structural and functional features of the two receptors.

II. p75 Nerve Growth Factor Receptor

Receptors for mammalian growth factors are integral membrane proteins designed to have multiple functions for communicating with other cells and for intracellular signal transduction. For NGF the interaction with cell surface receptors led to a remarkable set of reactions in responsive cells, including changes in gene expression, activation of neurotransmitter enzymes, and stimulation of neuronal processes. These many events led ultimately to the survival and maintenance of selective neuronal cell populations. The overall biological functions of NGF are cell survival and cell differentiation. As such, the receptor-signaling mechanisms must mediate and fulfill these functions.

The initial receptor identified for NGF was a membrane glycoprotein of apparent M_r of 75,000 daltons, p75^{NGFR} (Chao *et al.*, 1986; Radeke *et al.*, 1987). Two techniques contributed to the cloning of the p75 NGF receptor gene: the use of gene transfer and the ability to detect receptors using monoclonal antibodies. DNA-mediated gene transfer was used as an assay for the expression cloning of the p75^{NGFR} (Chao *et al.*, 1986; Radeke *et al.*, 1987). This method circumvented the requirement for mRNA enrichment or for protein purification and thus allowed for the isolation of a gene, the product of which was not easily accessible by biochemical means.

Because expression of cell surface receptors is not amenable to metabolic selection, transfer of the gene encoding the NGF receptor required co-transfection with genes carrying selectable markers. High molecular weight DNA was mixed with the purified herpes virus thymidine kinase gene to produce a calcium phosphate precipitate. This precipitate was used to transfect mouse fibroblasts deficient in thymidine kinase (tk$^-$). After 2 weeks in hypoxanthine, aminopterin, thymidine (HAT) selection, tk$^+$ colonies were screened for NGF receptor expression by an *in situ* rosette assay employing the ME20.4 monoclonal antibody (Chao *et al.*, 1986), which inhibited NGF binding to NGF receptors on human melanoma cells (Ross *et al.*, 1984), or by fluorescent activated cell sorting (Radeke *et al.*, 1987) using the 192-IgG monoclonal antibody against the rodent receptor (Chandler *et al.*, 1984). The isolation and availability of these antibodies were essential to the cloning and characterization of p75^{NGFR}.

Two different routes were taken to isolate the p75^{NGFR} gene following gene transfer and purification of the fibroblast transfectants. The first was to use a molecular tag to rescue the transforming sequences, and the second approach was to employ subtracted cDNA populations from transfected cells. Human NGF receptor sequences were isolated from a transfected mouse L cell line using human middle repetitive Alu sequences as a probe (Chao *et al.*, 1986). The rodent p75 NGF receptor gene was isolated following cDNA subtractive hybridization. Poly(A)$^+$ RNA from the L cell transfectant was converted to [^{32}P]cDNA, and the unique cDNA sequences were isolated after hybridization with RNA from Ltk$^-$ cells. This enriched cDNA population was used to identify a full-length rat p75^{NGFR} (Radeke *et al.*, 1987).

Complementary cDNAs representing the NGF receptor mRNA have been isolated for chicken and human species by screening cDNA libraries from chicken brain RNA (Large *et al.*, 1989; Heuer *et al.* 1990) and human A875 melanoma cells (Johnson *et al.*, 1986). The sequence analysis of these clones predicted a signal sequence of 28 amino acids, an extracellular domain containing four 40-amino acid repeats with six cysteine residues at conserved positions, followed by a region rich in threonine and serine residues, a single transmembrane domain, and a 155-amino acid cytoplasmic domain. The molecular weight of the fully processed receptor protein is approximately 75,000 daltons. The difference between the size of the mature receptor and the predicted size of the receptor protein (399 amino acids) is due to extensive O- and N-linked glycosylation. The extracellular sequence is also distinguished by its highly negative charge. The bulk of the negative charge is contributed by the cysteine-rich domain in which the net charge is -24, consistent with the acidic pI of the receptor (Grob *et al.*, 1985).

One of the most striking features of the p75 receptor protein sequence is the 28 cysteine residues, 24 of which are found within the first 160 residues of the receptor. Alignment of the cysteine residues indicated that they are contained in four 40-amino acid internal repeats that have many similar residues besides the cysteine residues. These internal repeats have led to the hypothesis that these units are evolutionarily related. Similar cysteine-rich repeats have been detected in the extracellular domains of a number of proteins, including cell surface proteins, such as the Fas antigen (Itoh *et al.*, 1991), two receptors for tumor necrosis factor, and B and T cell activation antigens, CDw40 and OX40 (Smith *et al.*, 1990). Such a diverse family of cell surface proteins sharing similar cysteine-rich repeats indicates that these domains probably share similar structural roles for ligand binding and for functional responses.

III. Proto-oncogene *Trk*

A second NGF receptor species of 135,000–158,000 daltons was originally identified by photoaffinity cross-linking studies that employed the reagent, hydroxysuccinimidyl-4-azido benzoate (HSAB) (Massague *et al.*, 1981; Hosang and Shooter, 1985). This cross-linked species was observed only on NGF-responsive cells, including PC12 cells and sympathetic neurons. HSAB was also capable of detecting p75^{NGFR} bound to [^{125}I]NGF (M_r = 90–100,000). Curiously, other cross-linking agents do not efficiently give rise to the 160-kDa species, presumably due to the different mechanism of cross-linking agents used and the lower levels of this cross-linked species (Green and Greene, 1986; Hempstead *et al.*, 1991).

The identity of the higher molecular weight cross-linked species was obscure until the proto-oncogene *trk* was found to be specifically and rapidly phosphorylated on tyrosine residues by NGF treatment of PC12 cells (Kaplan *et al.*, 1991a). Increased tyrosine phosphorylation of a number of cellular substrates had been previously observed in PC12 cells treated with NGF for short periods of time (Maher, 1988) and provided the initial evidence for protein tyrosine phosphorylation as a signal transduction mechanism for NGF. These important observations led to the identification of the *trk* proto-oncogene as a signaling receptor for NGF that is activated after ligand binding (Kaplan *et al.*, 1991b; Klein *et al.*, 1991). The finding that the proto-oncogene *trk* is an NGF receptor (Kaplan *et al.*, 1991a; Klein *et al.*, 1991) provided new insights into the mechanism by which NGF exerts its many effects upon responsive cells.

The *trk* gene was originally identified as a rearranged oncogene from a colon carcinoma (Martin-Zanca *et al.*, 1986); however, the normal cellular proto-oncogene is represented by a single membrane-spanning receptor molecule. The *trk* oncogene activity was first observed in gene transfer experiments using fibroblasts transfected with high molecular weight tumor DNA. The *trk* gene was discovered by its virtue of inducing focus formation in 3T3 cells and was found to be the result of a genomic rearrangement, wherein tropomyosin sequences were aberrantly fused onto the transmembrane and cytoplasmic domains of *trk* (hence, the name **t**ropomyosin **r**eceptor **k**inase), thus producing an oncogene with malignant properties. The oncogene was also cloned using Alu repetitive sequences to recover transforming sequences from 3T3 cells (Martin-Zanca *et al.*, 1986).

The cellular homology of *trk* contains an extracellular domain similar in size to other growth factor receptors. *Trk* encodes a receptor tyrosine kinase that exists as a transmembrane glycoprotein of M_r = 140,000 (p140^{prototrk}), capable of binding to NGF directly (Kaplan *et al.*, 1991b; Klein *et al.*, 1991). The extracellular sequence of p140^{prototrk} differs substantially from the p75 receptor ligand-binding domain. In particular, multiple cysteine repeats are absent in p140^{prototrk}. Carbodimide affinity cross-linking agents exist that can cross-link [^{125}I]NGF to p75^{NGFR} but are not effective in cross-linking ^{125}I-labeled NGF to p140^{prototrk} (Hempstead *et al.*, 1991). Therefore, different protein–protein interactions are made between NGF and each of the two receptors.

The product of the proto-oncogene product *trk* contains a tyrosine kinase catalytic activity in the cytoplasmic domain, which is most similar in sequence to the insulin receptor (Hanks *et al.*, 1988). A number of distinctive structural features distinguish the *trk* tyrosine kinase domain from that of other receptor tyrosine kinases. A short kinase insert domain is present in the middle of the tyrosine kinase domain, which has been observed in other tyrosine kinase receptors, such as for platelet-derived growth factor (PDGF). The C terminus is uncharacteristically short (15 amino acids) for all the *trk* family members, including *trkB* (Klein *et al.*, 1989) and *trkC* (Lamballe *et al.*, 1991), in direct contrast to other members of the receptor tyrosine kinase family. Other differences include a threonine substitution in place of a highly conserved alanine residue at amino acid position 647, an absence of a consensus helix-breaking proline at residue 766, and a conserved tryptophan residue at position 722. In addition, *trk* contains an unusual tyrosine redundancy in the middle of the kinase domain. The functional significance of these features in the *trk* tyrosine kinase domain is not yet known, but it is likely the differences in amino acid sequence function to direct *trk* phosphorylation activity to specific targets and substrates in neuronal cells.

Computer analysis of the extracellular domain of the *trk* tyrosine kinase has revealed a number of motifs, including three tandem leucine rich repeats of 24 amino acids, which are bounded by two clusters of cysteine residues (Schneider and Schweigei, 1991). These features have been found in the *toll* gene product of *Drosophila*, a receptor that is involved in cell-to-cell recognition. In addition, two immunoglobulin-like C2 repeats are found in the *trk* extracellular domain, further implicating a role of this receptor in cell adhesion events. These IgC_2-like domains are similar to those found in N-CAM, the PDGF receptor, and the carcinoembryonic antigen-related protein, raising the possibility that other cellular proteins or extracellular matrix proteins may interact with the *trk* family of receptors.

Several lines of evidence indicated that *trk* function is essential to NGF signal transduction. Introduction of the *trk* gene into mutant PC12 cells (PC12^{nnr}) completely unresponsive to NGF resulted in the recovery of morphological differentiation in these cells, showing that the expression of *trk* molecule is essential for maintaining the neuronal phenotype (Loeb *et al.*, 1991). In addition, a potent inhibitor of NGF action in PC12 cells, K-252a (Koizumi *et al.*, 1988), was shown to act specifically on p140^{prototrk} activity by blocking NGF-induced autophosphorylation of *trk* and also phosphorylation of a number of key cellular substrates (Berg *et al.*, 1992). K-252a did not affect phosphorylation events mediated by basic FGF and EGF, demonstrating that the initial signaling events initiated by NGF binding to p140^{prototrk} are distinctive during neuronal differentiation.

IV. Expression of Nerve Growth Factor Receptors

Although the *trk* gene was originally identified in colon carcinoma tumor, the mRNA for the *trk* proto-oncogene is expressed predominately in the nervous system, sensory neurons (Martin-Zanca *et al.*, 1990), sympathetic ganglia, and the basal forebrain (M. Chao and L. Parada, unpublished data). The finding that *trk* transcripts were observed by *in situ* hybridization in cranial sensory ganglia and dorsal root ganglia became the initial observation that led to the consideration of p140^{prototrk} as a signaling molecule involved in NGF action. The subsequent discovery of p140^{prototrk} in PC12 cells and the induction of *trk* autophosphorylation by NGF (Kaplan *et al.*, 1991a) verified its function in neurotrophin signal transduction.

Hence, there is a correlation between *trk* expression and the primary neuronal targets that are known to be responsive to NGF, which include PC12 cells, sensory neurons, sympathetic neurons, and basal forebrain cholinergic neurons. As well, the p75^{NGFR} is expressed in the same populations of cells (Buck *et al.*, 1987; Bothwell, 1990; Patil *et al.*, 1990). Therefore, in NGF-responsive cells, both receptors are expressed. However, the p75 receptor has been detected on a wide variety of both neuronal and nonneuronal derivatives of the neural crest (Ernfors *et al.*, 1988; Yan and Johnson, 1988; Thomson *et al.*, 1988; Large *et al.*, 1989), including sympathetic neurons (Frazier *et al.*, 1974) and Schwann cells (Taniuchi *et al.*, 1986). Also, tumor cell lines of neural crest origin, specifically melanoma (Fabricant *et al.*, 1977), neuroblastoma (Sonnenfeld and Ishii, 1982), neurofibromas (Ross *et al.*, 1984), and pheochromocytoma cells (Landreth and Shooter, 1980; Grob *et al.*, 1983; Hosang and Shooter, 1985; Green and Greene, 1986), display considerable NGF cell surface binding.

Classic target cell populations for NGF have been examined for developmentally regulated expression of the p75^{NGFR} mRNA *in vivo* (Buck *et al.*, 1987; Schatteman *et al.*, 1988; Ernfors *et al.*, 1988). In the rat sympathetic ganglia, which utilizes NGF throughout the entire life span, the steady-state levels of p75^{NGFR} mRNA increased with development. In contrast, dorsal root ganglia sensory neurons that require NGF for survival only during early developmental periods express lower levels of receptor mRNA levels after birth (Buck *et al.*, 1987). Receptor mRNA levels reach a peak in early chicken brain development at embryonic days 6–10 but decrease significantly at later embryonic ages (Escandon and Chao, 1989). Because NGF appears to be synthesized at a low, constant level in most tissues, regulation of receptor gene expression and numbers may determine which cells respond to NGF.

Perhaps the most interesting regions of p75 receptor expression have been in nonneuronal populations, such as in testis (Ayer-LeLievre *et al.*, 1988) and Schwann cells after peripheral nerve transection (Taniuchi *et al.*, 1986; Heumann *et al.*, 1987) or after culture *in vitro* (Lemke and Chao, 1988). The increased levels of transcripts and protein for both NGF and NGF receptor mRNA in Schwann cells after nerve lesion are not understood, but these events may function to facilitate nerve regeneration (Johnson *et al.*, 1988). The developmental expression of p75 NGF receptor mRNA in such areas as testis, muscle (Raivich *et al.*, 1987), and lymphoid tissues, such as thymus and spleen (Ernfors *et al.*, 1988), indicates that the actions of NGF may not be restricted only to cholinergic cells of the basal forebrain and cells derived from the neural crest. For example, NGF is known to bind to T cell populations (Thorpe *et al.*, 1987), influence the differentiation of eosinophils and basophils (Matsuda *et al.*, 1988), and have proliferative effects upon mast cells (Aloe and Levi-Montalcini, 1977). It is not yet known which of the two NGF receptors is responsible for these activities.

Thus, p75^{NGFR} displays a much wider pattern of expression compared with p140^{prototrk}. This widespread pattern of p75 receptor expression suggests additional roles for this receptor. An explanation for the diverse pattern of expression for the low-affinity NGF receptor can be accommodated by the finding that multiple, similar neurotrophic factors can directly bind p75^{NGFR} (Rodriguez-Tebar *et al.*, 1990; Rodriguez-Tebar, 1992). The binding of [^{125}I]BDNF and [^{125}I]NT-3 to fibroblast cells expressing high amounts of p75^{NGFR} is of a low-affinity ($K_d = 10^{-9}$ M) nature but appears to differ by kinetic parameters. These results indicate that the p75 receptor can act as a receptor for multiple neurotrophic factors and should be properly considered as a generic neurotrophin receptor. Furthermore, these results implicate the low-affinity p75 receptor as a receptor that potentially dictates specificity in neurotrophin interactions with the *trk* family of tyrosine kinases.

V. High-Affinity Nerve Growth Factor Binding Site

The finding of the *trk* tyrosine kinase as an NGF receptor provides a signaling mechanism of action for NGF. The discovery that a member of the receptor tyrosine kinase family is the signal transducer for NGF simultaneously solves a number of questions that had previously stymied the neurotrophic field, including the signaling mechanism for NGF and the way in which specificity of action is generated for the neurotrophic family of factors: NGF and its relatives, BDNF, and other neurotrophins NT-3, NT-4, and NT-5.

Related molecules to *trk*, such as *trkB* (Klein *et al.*, 1989) and *trkC* (Lamballe *et al.*, 1991), do not respond to NGF but interact with other neurotrophin factors, such as BDNF, NT-3, NT-4 (*trkB*), and NT-3 (*trkC*). Interestingly, NT-3 and Nt-5 can also bind *trk* and initiate *trk* autophosphorylation (Cordon-Cardo *et al.*, 1991; Berkemeier *et al.*, 1992). This apparent promiscuity or cross-talk among neurotrophin factors for the *trk* receptor tyrosine kinase members has been observed in fibroblast cells transfected with individual *trk* receptors. These *trk*-expressing fibroblast cells do not contain p75^{NGFR}; therefore, the p75 receptor is not required for the mitogenic activities of the *trk* family members (Glass *et al.*, 1991). It is likely that cells of more neuronal origin expressing both the p75 receptor and appropriate *trk* receptor will display far more specificity in response from the neurotrophin family members.

The requirement for both receptors has been established by binding studies in responsive cells. Equilbrium binding of [^{125}I]NGF to responsive cells revealed two distinct affinity states for the NGF receptor (Sutter *et al.*, 1979; Landreth and Shooter, 1980; Schechter and Bothwell, 1981). In most responsive cells, such as neurons and PC12 cells (Greene and Tischler, 1976), approximately 10–15% of the receptors display high-affinity binding with a K_d of 10^{-11} M, with the remainder of the receptors possessing a K_d of 10^{-9} M. The difference in equilibrium binding is accounted for by a 100-fold difference in the rate of dissociation of NGF. It is generally believed that the high-affinity receptor sites are internalized upon NGF binding and are absolutely required for the actions of NGF.

Human and rat p75^{NGFR} cDNAs introduced into mouse fibroblast cell lines only give rise to the low-affinity kinetic class of receptors (Chao *et al.*, 1986; Radeke *et al.*, 1987). The equilibrium-binding constant (10^{-9} M) is reflective of the extremely fast off-rate of [^{125}I]NGF from this receptor (Sutter *et al.*, 1979; Schechter and Bothwell, 1981). Low-affinity receptors in most cell lines are not responsive to NGF. In p75^{NGFR} transfected cells, as many as 500,000 receptors per cell can be observed that are capable of binding NGF; however, none of the cell lines displays any detectable responses with NGF treatment (Hempstead *et al.*, 1988, 1989).

Both p140^{prototrk} and p75^{NGFR} bind [^{125}I]NGF with a similar low affinity (10^{-9} M K_d) in equilibrium-binding experiments (Kaplan *et al.*, 1991b; Hempstead *et al.*, 1991; Klein *et al.*, 1991). Reconstitution experiments by membrane fusion and transfection of cloned cDNAs for *trk* and p75^{NGFR} in heterologous cells indicated that high-affinity NGF binding ($K_d = 10^{-11}$ M) required coexpression and binding to both the p75 NGF receptor and the tyrosine kinase *trk* gene product (Hempstead *et al.*, 1991). These reconstitution studies with the two cellular receptors for NGF imply that both receptors are required for NGF-induced differentiation and cell survival.

NGF-induced *trk* tyrosine kinase activity was sufficient for cell proliferation in fibroblasts (Cordon-Cardo *et al.*, 1991), suggesting that proliferative effects mediated by NGF may be carried out solely by interaction with p140^{prototrk}. The major biological function of NGF in the nervous system, however, is in mediating cell survival and maintenance in postmitotic neuronal populations (Levi-Montalcini, 1987). Hence, the *cellular context* that the *trk* tyrosine kinase family is found in is probably a major parameter in determining biological functions of the neurotrophic factor family. Functional responses have been documented by *trk* in the absence of p75^{NGFR} receptors (Cordon-Cardo *et al.*, 1991; Birren *et al*, 1992) and when NGF binding to p75^{NGFR} is compromised (Weskamp and Reichardt, 1991; Ibanez *et al*, 1992) indicating additional mechanisms may exist for neurotrophin receptor action depending upon the ratio of receptors.

Introduction of the p75 receptor cDNA into cells derived from neuronal origins gave rise to both high- and low-affinity receptors (Hempstead et al., 1989; Matsushima and Bogenmann, 1990; Pleasure et al., 1990) as assessed by equilibrium binding, and led to certain functional responses, such as the induction of immediate early gene expression. Therefore, the p75 receptor gene can participate in both kinetic forms of the receptor in the appropriate cell lines that express the trk tyrosine kinase (Hempstead et al., 1991). This conclusion was consistent with previous cross-linking studies of the NGF receptor in PC12 cells, suggesting that both kinetic classes are represented by the same p75 receptor (Green and Greene, 1986).

From affinity cross-linking with different reagents, the two receptors bind to NGF through different protein–protein contacts. This conclusion is supported by kinetic studies, in which NGF receptors display different interactions with [^{125}I]NGF, with respect to association and dissociation rates. The kinetic rates of [^{125}I]NGF binding have been found to differ dramatically for the two receptors, with p140^{prototrk} exhibiting a much slower on- and off-rate than p75^{NGFR}. The faster association and dissociation of p75 and the much slower rates measured for p140^{prototrk} accounted for the low-affinity equilibrium-binding behavior for these receptors (B. L. Hempstead et al., unpublished). Therefore, the two receptors can be distinguished by kinetic parameters, rather than by equilibrium binding, as assessed by Scatchard analysis. Other studies have suggested that p140^{prototrk} exhibited an inherent higher affinity for NGF binding than the p75 receptor (Meakin and Shooter, 1991; Klein et al., 1991). Resolution of the nature of the functional NGF receptor complex will be resolved by correlating the action of the two receptors with appropriate neutrophic responses and determining which downstream signaling events are specific for NGF.

Two different receptors interact with NGF, leading to cell survival and cell differentiation and maintenance of neurons. The finding that NGF binding to trk can induce cell proliferation suggests new roles for neurotrophic factors. The product of the proto-oncogene trk is the primary signaling NGF receptor, the function of which is to increase tyrosine phosphorylation of cellular proteins. Sensory deficits in mutant mice missing the p75^{NGFR} gene suggest that this receptor has an essential role in neuronal development (Lee et al., 1992). The low-affinity p75 receptor is believed to participate in retrograde transport (Johnson et al., 1987) and in facilitating the trk tyrosin kinase activity, either by specifying substrates or directly modulating the phosphorylation of specific targets. Other potential functions for the p75 receptor include a direct signaling function (Yan et al., 1991) and binding to all the neurotrophic factors (Rodriguez-Tebar et al., 1990; 1992) to directly determine which neurotrophic factor will interact with its cognate trk tyrosine kinase. Hence, the p75 receptor may act as a common subunit, interacting with the trk family members. Together the two receptors for NGF represent a unique growth factor receptor mechanism for signal transduction in the differentiating nervous system.

References

Aloe, L., and Levi-Montalcini, R. (1977). *Brain Res.* **133**, 356–366.

Ayer-Lelievre, C., Olson, L., Ebendal, T., Hallbook, F., and Persson, H. (1989). *Proc. Natl. Acad. Sci. U.S.A.* **85**, 2628–2632.

Barde, Y. A., Edgar, D., and Thoenen, H. (1982). *EMBO J.* **1**, 549–553.

Berg, M. M., Sternberg, D., Parada, L. F., and Chao, M. V. (1992). *J. Biol. Chem.* **267**, 13–16.

Berkemeier, L. R., Winslow, J. W., Kaplan, D. R., Nikolics, K., Goedell, D. V., and Rosenthal, A. *Neuron* **7**, 857–866.

Birren, S. J., Verdi, J. M. and Anderson, D. J. (1992) *Science* **257**, 395–397.

Bothwell, M. A. (1990). *Curr. Topics Microbiol. Immunol.* **165**, 55–70.

Bothwell, M. (1991). *Cell* **65**, 915–918.

Buck, C. R., Martinez, H. M., Black, I. B., and Chao, M. V. (1987). *Proc. Natl. Acad. Sci. U.S.A.* **84**, 3060–3063.

Chandler, C., Parsons, L. M., Hosang, M., and Shooter, E. M. (1984). *J. Biol. Chem.* **259**, 6882–6889.

Chao, M. V., Bothwell, M. A., Ross, M. A., Koprowski, H., Lanahan, A. A., Buck, C. R., and Sehgal, A. (1986). *Science* **232**, 418–421.

Cordon-Cardo, C., Tapley, P., Jing, S., Nanduri, V., O'Rourke, E., Lamballe, F., Kovary, K., Klein, K., Jones, K. R., Reichardt, L. F., and Barbacid, M. (1991). *Cell* **66**, 173–183.

Ernfors, P., Hallbook, F., Ebendal, T., Shooter, E. M., Radeke, M. J., Misko, T. P., and Persson, H. (1988). *Neuron* **1**, 983–996.

Escandon, E., and Chao, M. V. (1989). *Dev. Brain Res.* **47**, 187–196.

Fabricant, R. F., DeLarco, J. E., and Todaro, G. J. (1977). *Proc. Natl. Acad. Sci. U.S.A.* **74**, 565–569.

Frazier, W. A., Boyd, L. F., Pulliman, L. W., Szutowicz, A., and Bradshaw, R. A. (1974). *J. Biol. Chem.* **249**, 5918–5923.

Glass, D. J., Nye, S. H., Hantzopoulos, P., Macchi, M. J., Squinto, S. P., Goldfarb, M., and Yancopoulos, G. D. (1991). *Cell* **66**, 405–413.

Green, S., and Greene, L. A. (1986). *J. Biol. Chem.* **261**, 15316–15326.

Greene, L. A., and Tischler, A. S. (1976). *Proc. Natl. Acad. Sci. U.S.A.* **73**, 2424–2428.

Grob, P. M., Berlot, C. H., and Bothwell, M. A. (1983). *Proc. Natl. Acad. Sci. U.S.A.* **80**, 6819–6823.

Grob, P. M., Ross, A. H., Koprowski, H., and Bothwell, M. A. (1985). *J. Biol. Chem.* **260**, 8044–8049.

Hallbook, F., Ibanez, C. F., and Persson, H. (1991). *Neuron* **6**, 845–858.

Hanks, S. K., Quinn, A. M., and Hunter, T. (1988). *Science* **241**, 42–52.

Hempstead, B. L., Patil, N., Olson, K., and Chao, M. V. (1988). *Cold Spring Harbor Symp. Quant. Biol.* **53**, 477–485.

Hempstead, B. L., Schleifer, L. S., and Chao, M. V. (1989). *Science* **243**, 373–375.

Hempstead, B. L., Martin-Zanca, D., Kaplan, D., Parada, L. F., and Chao, M. V. (1991). *Nature* **350**, 678–683.

Heuer, J. G., Fatemie-Nainie, S., Wheeler, E. F., and Bothwell, M. (1990). *Dev. Biol.* **137**, 287–304.

Heumann, R., Lindholm, D., Bandtlow, C., Meyer, M., Radeke, M. J., Misko, T. P., Shooter, E., and Thoenen, H. (1987). *Proc. Natl. Acad. Sci. U.S.A.* **84**, 8735–8739.

Hohn, A., Leibrock, J., Bailey, K., and Barde, Y. A. (1990). *Nature* **334**, 339–341.

Hosang, M., and Shooter, E. M. (1985). *J. Biol. Chem.* **260**, 655–662.

Ibanez, C. F., Ebendal, T., Barbany, G., Murray-Rust, J., Blundell, T. L. and Persson, H. (1992). *Cell* **69**, 329–341.

Itoh, N., Yonehara, S., Ishii, A., Yonehara, M., Mizushima, S. I., Sameshima, M., Hase, A., Seto, Y., and Nagata, S. (1991). *Cell* **66**, 233–243.

Johnson, D., Lanahan, A., Buck, C. R., Sehgal, A., Morgan, C., Mercer, E., Bothwell, M., and Chao, M. (1986). *Cell* **47**, 545–554.

Johnson, E. M., Taniuchi, M., Clark, H. B., Springer, J. E., Koh, S., Tayrien, M. W. and Loy, R. (1987). *J. Neurosci.* **7**, 923–929.

Kaplan, D. R., Martin-Zanca, D., and Parada, L. F. (1991a). *Nature* **350**, 158–160.

Kaplan, D. R., Hempstead, B. L., Martin-Zanca, D., Chao, M. V., and Parada, L. F. (1991b). *Science* **252**, 554–558.

Klein, R., Parada, L. F., Coulier, F., and Barbacid, M. (1989). *EMBO J.* **8**, 3701–3709.

Klein, R., Jing, S., Nanduri, V., O'Rourke, E., and Barbacid, M. (1991). *Cell* **65**, 189–197.

Koizumi, S., Contreras, M. L., Matsuda, Y., Hama, T., Lazarovici, P., and Guroff, G. (1988). *J. Neurosci. Res.* **8**, 715–721.

Korsching, S. (1986). *Trends Neurosci.* **5**, 570–573.

Lamballe, F., Klein, R., and Barbacid, M. (1991). *Cell* **66**, 967–979.

Landreth, G. E., and Shooter, E. M. (1980). *Proc. Natl. Acad. Sci. U.S.A.* **77**, 4751–4755.

Large, T. H., Weskamp, G., Helder, J. C., Radeke, M. J., Misko, T. P., Shooter, E. M., and Reichardt, L. F. (1989). *Neuron* **2**, 1123–1134.

Lee, K. F., Li, E., Huber, L. J., Landis, S. C., Sharpe, A. H., Chao, M. V. and Jaenisch, R. (1992). *Cell* **69**, 737–749.

Leibrock, J., Lottspeich, F., Hohn, A., Hofer, M., Hengerer, D., Masiakowski, P., Thoenen, H., and Barde, Y. A. (1989). *Nature* **341**, 149–152.

Lemke, G., and Chao, M. V. (1988). *Development* **102**, 499–504.

Levi-Montalcini, R. (1987). *Science* **237**, 1154–1164.

Levi-Montalcini, R., Aloe, L., and Alleva, E. (1990). *Prog. Neuro. Endocrinimmunol.* **3**, 1–10.

Loeb, D., Martin-Zanca, D., Chao, M. V., Parada, L. F., and Greene, L. A. (1991). *Cell* **66**, 961–966.

Maher, P. (1988). *Proc. Natl. Acad. Sci. U.S.A.* **85**, 6788–6791.

Maisonpierre, P. C., Belluscio, L., Squinto, S., Ip, N. Y., Furth, M. E., Lindsay, R. M., and Yancopoulos, G. D. (1990). *Science* **247**, 1446–1451.

Martin-Zanca, D., Hughes, S. H., and Barbacid, M. (1986). *Nature* **319**, 743–777.

Martin-Zanca, D., Barbacid, M., and Parada, L. F. (1990). *Genes Dev.* **4**, 683–688.

Massague, J., Guillette, J. J., Czech, M. P., Morgan, C. J., and Bradshaw, R. A. (1981). *J. Biol. Chem.* **256**, 9419–9424.

Matsuda, H., Coughlin, M. D., Bienenstock, J., and Denburg, J. A. (1988). *Proc. Natl. Acad. Sci. U.S.A.* **85**, 6508–6512.

Matsushima, H., and Bogenmann, E. (1990). *Mol. Cell. Biol.* **10**, 5015–5020.

Meakin, S. O., and Shooter, E. M. (1991). *Neuron* **6**, 153–163.

Patil, N., Lacy, E., and Chao, M. V. (1990). *Neuron* **4**, 437–447.

Pleasure, S. J., Reddy, U. R., Venkatakrishnan, G., Roy, A. K., Chen, J., Ross, A. H., Trojanowski, J. Q., Pleasure, D. E., and Lee, V. M. (1990). *Proc. Natl. Acad. Sci. U.S.A.* **87**, 8496–8500.

Radeke, M. J., Misko, T., Hsu, C., Herzenberg, L. A., and Shooter, E. M. (1987). *Nature* **325**, 593–597.

Raivich, G., Zimmermann, A., and Sutter, A. (1987). *J. Comp. Neurol.* **256**, 229–245.

Rodriguez-Tebar, A., Dechant, G., and Barde, Y. A. (1990). *Neuron* **4**, 187–192.

Rodriguez-Tebar, A., Dechant, G., Gotz, R., and Barde, Y. A. (1992). *EMBO J.* **11**, 917–922.

Ross, A. H., Grob, P., Bothwell, M. A., Elder, D. E., Ernst, C. S., Marano, N., Ghrist, B. F. D., Slemp, C. C., Herlyn, M., Atkinson, B., and Koprowski, H. (1984). *Proc. Natl. Acad. Sci. U.S.A.* **81**, 6681–6685.

Schatteman, G. C., Gibbs, L., Lanahan, A. A., Claude, P., and Bothwell, M. (1988). *J. Neurosci.* **8**, 860–873.

Schechter, A. L., and Bothwell, M. A. (1981). *Cell* **24**, 867–874.

Schneider, R., and Schweiger, M. (1991). *Oncogene* **6**, 1807–1811.

Smith, C. A., Davis, T., Anderson, D., Solam, L., Beckmann, M. P., Jerzy, R., Dower, S. K., Cosman, D., and Goodwin, R. G. (1990). *Science* **248**, 1019–1023.

Sonnenfeld, K. H., and Ishii, D. N. (1982). *J. Neurosci. Res.* **8**, 375–391.

Sutter, A., Riopelle, R. J., Harris-Warrick, R. M., and Shooter, E. M. (1979). *J. Biol. Chem.* **254**, 5972–5982.

Taniuchi, M., Clark, H. B., and Johnson, E. M. (1986). *Proc. Natl. Acad. Sci. U.S.A.* **83**, 4094–4098.

Thoenen, H., and Barde, Y. A. (1980). *Physiol. Rev.* **60**, 1284–1335.

Thoenen, H., Bandtlow, C., and Heumann, R. (1987). *Rev. Physiol. Biochem. Pharmacol.* **109**, 145–178.

Thomson, T. M., Rettig, W. J., Chesa, P. G., Green, S. H., Mena, A. C., and Old, L. J. (1988). *Exp. Cell Res.* **174**, 533–539.

Thorpe, L. W., Stach, R. W., Hashim, G. A., Marchetti, D., and Perez-Polo, J. R. (1987). *J. Neurosci. Res.* **17**, 128–134.

Weskamp, G. and Reichardt, L. F. (1991). *Neuron* **6**, 654–663.

Yan, Q., and Johnson, E. M. (1988). *J. Neurosci.* **8**, 3481–3498.

Yan, H., Schlessinger, J., and Chao, M. V. (1991). *Science* **252**, 561–563.

Asialoglycoprotein Receptor

Iris Geffen and Martin Spiess

Department of Biochemistry, Biocenter, University of Basel, Basel, Switzerland

I. Introduction

A hepatic receptor and uptake system for desialylated glycoproteins was initially discovered by Ashwell, Morell and their co-workers in their seminal studies on the metabolism of serum glycoproteins in mammals (reviewed by Ashwell and Morell, 1974; Ashwell and Harford, 1982). The half-life in the circulation of proteins carrying N-linked oligosaccharides was found to be drastically reduced when the terminal sialic acid residues had been enzymatically removed. Such asialoglycoproteins (ASGPs) accumulate and are degraded inside parenchymal cells of the liver. Hepatic recognition of ASGPs is dependent on the normally penultimate galactose residues, which become exposed upon desialylation; asialo-agalactoglycoproteins exposing terminal N-acetyl glucosamine units are again slowly cleared from the circulation. The proteins in the hepatocyte membrane that are responsible for the binding and internalization of ASGPs were isolated by affinity chromatography. They have been alternatively called ASGP receptors to indicate the initially used and the presumed natural substrates, galactose (or galactose/N-acetyl galactosamine) receptors to emphasize the molecular specificity, or simply hepatic lectins, a name that reflects the fact that these proteins were among the first known carbohydrate-binding proteins of mammalian origin.

Since its discovery, the ASGP receptor has become one of the best characterized model systems for the process of receptor-mediated endocytosis, and its analysis in perfused rat liver, isolated rat hepatocytes, and HepG2 cells (a human hepatoma cell line expressing the native receptor) has greatly contributed to the identification and the understanding of endosomal compartments in the cell and of intracellular membrane traffic. In contrast to transporters of small molecules, such as ions, monosaccharides, and amino acids, endocytic receptors do not deliver their ligands (which invariably are macromolecules) into the cytoplasm. Instead, after binding their ligands on the cell surface, they cluster into specialized plasma membrane domains characteristically coated with clathrin on the cytoplasmic side. By invagination of these areas as coated pits and pinching off as coated vesicles, receptor–ligand complexes are internalized as part of the membrane. Upon fusion of these vesicles with endosomal compartments, acidic conditions cause dissociation of the ligand from the receptor. Typical transport receptors, such as the ASGP receptor, efficiently return to the plasma membrane via recycling vesicles. The majority of the ligand molecules are transported to lysosomes where they are hydrolyzed. Only as degradation products do the ligands finally reach the cytoplasm.

The aim of this chapter is to summarize what is known about the intracellular traffic of the ASGP receptor and its ligands and to integrate this information into the current understanding of intracellular compartments. cDNA cloning and sequence analysis have provided insights into the receptor structure, and specific signals for the intracellular sorting are beginning to be identified.

For many years functional ASGP receptors were believed to be expressed exclusively in hepatocytes. However, binding and uptake of ASGPs have also been discovered in rat peritoneal macrophages. Analysis of the responsible protein revealed the existence of a distinct but homologous receptor in these cells. The macrophage ASGP-binding protein, as it was named, is compared with the hepatic ASGP receptor in Section VI.

II. Physiological Functions

The ASGP receptor contains carbohydrate-binding sites specific for terminal nonreducing galactose and N-acetyl galactosamine residues (Schwartz, 1984b). The interaction of individual residues of this type with the receptor protein is of fairly low affinity with dissociation constants on the order of $10^{-3}\ M$. The presence of two terminal galactoses in biantennary oligosaccharides and of three in triantennary oligosaccharides dramatically increases the binding affinity and yields dissociation constants of $\sim 10^{-6}$ and $\sim 5 \times 10^{-9}$, respectively (Lee et al., 1983). A fourth galactose in a tetra-antennary structure further increases the affinity only fivefold. These observations indicate that several galactose-binding sites are combined within each high-affinity ASGP-binding site (Section III,C). Besides the number of terminal galactoses, their geometry within the ligand molecule is important for the affinity of receptor binding. Among the best of a large number of natural and synthetic ligands that have been characterized (Lee et al., 1983; Hardy et al., 1985; Townsend et al., 1986; Schwartz, 1984b) are the desialylated forms of the typical tri- and tetra-antennary N-linked oligosaccharides found in most serum glycoproteins.

This suggests a physiological function of the ASGP receptor in the turnover of serum glycoproteins. The ASGP receptor, in combination with a desialylating activity in the serum, appears to provide a degradative pathway in glycoprotein homeostasis. In birds, the serum proteins of which are largely unsialylated (Regoeczi *et al.*, 1975), there is an analogous hepatic receptor specific for agalacto-glycoproteins, that is, glycoproteins with terminal N-acetyl glucosamine residues (Lunney and Ashwell, 1976; Drickamer, 1981).

It was further observed that ASGPs, poly-galactosylated albumin, and antibodies against the ASGP receptor substantially reduced the binding of chylomicron remnants to HepG2 cells (Windler *et al.*, 1991). It was concluded that the ASGP receptor may contribute to the hepatic uptake of remnants by binding apolipoprotein E. Consistent with this transport mechanism is the finding that apolipoprotein E is secreted as a sialoglycoprotein and is subsequently desialylated in the plasma (Zannis *et al.*, 1984).

A general function of the ASGP receptor in glycoprotein clearance could explain why hepatocytes are not impaired in I cell disease patients (Kornfeld, 1987). Because the addition of the mannose-6-phosphate recognition marker for lysosomal sorting is defective in these patients, lysosomal enzymes are secreted. As a result, most cells accumulate in their lysosomes massive amounts of extracellular material that is insufficiently degraded. Because the secreted lysosomal enzymes would eventually be desialylated and then taken up by the ASGP receptor, hepatic lysosomes should always have the full complement of lysosomal hydrolases.

In addition to general glycoprotein clearance, the ASGP receptor may also be important for the uptake of specific ligand molecules in the liver. One example is immunoglobulin A (IgA; Brown and Kloppel, 1989; Daniels *et al.*, 1989). In several species, including humans, which express little polymeric immunoglobulin receptor in the serosal membrane, IgA was shown to be endocytosed into hepatocytes to a large extent via the ASGP receptor. The oligosaccharides of IgA contain terminal galactose and N-acetyl galactosamine residues (Kerr, 1990). Other glycoproteins that use the hepatic transport system for ASGPs may still be discovered. Moreover, artificial ligands for the ASGP receptor are being created for the specific delivery of drugs, DNA, and other molecules to hepatocytes (e.g., Dragsten *et al.*, 1987; Ishihara *et al.*, 1990; Weissleder *et al.*, 1990; Wilson *et al.*, 1992).

III. Structure of the Asialoglycoprotein Receptor Proteins

A. A General Protein Structure

The receptor proteins are integral membrane proteins. Upon solubilization with detergents, they can be purified by affinity chromatography on immobilized ASGPs or galactose: because ligand binding is dependent on the pH and on Ca^{2+}, the receptor proteins are adsorbed to the affinity resin at pH 7.8 in the presence of 20 mM Ca^{2+} and eluted at pH 5.1 without Ca^{2+}. ASGP receptor proteins have been isolated from rabbit (Kawasaki and Ashwell, 1976), human (Baenziger and Maynard, 1980), rat (Drickamer et al., 1984), and mouse liver (Hong et al., 1988) as well as from the human hepatoma cell line HepG2 (Schwartz and Rup, 1983; Bischoff and Lodish, 1987). The receptor preparations contain two (in rat three) distinct but immunologically related proteins in the range of 40–60 kDa, one polypeptide being roughly three times as abundant as the other(s).

The ASGP receptor proteins are themselves glycoproteins carrying two or three N-linked oligosaccharides, which are co-translationally added in the endoplasmic reticulum (ER). As the proteins are transported from the ER through the Golgi compartments to the plasma membrane, the oligosaccharide moieties, which are originally of the high-mannose type, undergo the typical modifications (trimming and addition of monosaccharides) to the complex type within 45–60 minutes (in HepG2 cells; Schwartz and Rup, 1983; Bischoff and Lodish, 1987). However, inhibition of glycosylation or of maturation by the inhibitors tunicamycin and swainsonine, respectively, does not affect protein stability or receptor function (Breitfeld et al., 1984; Hsueh et al., 1986).

Partial sequencing of the rat proteins (Drickamer et al., 1984) and the cloning of the cDNAs of the rat (Holland et al., 1984; McPhaul and Berg, 1987; Halberg et al., 1987) and human receptors (Spiess et al., 1985; Spiess and Lodish, 1985) confirmed the existence of two homologous genes encoding the major form of the hepatic lectin, HL-1, and the minor form(s), HL-2, respectively. The two minor rat sequences originate from the same gene and differ only in their carbohydrate structures (Halberg et al., 1987). In the literature the rat receptors are usually referred to as RHL-1 and RHL-2/3, and the human receptors as H1 and H2.

The cDNA and deduced protein sequences of the two human and the two rat receptors are highly homologous over almost the entire length: 39% of the amino acids are identical in all four polypeptides. With 80% identities the two HL-1 sequences are more closely related to each other than to the HL-2 sequences, with which they have only 50–58% of the residues in common. The HL-2 sequences are likewise more homologous to each other, but with only 62% identical residues they appear to be less conserved than the HL-1 sequences. The two receptor genes must have been generated by gene duplication before primates and rodents diverged in evolution. In the mouse the two genes were shown to be closely linked on chromosome 11 (Sanford et al., 1988). The different receptor proteins appear simultaneously in the developing rat (Petell and Doyle, 1985), suggesting that the two genes may be coordinately regulated by common cis-acting elements.

The overall structure of the ASGP receptor proteins, as deduced from the sequence data and from biochemical analyses, is summarized in Figure 1. The

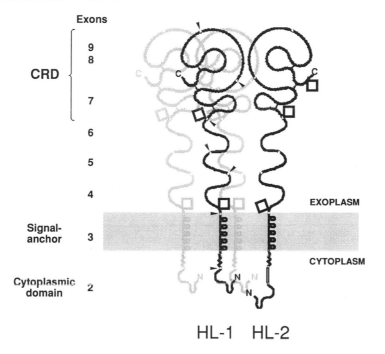

FIG. 1   Structure of the hepatic ASGP receptor. The domain structure of the HL-1 and HL-2 polypeptides is schematically illustrated. Exon boundaries (as deduced from the RHL-1 gene structure; Leung *et al.*, 1985) are indicated in HL-1 by arrowheads, and the exons are numbered on the left. N and C termini of the polypeptides, disulfide bridges, and N-linked oligosaccharides (squares) are indicated. To illustrate the oligomeric structure of the receptor complex (Fig. 2), two additional copies of HL-1 are shown in grey.

polypeptide chains consist of approximately 300 amino acids and span the membrane once in a Type II orientation: a single transmembrane domain of approximately 20 hydrophobic residues separtes a small amino-terminal portion protruding into the cytoplasm (40 residues in HL-1, 58 residues in HL-2) from a large carboxy-terminal domain (~230 residues) exposed on the exoplasmic side of the membrane (Chiacchia and Drickamer, 1984). The carboxy-terminal half of the polypeptide constitutes the carbohydrate recognition domain (CRD): a clostripain fragment of RHL-1 consisting of residues 134–283 was shown to be sufficient for galactose binding (Hsueh *et al.*, 1986). The most striking difference between the two receptor proteins is an 18-residue insertion at position 24 in the cytoplasmic portion of HL-2 that does not have a counterpart in HL-1.

The general organization of the receptor protein is reflected in the structure of the gene as analyzed for RHL-1 (Leung *et al.*, 1985). Whereas exon 1 covers 5'

TABLE I
Domain Organization of C-Type Lectins

Group[a]	Introns (S–S)	Domain Organization[b]	Proteins	References[c]
I				
Proteoglycans	+	(signal) – GAG – **CRD**	Cartilage proteoglycans	Sai et al., 1986; Doege et al., 1986; Oldberg et al., 1987; Tanaka et al., 1988
	(3)	(signal) – link – GAG – EGF – **CRD** – CB	Fibroblast proteoglycan (Versican)	Zimmermann and Ruoslahti, 1989
II				
Type II receptors	+	cyt – TM – x – **CRD**	**Hepatic ASGP receptors**	Spies and Lodish, 1985; Halberg et al., 1987; Leung et al., 1985*
	(3)		Macrophage ASGP binding protein	Ii et al., 1990
			Kupffer cell receptor	Hoyle and Hill, 1991*
			IgE receptor	Suter et al., 1987*
			Natural killer cell receptor	Giorda et al., 1990
			Chicken hepatic lectin	Bezouska et al., 1991
		cyt – TM – x – $[\mathbf{CRD}]_8$	Mannose receptor	Taylor et al., 1990
III				
CRDs with collagen domains	–	(signal) – coll – **CRD**	Pulmonary surfactant apoproteins	White et al., 1985*; Benson et al., 1985; Sano et al., 1987; Boggaram et al., 1988; Rust et al., 1991
	(2)		Conglutinin	Lee et al., 1991
			Mannose-binding proteins	Drickamer et al., 1986; Drickamer and McCreary, 1987*; Taylor et al., 1989*
IV				
Cell adhesion molecules	–	(signal) – **CRD** – EGF – O-glyc – TM – cyt	Thrombomodulin	Suzuki et al., 1987
	(2)	(signal) – **CRD** – EGF – $[\mathrm{CB}]_n$ – TM – cyt	ELAM-1	Collins et al., 1991*
			GMP-140	Johnston et al., 1990*
			Lymph node homing receptor	Ord et al., 1990*; Dowbenko et al., 1991*
V				
Separate CRDs	?	(signal) – **CRD**	Acorn barnacle lectin	Muramoto and Kamiya, 1986
	(3)		Echinoidin	Giga et al., 1987
			Pancreatic stone protein	De Caro et al., 1987
			Tetranectin	Fuhlendorff et al., 1987
			Tunicate lectin	Suzuki et al., 1990
			Rattlesnake venom lectin	Hirabayashi et al., 1991
			Factor IX/X binding proteins	Atoda et al., 1991
			Fly hemolymph lectin	H. Takahashi et al., 1985
	(2)		Snake phospholipase A2 inhibitors	Inoue et al., 1991

[a]Numbering according to Bezouska et al., 1991 for groups I–IV.

[b]Abbreviations: CRD, C-type carbohydrate-recognition domain; CB, complement binding domain; cyt, cytoplasmic domain; EGF, EGF-like repeats; GAG, glycosaminoglycan-attachment sequence; link, link protein-like sequence; O-glyc, O-glycosylated domain; (signal), cleaved signal sequence; TM, transmembrane domain; x, other domain(s).

[c]References with an asterisk include information on the gene structure.

untranslated sequences, most of the cytoplasmic domain is encoded by exon 2. The 18-codon insertion in HL-2 coincides exactly with the position of intron 2, suggesting that its presence in HL-2 (or its absence in HL-1) is the result of an aberrant splicing event. Exon 3 corresponds to the membrane-spanning segment. Exons 4–6 constitute the glycosylated peptide that extends to the CRD, which is itself encoded by exons 7–9.

Based on cloned cDNA sequences, two forms of H2 mRNAs have been discovered which differ only in the presence (H2a) or absence (H2b) of a segment of 15 base pairs immediately carboxy-terminal to the transmembrane domain (Spiess and Lodish, 1985). This was shown to be the result of alternative splicing of intron 3 (Lederkremer and Lodish, 1991). About 8% of H2 mRNA in HepG2 cells is of the H2a type. The function of the additional five residues is not known. However, when expressed in NIH 3T3 fibroblasts, H2a protein is retained and degraded in the ER or a related pre-Golgi compartment (Section III,C).

B. A Family of Ca^{2+}-Dependent Lectins

The ASGP receptor proteins are members of a rapidly growing family of proteins that possess homologous CRDs (summarized in Table I). The hallmark of these lectin domains is the absolute requirement for Ca^{2+} for carbohydrate binding. They have been classified as C-type animal lectins, as opposed to the S-type lectins, which depend on free thiol groups (Drickamer, 1988) and to other lectins that belong to neither group (e.g., the mannose-6-phosphate receptors; Thiel and Reid, 1989). We currently know of more than 30 proteins with C-type CRDs. Sequence comparison of CRDs reveals several invariant or strongly conserved residues in a characteristic pattern (Spiess, 1990). Among the invariant amino acids are four cysteines that were shown to form disulfide bridges in the acorn barnacle lectin, echinoidin, tetranectin, rattle snake lectin, and the human mannose-binding protein. In a subclass of CRDs, which includes the ASGP receptors, an additional pair of cysteines near the amino-terminal end of the domain is conserved forming an additional disulfide bond.

With the exception of the macrophage ASGP-binding protein, sequence homology to the ASGP receptors is restricted to the CRD. However, several of the C-type lectins are structurally similar Type II membrane proteins with a single transmembrane segment and a short amino-terminal cytoplasmic domain (Table I): chicken hepatic lectin, lymphocyte IgE receptor, Kupffer cell receptor, mannose receptor, and natural killer cell receptor. In another group of this lectin family, the protein is anchored in the membrane by a transmembrane domain near the carboxy terminus (Type I membrane proteins) and the CRD is combined with EGF-like repeats and complement-binding domains. Other members are soluble proteins and their CRDs are combined with collagen-like domains or with glycan attachment sequences, or the CRD constitutes essentially the entire protein. The C-type lectins thus could be described as mosaic proteins composed of several functional units that have been combined in evolution by exon shuffling from different ancestral genes.

Sequence comparison of the CRDs revealed that the grouping according to different functional contexts also reflects the evolutionary descent of these proteins (Bezouska *et al.*, 1991): The CRDs of groups I–IV, as summarized in Table I (the proteins summarized here in group V had not been analyzed), are most closely related to other members of the same group, indicating that the CRDs of each group derived from a single ancestral gene. Based on sequence similarly, CRDs of proteoglycans and type II receptors (groups I and II) are more closely related to each other than to the other groups (Bezouska *et al.*, 1991). Consistently, almost all members of groups I and II contain three pairs of invariant cysteines as opposed to only two pairs in the proteins of group III and IV. How the precursor genes of groups I/II, III, and IV evolved from their common ancestral CRD gene cannot be derived with certainty from the current sequence data. Most of the single domain proteins, which we have tentatively summarized in group V, contain the three pairs of cysteines typical for groups I and II; their position within the evolutionary tree remains to be analyzed in detail.

The gene structures that have been elucidated for currently 12 C-type lectins support the dendrogram outlined above. The CRDs of the type II receptors RHL-1, IgE receptor, Kupffer cell receptor, and the chicken hepatic lectin are encoded by three exons with exactly conserved boundary positions. Similarly, the CRD of the chicken cartilage proteoglycan is encoded by three exons, but the intron positions differ from those in the group II genes by up to four codons. In contrast, the CRD sequences of group III and group IV proteins are contained in a single exon. Whether the ancestral gene of C-type lectin domains contained introns which were lost in a precursor of groups II and IV, or whether it lacked introns which were later acquired by an immediate precursor of group I and II CRDs cannot be decided at the moment (see Bezouska *et al.*, 1991, for discussion).

An important feature of many endocytic receptors is the pH dependence of ligand binding, which is essential for the efficient release of the ligand in acidic endosomes (Mellman *et al.*, 1986). The carbohydrate-binding activities of the ASGP receptor (Breitfeld *et al.*, 1985), the chicken hepatic lectin (Loeb and Drickamer, 1988), and the Kupffer cell receptor (Lehrman *et al.*, 1986) have been shown to be gradually lost at pH below 6.5; pH-dependent conformational changes were observed for the rabbit ASGP receptor (DiPaola and Maxfield, 1984) and the chicken hepatic lectin (Loeb and Drickamer, 1988). Because lowering the pH also dramatically reduces the affinity of the CRD for Ca^{2+} (as shown for the chicken hepatic lectin; Loeb and Drickamer, 1988), the acidic environment of endosomes causes C-type receptors to release both Ca^{2+} and the ligand. Ca^{2+}-binding studies performed with the rabbit ASGP receptor and the chicken hepatic lectin indicate that two molecules of Ca^{2+} are bound per CRD (Andersen *et al.*, 1982; Loeb and Drickamer, 1988).

The CRD of one member of C-type lectins, of the rat mannose-binding protein A, has recently been crystallized (Weis *et al.*, 1991a) and its structure determined (Weis *et al.*, 1991b). The polypeptide adopts a previously undescribed fold of approximate dimensions 40 by 25 by 25 $Å^3$. Loops and extended regions account for more than half of the CRD structure; 29% of the residues are contained in five β strands and 18% in two helices. Two Ca^{2+} ions form an integral part of the structure by pinning together several loop regions near the "top" of the molecule. The structure around the cation sites bears no resemblance to known Ca^{2+}-binding motifs. The two ions are linked by two invariant acidic amino acids which are likely to contribute to the loss of ligand binding at mildly acidic pH by reducing the affinity for Ca^{2+} as they are protonated. The site for carbohydrate binding has not been identified yet.

C. State of Oligomerization

The functional ASGP receptor is a hetero-oligomeric complex composed of both subunits HL-1 and HL-2. The affinity of the ASGP receptor for mono-, di-, tri-, and tetra-antennary asialo-oligosaccharides increases with dissociation constants of $\sim 10^{-3}$, $\sim 10^{-6}$, $\sim 5 \times 10^{-9}$, and $\sim 10^{-9}$ M, respectively (Lee *et al.*, 1983), suggesting a close arrangement of at least three, and probably four or more, galactose-binding sites within the span of a ligand oligosaccharide. In gel filtration and sedimentation equilibrium analysis in detergent solution, the receptor behaves as a complex of approximately 260 kDa (Kawasaki and Ashwell, 1976; Andersen *et al.*, 1982), setting the upper limit for the size of the complex at a hexamer. Radiation inactivation experiments suggested a target size for high-affinity ligand binding of 70–140 kDa (Steer *et al.*, 1981; Schwartz *et al.*, 1984).

Expression of the cDNAs of both HL-1 and HL-2 is required to generate high-affinity ASGP-binding sites in transfected HTC rat hepatoma cells and fibroblast cells (McPhaul and Berg, 1986; Shia and Lodish, 1989), even though each subunit contains a functional CRD. When receptor degradation was induced by cross-linking with antibodies against H1 or H2 in HepG2 cells, both polypeptides were internalized and degraded simultaneously (Bischoff *et al.*, 1988). Similarly, antisera specific for each subunit co-precipitated the other receptor species from surface-iodinated rat hepatocytes (Sawyer *et al.*, 1988). In fluorescence photobleaching recovery experiments, antibodies against one subunit also immobilized the other subunit on the surface of HepG2 cells (Henis *et al.*, 1990). Chemical cross-linking of HepG2 cells produced covalent dimers and trimers of H1 as well as mixed trimers (Bischoff *et al.*, 1988). Cross-linking of H1 also

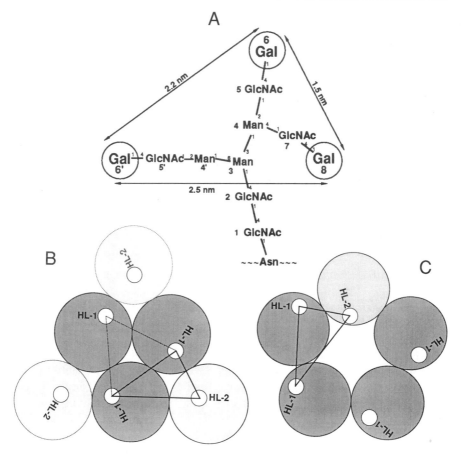

FIG. 2 Oligomeric organization of the ASGP receptor. Two models for the hetero-oligomeric structure of the ASGP receptor are shown (B and C) in relation to the dimensions of an optimal triantennary ligand oligosaccharide (A). Individual galactose-binding sites are indicated as white
circles.

yielded dimers and trimers in transfected fibroblasts lacking H2 (Shia and
Lodish, 1989), suggesting that HL-1 is oligomerizing at least to a homotrimer
$(HL-1)_3$, which in the presence of H2 serves as a core for the formation of hetero-oligomers $(HL-1)_3(HL-2)_x$. The subunit ratio of 3:1 in HepG2 cells (Henis *et
al.*, 1990) together with the molecular weight estimates mentioned herein indicates a stoichiometry of $(HL-1)_3(HL-2)_1$, as illustrated in Figure 2B (solid circles). The transfected fibroblast cell line 1-7-1 expresses equal amounts of the
two subunits on the surface. Based on lateral diffusion measurements, all H1
subunits are associated with H2 (i.e., can be immobilized by cross-linking with

anti-H2 antibodies), but only two-thirds of H2 molecules are associated with H1 (Henis *et al.*, 1990). This suggest a different (average) stoichiometry in 1-7-1 cells of $(HL-1)_3(HL-2)_2$ (Fig. 2B).

However, a 4:1 ratio of subunits for the receptor in HepG2 cells would still be within the limits of experimental error. A heteropentameric structure is thus possible, with a possible stoichiometry of $(HL-1)_4(HL-2)_1$ in HepG2 cells (Fig. 2C) and of $(HL-1)_3(HL-2)_2$ in 1-7-1 cells (Y. Henis, personal communication). Similar pentameric structures exist in other transmembrane proteins, such as the nicotinic acetylcholine receptor (a heteropentamer of the composition $\alpha_2\beta\gamma\delta$, in which the subunits are sufficiently homologous to yield a quasisymmetrical quarternary structure; Changeux *et al.*, 1990).

These models must also be evaluated with regard to the structure of the ligand (Lodish, 1991). Extensive studies using natural and synthetic oligosaccharides have demonstrated the importance of the geometric arrangement of the terminal galactoses for the affinity of binding to the receptor (Lee *et al.*, 1983, 1984; Hardy *et al.*, 1985; Townsend *et al.*, 1986). Among the triantennary ligands, the naturally occurring oligosaccharide shown in Figure 2A was found to have the highest affinity. Based on this structure, galactose-binding sites would be predicted to be positioned at the corners of a triangle with sides of 1.5, 2.2, and 2.5 nm (Lee *et al.*, 1984). Molecules with intergalactose distances shorter than these lengths were invariably poor ligands. Among the other ligands, those with the most flexible structures had the highest affinities.

To define the geometry of ligand binding, Rice *et al.*, (1990) used derivatives of a triantennary glycopeptide (Fig. 2A), in which a photolyzable reagent was coupled to carbon-6 of either galactose 6, 6', or 8. Upon binding to rat hepatocytes and photolysis, the ligands derivatized at galactose 6 or 6' specifically labeled RHL-1, whereas the ligand derivatized at galactose 8 labeled only RHL-2/3. Based on these data, a model for the arrangement of the receptor subunits in the hetero-oligomeric complex has been proposed by Lodish (1991) (Fig. 2B). Because the binding sites on three equivalent subunits HL-1 would form an equilateral triangle, a differently positioned subunit HL-2 could create the required geometry of galactose-binding sites. Such an asymmetric arrangement of binding sites could also be achieved by a heteropentamer (Y. Henis, personal communication; Fig. 2C).

The only known function of subunit HL-2 is to create high-affinity ligand-binding sites as a complex with subunit HL-1. In the absence of HL-2, HL-1 was shown to oligomerize, to be transported to the cell surface (Shia and Lodish, 1989), and to constitutively endocytose and recycle (Geffen *et al.*, 1989; Braiterman *et al.*, 1989). HL-1 thus contains all the signals necessary for correct intracellular transport of the receptor.

In contrast, only 20–40% of H2b (the major splicing variant of subunit H2), when expressed in transfected fibroblasts in the absence of H1, reached the cell surface (Lederkremer and Lodish, 1991). These H2b molecules, unlike H1, lack the ability to undergo efficient internalization and are mostly accumulated at the cell surface (Fuhrer *et al.*, unpublished). The other variant, H2a, was found to have a very short half-life and to be retained intracellularly almost completely in a high-mannose glycosylated form (Shia and Lodish, 1989). The only difference between the two variants, which is responsible for these different cellular fates, is a 5 residue insertion near the extracytoplasmic face of the membrane in H2a. As a first step in the degradation of H2a and of a large fraction of H2b, a 35 kDa fragment corresponding to the extracytoplasmic portion of the polypeptide is generated (Amara *et al.*, 1989). This cleavage is insensitive to lysosomotropic agents, but can be inhibited by TLCK and TPCK (Wikström and Lodish, 1991, 1992). It appears to occur in the ER, since the fragment lacks any Golgi-associated carbohydrate modifications and accumulates in the rough ER of cycloheximide-treated cells. Interestingly, in HepG2 cells a portion of the cleaved luminal portion is secreted to the extracellular medium (G. Lederkremer, personal communication). Moreover, preliminary data indicates that while the secreted H2b molecules are glycosylated at all three N-glycosylation sites, the intracellularly retained ones are underglycosylated (G. Lederkremer, personal communication). Retained precursor forms of H2b also appear to be underglycosylated, while all H2b polypeptides processed by the Golgi are fully glycosylated. For a variety of integral membrane proteins it has been shown that assembly into an oligomeric structure takes place in the ER and is a prerequisite for transport out of the ER (Hurtley and Helenius, 1989). It is therefore likely that folding or oligomerization of H2 (particularly of the H2a isoform) occurs inefficiently in the absence of H1 and full glycosylation is probably linked to proper folding. When co-expressed with subunit H1, H2 is rescued from retention and degradation in the ER by incorporation into hetero-oligomers.

IV. Intracellular Traffic of Receptor and Ligand

Receptor-mediated endocytosis involves several steps of sorting and segregation of receptors from other membrane proteins and from soluble proteins within intracellular compartments as well as efficient transport between organelles. The intracellular distribution of the receptor is thus intimately related to its function. Indeed, many of the concepts in cell biology come from kinetic biochemical and morphological studies on a few specialized receptor transport systems among which is the ASGP uptake in hepatocytes. Comparative studies of different receptors in the same cell have helped to define intracellular pathways and distinct endosomal compartments. The current model of the compartments and pathways used by the ASGP receptor and its ligand is schematically summarized in Figure 3.

A. Surface Events

The initial step in endocytosis is the clustering of receptors into specialized clathrin-coated domains of the cell surface. Coated pits are distinguished by a

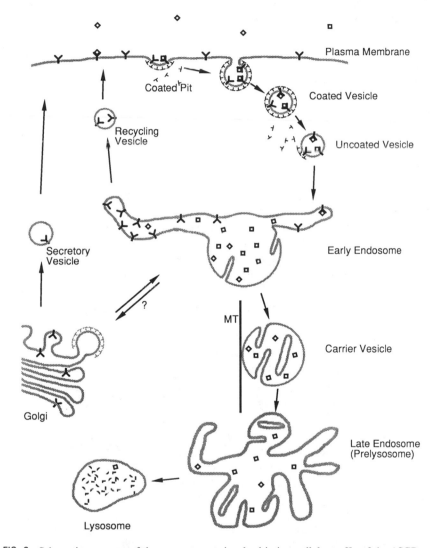

FIG. 3 Schematic summary of the compartments involved in intracellular traffic of the ASGP receptor and its ligands. ASGP receptors (Y), ligand molecules (squares), clathrin (⊥), and microtubules (MT).

characteristic coat of ~20 nm on the cytoplasmic face of the plasma membrane and have been observed in all cell types except erythrocytes. A single coated pit covers approximately 0.1 μm^2. The coated area typically occupies ~2% of the cell surface (Anderson et al., 1976) but can range between 0.4% and 3.4% depending on the cell type (Van Deurs et al., 1990).

It has been suggested that receptors are concentrated in coated pits, whereas other surface proteins are passively excluded (Bretscher, 1984; Roth et al., 1986). According to this view coated pits can be considered to function as molecular filters that allow internalization of endocytic proteins, leaving resident plasma membrane proteins behind. Active exclusion of a membrane protein from coated pits by interaction with cytoskeletal elements is another possibility, as has been suggested for the macrophage F_c receptor isoform B1 (Miettinen et al., 1989). The extent of clustering into coated pits is directly related to the initial rates of receptor internalization (Miettinen et al., 1989; Fuhrer et al., 1991).

The ASGP receptor, like other receptors primarily involved in the uptake of ligand molecules into the cell (such as the low density lipoprotein [LDL], transferrin, and mannose-6-phosphate receptors) is clustered in coated pits constitutively, that is, independently of ligand (Geffen et al., 1989). As a result, a considerable fraction of the receptor is always found intracellulary (Geuze et al., 1984b). In contrast, transcytotic receptors, such as those for polymeric immunoglobulins (poly-Ig) and F_c, and signal-transducing (hormone) receptors, such as those for epidermal growth factor (EGF) and insulin, are efficiently internalized only upon ligand binding. Binding of ligand to the ASGP receptor nevertheless induces an increased rate of internalization (by a factor of ~2) and a subsequent reduction of the plasma membrane receptor pool (Schwartz et al., 1984; Zijderhand-Bleekemolen et al., 1987). Whether this increase in endocytosis rate is due to receptor cross-linking by the ligand or to a conformational change in the receptor that increases the affinity for coated pits is presently unknown.

Coated pit formation was suggested on theoretical grounds to result from a process in which receptors and coat proteins are added simultaneously at the edge of a growing lattice (Bretscher and Pearse, 1984). This notion is supported by the finding that the level of expression of transferrin receptor in transfected cells was correlated with the number of coated pits on the cell surface (Iacopetta et al., 1988). The average assembly time of these structures was determined to be 1 minute (Griffiths et al., 1989). Invagination of the coated membrane domain was proposed to occur spontaneously as a result of a rearrangement of the clathrin triskelions (transformation of triskelion hexagons into pentagons; Heuser, 1989). In an alternative model the curvature is built into the clathrin lattice (Pearse and Crowther, 1987). Pinching off of vesicles occurs with an estimated half-time of less than 1 minute (Anderson et al., 1977) and is inhibited by low intracellular K^+ concentrations, low cytosolic pH, and hypertonic extracellular conditions (Larkin et al., 1986; Heuser, 1989).

B. Endosomal Compartments

Different receptors are segregated from other proteins at the plasma membrane, and they are found together in the same coated pits and vesicles. After vesicle formation the clathrin coat is removed by an uncoating ATPase (Rothman and Schmid, 1986), and the uncoated vesicles are believed to fuse with an endosomal compartment. Morphological observations and subcellular fractionation experiments suggest that at least two prelysosomal compartments, early and late endosomes, are involved in intracellular transport of receptors and ligands. These compartments are characteristically acidified by an ATP-driven proton pump to pH 6.0–6.2 in early endosomes, pH ~5.5 in late endosomes, and lower still in lysosomes (Kornfeld and Mellman, 1989).

The endosomal compartments are major sorting organelles. In early endosomes ASGPs and other ligands undergo acid-induced dissociation from their recycling receptors and are routed via late endosomes to a terminal endocytic compartment, the dense lysosomes. The receptors are rapidly returned from the early endosomes to the plasma membrane by recycling vesicles. The transferrin receptor uses a modification of this scheme: iron is released from the internalized transferrin in the early endosomes and apotransferrin still bound to the receptor is sorted into the recycling vesicles (Dautry-Varsat et al., 1983). An alternative pattern is exemplified by the EGF receptor system in which both ligand and receptor are delivered from the plasma membrane to early endosomes, then to late endosomes, and finally to lysosomes for degradation (Yarden and Ullrich, 1988; Schlessinger, 1988). Transcytotic receptors, such as the poly-Ig receptor, are packaged in early endosomes together with their ligands into yet another class of vesicles, the destination of which is the opposite plasma membrane domain (Mostov and Simister, 1985). The mannose-6-phosphate receptor, which is responsible for the correct delivery of newly synthesized and secreted lysosomal enzymes to lysosomes, is transported via vesicles between early endosomes, late endosomes, the trans-Golgi network (TGN), and the cell surface (Kornfeld and Mellman, 1989).

1. Sorting in Early Endosomes

Kinetic studies indicate that the early endosomes at the cell periphery are the first to receive internalized molecules. Within a few minutes at 37°C they are loaded with fluid-phase markers, of which ~50% is recycled back to the extracellular medium (Steinman et al., 1983; Swanson and Silverstein, 1988). Surface receptors, such as those for LDL and ASGP (Goldstein et al., 1985; Breitfeld et al., 1985), are also internalized to this compartment within a few minutes at 37°C. Receptors are recycled back to the cell surface with a half-time of 5–7 minutes. Internalized ligand ASGPs, however, are largely (50–75%)

retained inside the cell (Weigel and Oka, 1984; Simmons and Schwartz, 1984), reflecting acid-induced dissociation and segregation of ligands and receptors.

The endosomal compartment containing both ASGP receptor and ligand was morphologically identified by double-label immunoelectron microscopy of rat liver cells after 60 minutes of continuous infusion of asialo-orosomucoid (ASOR) (Geuze et al., 1983b). Receptor and ligand were found segregated in organelles with a tubulovesicular morphology. Interestingly, ligand was abundant within the vesicular portion, whereas receptors were concentrated in connected tubular extensions, which may give rise to the recycling vesicles. This compartment was therefore also named the "compartment for the uncoupling of receptors and ligands" (CURL; Geuze et al., 1983b). Entry of ligand–receptor complexes into this compartment seemed to occur from tubular regions as judged from sequential morphological studies in Hep G2 cells (Geuze et al., 1983a).

Segregation of receptors with different final destinations was also shown to take place in this compartment. The receptors for ASGP, mannose-6-phosphate, and poly-Ig were found co-localized within coated pits at the sinusoidal plasma membrane of hepatocytes and in coated vesicles. In early endosomes poly-Ig receptor was enriched in tubular regions separate from domains in which mannose-6-phosphate and ASGP receptors were co-distributed (Geuze et al., 1984b). As expected, poly-Ig remained attached to its receptor for delivery to the apical domain. That the biochemistry defined early endosomes and morphologically defined CURL correspond to the same compartment (or at least partly overlap) is supported by co-localization of transferrin and ASOR immediately after internalization but before their rapid segregation into different membrane fractions assessed by diamino benzidine (DAB) density shift experiments (Stoorvogel et al., 1987, 1989).

Different models have been proposed to explain ligand–receptor segregation and sorting. The inverse relationship between the enclosed volume and the receptor content observed within tubular and vesicular regions of early endosomes could result in less membrane protein and more content solutes in the vesicles detaching for the degradative pathway than in the recycling vesicles budding off from the tubules (Rome, 1985). In addition to this proposed passive sorting, morphological data on early endosomes in hepatoma cells reveal a higher concentration of ASGP receptors in tubules than in vesicles. These observations suggest the existence of barriers to receptor diffusion at the junctions of tubules and vesicles (Geuze et al., 1987). The mechanism of this selective lateral segregation remains unknown.

Sorting efficiency could be a function of the affinity of a ligand for its receptor and of how freely and rapidly a receptor enters the tubules. This model proposes a single efficient step of sorting (Linderman and Lauffenburger, 1988). Alternatively, sorting may be accomplished in a more continuous fashion by many iterations of a less efficient sorting process. This possibility is supported by the finding that, as internalization proceeds, ligands accumulate in early endosomes while receptors reach a low steady-state level. By the use of digital image analysis to quantify accumulation of fluorescent ligands in endosomes during continuous endocytosis, it has been shown that LDL accumulates in endosomes that rapidly export apotransferrin but continue to fuse with other endosomes (Dunn *et al.*, 1989). As predicted from this model, the efficiency of receptor recycling is much higher than that of sorting of ligands to lysosomes: 25–50% of internalized ASOR is released intact from the cells (Weigel and Oka, 1984; Simmons and Schwartz, 1984). Other studies both *in vivo* (Salzman and Maxfield, 1989) and *in vitro* (Gruenberg and Howell, 1989) also support this concept.

2. Transport of Ligand to Late Endosomes and Lysosomes

While the ASGP receptors recycle from early endosomes to the cell surface (Section IV,C), the released ligands continue along the endocytic pathway to lysosomes. After 30 minutes of ligand uptake at 37°C, the majority of intracellular ligand is localized to lysosomes in HepG2 cells (Geuze *et al.*, 1983a), and degradation products of [^{125}I]ASOR begin to appear in the medium after 1 hour (Schwartz *et al.*, 1982).

Distinct endosomal compartments that successively receive internalized ASOR were identified by subcellular fractionation of perfused rat liver. After 2.5 minutes at 37°C, [^{125}I]iodinated ASOR was detectable in receptor-positive endosomes, whereas after 15 minutes the ligand accumulated in receptor-negative endosomes (Mueller and Hubbard, 1986). At that time no degradation occurred, indicating that the late endosomal, receptor-negative compartment is distinct from lysosomes. Two functionally and biochemically distinct endosomal compartments were isolated from rat hepatocytes on sucrose density gradients after continuous incubation with [^{125}I]ASOR at 37°C. When cells were incubated in high K$^+$ buffer, internalized ASOR was not delivered to lysosomes but accumulated in a less dense prelysosomal compartment. Organelles containing both receptor and ligand were of even lower density and correspond to early endosomes, whereas the intermediate-density, receptor-negative organelles were proposed to be a prelysosomal transit compartment (Baenziger and Fiete, 1986) with the appearance of multivesicular bodies (Geuze *et al.*, 1983a). Similarly, the EGF receptor was shown to become incorporated into vesicles budding from early endosomes and to congregate within such multivesicular bodies, eventually to be delivered to lysosomes (Felder *et al.*, 1990).

The observation that reduced temperatures of 16–18°C block degradation but not internalization was instrumental in defining endosomes. While ASOR was still taken up at these temperatures into the early, receptor-positive endosomes, it was not segregated from the receptor and did not reach receptor-negative endosomes (Wolkoff *et al.*, 1984; Mueller and Hubbard, 1986; Casciola-Rosen and Hubbard, 1991).

Based on immunoelectron-microscopic localization of a variety of markers in normal rat kidney (NRK) cells, a polymorphic compartment in the perinuclear region was postulated to be a late endosomal organelle where endocytosed material and newly synthesized lysosomal enzymes converge (Griffiths *et al.*, 1988). This compartment contained mannose-6-phosphate receptor, lysosmal enzymes (cathepsin D), and lysosomal membrane proteins (lgp 120) and most likely corresponds to the multivesicular bodies described in hepatocytes and to perinuclear CURL structures described earlier (Geuze *et al.*, 1983b). Disruption of microtubules by nocodazole in baby hamster kidney (BHK) cells led to the accumulation of a transport intermediate. Under these conditions internalized horseradish peroxidase is found in large spherical vesicles (0.4–0.7 μm), which presumably are involved in microtubule-dependent translocation from early to late endosomes (Gruenberg *et al.*, 1989). These data are consistent with reports that disruption of the microtubule network slows down degradation of internalized ASOR in rat hepatocytes (Oka and Weigel, 1983; Wolkoff *et al.*, 1984).

These results suggest that solutes and membrane components are sorted into vesicles to be transported from one separate compartment to the next, that is, from early to late endosomes. An alternative view is that early and late endosomes exist as a continuous reticulum. This model is based on video-enhanced fluorescence microscopy of living cells after internalization of fluorescently labeled EGF and transferrin (Hopkins *et al.*, 1990). Transferrin was shown to move through a branched reticulum with tubular elements. Inside these tubules EGF appeared to become concentrated in boluses, first at the perimeter and then in membrane invaginations, which seemed to pinch off and congregate within the lumen. Electron microscopy suggested that these boluses are multivesicular bodies. These observations and interpretations conflict with the clear biochemical separation of early and late endosomes (Mueller and Hubbard, 1986; Stoorvogel *et al.*, 1987; Baenziger and Fiete, 1986; Schmid *et al.*, 1988) and with experiments showing that the contents of these two compartments do not mix (Parton *et al.*, 1989). It is also difficult to explain the existence of a pH gradient along the endocytic path because such a gradient would be expected to collapse rapidly in a continuous network. It is conceivable that the extensive interconnected network corresponds to the complex polymorphic structure containing tubular and vesicular regions characteristic of early endosomes (Griffiths *et al.*, 1989).

In this context two models for the biogenesis of the endocytic compartments have been proposed (Griffiths and Gruenberg, 1991; Murphy, 1991). The maturation model assumes that endosomes mature by acquring new membrane components and hydrolytic enzymes from fusing vesicles (e.g., from the plasma membrane or the Golgi) and by releasing vesicles (e.g., recycling or transcytotic vesicles). In the process they gradually assume the characteristics of lysosomes. The second model, the preexisting organelles model, postulates the existence of distinct, stable compartments that exchange material by specific transport vesicles. With the available of *in vitro* systems for studying highly enriched endosomal preparations, the biogenesis and the molecular principles by which these organelles mediate intracellular transport and sorting may be further elucidated (Section IV,E).

C. Receptor Recycling

Recycling of receptors back to the cell surface occurs with high efficiency. From the kinetics of internalization and degradation it can be estimated that each ASGP receptor recycles ~250 times (Schwartz and Rup, 1983). However, little is known about the mechanism of this process. The ASGP receptor has been demonstrated to recycle constitutively, that is, independently of ligand occupancy. Morphological studies in rat liver and in HepG2 cells showed that at steady state only ~35% of the receptor is on the cell surface ((Geuze *et al.*, 1984b; Zijderhand-Bleekemolen *et al.*, 1987). This surface pool was depleted in a dose-dependent and reversible manner by the addition of lysosomotropic amines, which had no effect on endocytosis but blocked recycling (Schwartz *et al.*, 1984). The intracellular receptor pool in HepG2 cells and in transfected fibroblasts expressing the ASGP receptor was biochemically shown to recycle (Geffen *et al.*, 1989). During a 30-minute incubation at 37°C in the presence of external protease, all receptor molecules reached the cell surface and were degraded.

Although the majority of the ASGP receptor molecules are found on the cell surface and in early endosomes, it has been reported that in hepatocytes and in HepG2 cells ~20% of ASGP receptors reside in the Golgi (Geuze *et al.*, 1984a; Zijderhand-Bleekemolen *et al.*, 1987). Considering a maximum of 2 hours for newly synthesized receptors to arrive at the cell surface and a half-life of ~30 hours, one would expect to find only 2–3% in the Golgi, suggesting that these receptors in the Golgi complex are not in transit along the biosynthetic route but rather recycling. Cycloheximide, which blocks protein synthesis without affecting intracellular transport or ligand uptake and degradation, caused depletion of poly-Ig receptor and 5'-nucleotidase from the Golgi but did not affect the Golgi pools of the ASGP and mannose-6-phosphate receptors (Geuze *et al.*, 1984a; Van den Bosch *et al.*, 1986). Mannose-6-phosphate receptor, after desialylation at the cell surface, was shown to be resialylated in the absence of protein synthesis (Duncan and Kornfeld, 1988; Goda and Pfeffer, 1988; Jin *et al.*, 1989). Because sialyl transferase is localized to the TGN (Roth *et al.*, 1985), it seems that the molecules have cycled through this compartment. Similarly, when sialic acid was removed from the transferrin receptor at the plasma membrane, a slow return of resialylated receptors to the cell surface was observed (Snider and Rogers, 1985). Kinetic studies of ricin uptake combined with ultrastructural immunocytochemistry in BHK-21 cells showed that ricin was internalized through coated pits into endosomes and that after 60 minutes at 37°C 4–6% was found in the Golgi (Van Deurs *et al.*, 1988). Because ricin binds glycoproteins and glycolipids with terminal galactosyl residues on the cell surface, this implies that some internalized plasma membrane molecules are routed to the TGN. Studies on surface glycoproteins in K562 and HepG2 cells led to the conclusion that receptors recycle only to the TGN and not to the *cis*-medial Golgi, the site of action of mannosidase I (Neefjes *et al.*, 1988). Transport of membrane glycoproteins through the TGN was argued not to be a general phenomenon based on resialylation studies in K562 cells (Reichner *et al.*, 1988). It rather seems that this extended recycling route through the TGN is a selective process that is only taken by certain endocytic proteins. The function of such a route for the ASGP receptor is not known at present.

As discussed, a significant fraction of internalized ligand is not degraded but is released back into the extracellular medium, a process called diacytosis (Regoeczi *et al.*, 1982; Weigel and Oka, 1984; Simmons and Schwartz, 1984). It has been suggested that recycling of ligand occurs via a minor pathway involving a preacidic compartment (Clarke *et al.*, 1987). Released ligand can rebind and undergo another round of endocytosis.

D. Modulation of Receptor Traffic

The ASGP receptor undergoes constitutive endocytosis and recycling. However, intracellular transport and distribution of the ASGP receptor is affected by ligand and by various pharmacological agents. Binding of ligands, such as ASOR, causes a specific down-regulation of cell surface receptors. In HepG2 cells it was observed that saturating concentrations of ASOR in the incubation medium caused a twofold increase in the number of intracellular receptors without affecting the total number of receptors (Schwartz *et al.*, 1982). A ~50% loss of surface binding sites occurs within 5 minutes at 37°C in the presence of 2 μg/ml ASOR and persists as long as ligand is present (I. Geffen *et al.*, unpublished results). The initial rate of receptor internalization at 37°C was shown to be increased by roughly 35–50% after preincubation with ASOR at 4°C (Schwartz *et al.*, 1984). This change in endocytosis rate could account for the observed down-regulation. By quantitative immunoelectron microscopy, a decrease of the surface receptor pool from 34% to 20% and an increase of the endosome pool from 37% to 50% was observed after incubation with ASOR, whereas the Golgi pool of approximately 20% remained unchanged (Zijderhand-Bleekemolen *et al.*, 1987).

In addition to the short-term effects, the presenced of excess levels of ASGPs in the culture medium for an extended period of 2–4 days had a modulating effect on receptor activity (Steer *et al.*, 1987). Prolonged exposure of HepG2 cells to ligand caused a marked decrease in ASOR binding without a change in affinity but also without a change in surface antibody-binding sites. This diminished binding activity was shown to correlate with a reduced sialic acid content in surface molecules and could be imitated by a minimal exposure of cells to neuraminidase (Weiss and Ashwell, 1989). Reduced surface sialylation could result in the blocking of binding sites by hyposialylated endogenous surface proteins, thereby limiting the available sites for exogenously added ligands.

Many cell surface proteins are internalized and degraded upon incubation of the cells with specific antibodies. When anti-ASGP receptor antibodies were added to HepG2 cells, a rapid and specific loss of receptor was observed (Schwartz et al., 1986; Bischoff et al., 1988). Antibody-induced degradation was similarly shown for the mannose-6-phosphate, LDL, and epidermal growth factor (EGF) receptors (Gartung et al., 1985; Anderson et al., 1982; Schreiber et al., 1983). It was thought to result from extensive cross-linking of the receptors at the cell surface because Fab fragments did not have this effect. Interestingly, in the case of the ASGP receptor, degradation was reported to take place in a prelysosomal compartment (Schwartz et al., 1986): the half-life for degradation was only 30 minutes, which is not long enough for surface molecules to reach lysosomes, degradation was insensitive to inhibitors of lysosomal proteases, and still occurred at 18°C, which blocks delivery to lysosomes. This supports the notion that some hydrolytic enzymes may be targeted to an early endocytic compartment. Previously it had been shown that hydrolysis of endocytosed molecules in 3T3 fibroblasts by cathepsin B began only a few minutes after uptake at 37°C and also occurred at 17°C (Roederer et al., 1987). Proteolytic activity capable of processing internalized EGF was identified in early endosomes of perfused rat liver in vivo and in isolated endosomes (Renfrew and Hubbard, 1991).

Lysosomotropic amines, such as weak bases and carboxylic ionophores, which neutralize the pH in acidic organelles, cause a dose-dependent and reversible loss of surface ASGP receptors in HepG2 cells (Schwartz et al., 1984). Incubation of cells with ammonium chloride, chloroquine, or primaquine did not interfere with internalization of bound $[^{125}I]ASOR$ but inhibited ligand degradation and return of receptors to the cell surface. Morphological analysis of receptor redistribution after incubation with 0.3 mM primaquine showed that the cell surface was almost completely depleted of receptors, whereas they were found to be enriched in the TGN (Zijderhand-Bleekemolen et al., 1987). It is possible that primaquine, in addition to inhibiting receptor recycling from endosomes, also induces transfer of receptors to the TGN. In that case recycling to the surface is blocked both from endosomes and from the TGN.

Tumor-promoting phorbol esters have been shown to have a marked effect on the intracellular distribution of many surface receptors, among them the transferrin, ASGP, and mannose-6-phosphate receptors (Klausner et al., 1984; May et al., 1984; Fallon and Schwartz, 1986; Braulke et al., 1990). Phorbol esters stimulate protein kinase C and thus trigger hyperphosphorylation of cell surface receptors at serine and threoninie residues, an observation that led to suggest that this modification may be important for regulating receptor traffic (Section V,B,2). However, the intracellular redistribution of both the transferrin and ASGP receptors caused by phorbol esters was shown to be independent of receptor phosphorylation (Davis and Meisner, 1987; McGraw *et al.*, 1988; I. Geffen and Spiess, 1992).

E. Cell-Free Systems

Recent studies in cell-free systems are beginning to reveal the biochemical requirements and molecular mechanisms responsible for the sorting and transport steps involved in receptor-mediated endocytosis (Gruenberg and Howell, 1989). Conversion of shallow coated pits to the deeply invaginated form is temperature-dependent, not requiring cytosol or adenosine triphosphate (ATP) (Smythe *et al.*, 1989). All steps of vesicle formation and membrane fusion events at different stages of intracellular transport were found to be dependent on temperature, ATP, and cytosol. Some of the necessary cytosolic components have been identified. An N-ethylmaleimide-sensitive factor (NSF), a 76-kDa tetrameric protein that was first shown to be required for vesicular fusion at the Golgi stack (Block *et al.*, 1988), and GTP-binding proteins have been isolated and are necessary for all vesicular transport events so far reconstituted (Balch, 1990). Both small, ras-like, GTP-binding proteins (reviewed by Balch, 1990), as well as trimeric G-proteins (Barr *et al.*, 1991; Colombo *et al.*, 1992) have been found to have a role in regulating membrane traffic. Using [^{125}I]asialofetuin as a marker, transport from late endosomes to lysosomes was reconstituted *in vitro* and was shown to depend on the same factors required for other transport steps, namely, ATP, temperature, and cytosol (Mullock *et al.*, 1989).

Many aspects of membrane traffic in the endocytic pathway remain poorly understood and controversial. The possibility to reconstitute individual steps in cell-free systems, which can be experimentally manipulated *in vitro*, will provide the means to elucidate the molecular mechanisms involved and to mechanistically identify distinct compartments.

V. Sorting Signals

Intracellular traffic of the receptor during both biosynthesis and endocytosis is dependent on information within the receptor structure. Signals for targeting to the plasma membrane and for cycling through intracellular compartments must be recognized by sorting receptors or other components of a cellular machinery that decodes these signals. There are at least four sorting steps in the pathway of the ASGP receptor: specific insertion into the ER membrane, targeting to the basolateral plasma membrane domain, clustering into plasma membrane-coated pits for endocytosis, and segregation into recycling vesicles in early endosomes.

A. Insertion into the Endoplasmic Reticulum

The ASGP receptor proteins are Type II membrane proteins, meaning that they span the membrane once, oriented with the amino terminus in the cytoplasm and the carboxy terminus exposed to the exoplasmic phase. They lack a cleaved amino-terminal signal sequence for sorting to the ER; ASGP receptor proteins are not proteolytically processed, except for the removal of the initiator methionine. Analyzing the membrane insertion of deletion and fusion constructs derived from human and rat HL-1 proteins, it was demonstrated that the transmembrane domain serves the function of an internal signal sequence that is necessary and sufficient to target the protein to the ER membrane and to initiate insertion (Spiess and Lodish, 1986; Holland and Drickamer, 1986). The functional characteristics of this domain are similar to those of amino-terminal cleaved signals: insertion occurs only co-translationally and requires signal recognition particle (SRP; a ribonucleoprotein particle that acts as a sorting receptor; Walter et al., 1984) and the SRP receptor in the ER membrane. Insertion of the transmembrane domain initiates translocation of the carboxy-terminal portion of the protein. Like cleaved signals, this internal signal sequence consists of an apolar, helical segment that is preceded by a positively charged region. With approximately 20 hydrophobic residues (the number required to span the lipid bilayer in an α-helical conformation), the apolar portion is longer than in typical cleaved signal peptides where it consists of 7–12 hydrophobic residues (Gierasch, 1989). However, by deletion analysis it was shown that the hydrophobic segment can be shortened by almost 50% without completely abolishing insertion activity (Spiess and Handschin, 1987).

Because the ER insertion signal of the ASGP receptor proteins is internal, the amino-terminal hydrophilic domain, as it folds in the cytoplasm, must not prevent the signal sequence from being sterically accessible for SRP. A proline-rich segment immediately preceding the transmembrane domain may serve the purpose of separating the amino-terminal domain from the signal. The proline-rich segment is encoded in the same exon as the hydrophobic sequence and thus might be part of the same functional unit. Unlike most amino-terminal signal peptides, internal signals lack a signal peptidase cleavage site at their exoplasmic end. However, a cryptic cleavage site exists after Gly-60 of H1 becomes activated upon deletion of the amino-terminal portion up to residue 37 (Schmid and Spiess, 1988). Most likely, the absence of this portion affects the position of the transmembrane segment in the membrane and thus alters the accessibility of the potential cleavage site to signal peptidase.

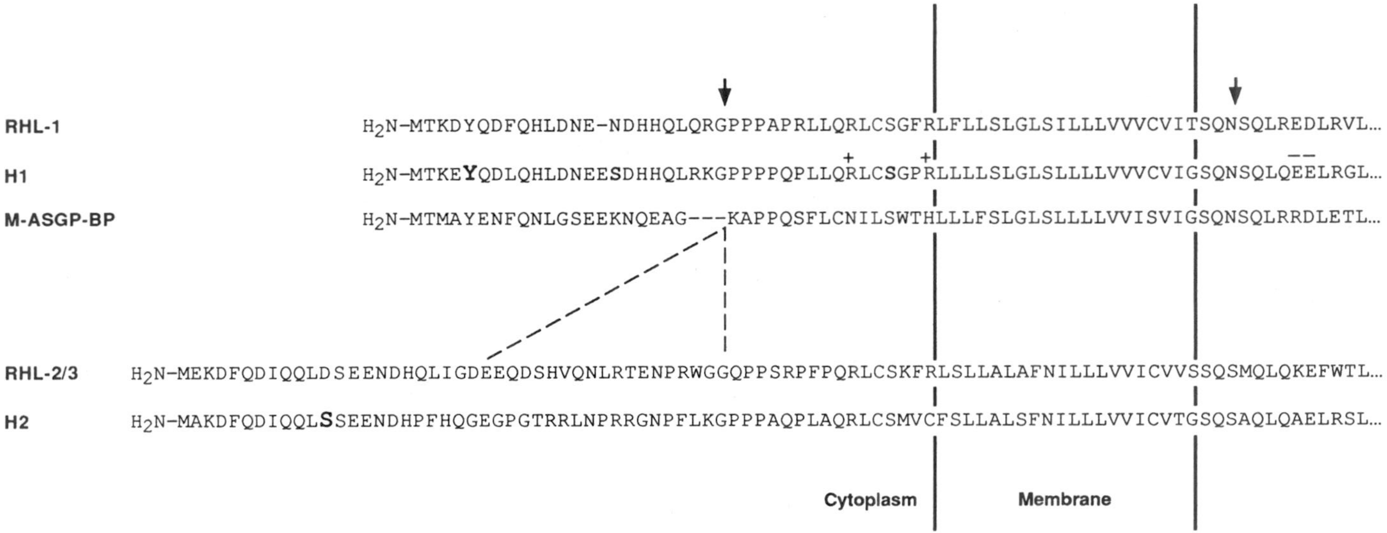

RHL-1   H$_2$N–MTKDYQDFQHLDNE–NDHHQLQRGPPPAPRLLQRLCSGFRLFLLSLGLSILLLVVVCVITSQNSQLREDLRVL...

H1   H$_2$N–MTKE**Y**QDLQHLDNEESDHHQLRKGPPPQPLLQRLC**S**GPRLLLSLGLSLLLLVVVCVIGSQNSQLQEELRGL...

M-ASGP-BP   H$_2$N–MTMAYENFQNLGSEEKNQEAG---KAPPQSFLCNILSWTHLLLFSLGLSLLLLVISVIGSQNSQLRRDLETL...

RHL-2/3   H$_2$N–MEKDFQDIQQLDSEENDHQLIGDEEQDSHVQNLRTENPRWGGQPPSRPFPQRLCSKFRLSLLALAFNILLLVVICVVSSQSMQLQKEFWTL...

H2   H$_2$N–MAKDFQDIQQL**S**SEENDHPFHQGEGPGTRRLNPRRGNPFLKGPPPAQLAQRLCSMVCFSLLALSFNILLLVICVTGSQSAQLQAELRSL...

Cytoplasm | Membrane

FIG. 4 Amino-terminal sequences of the hepatic ASGP receptor proteins and the macrophage ASGP-BP. The sequences were aligned with broken lines indicating the 18-amino acid insertion in the HL-2 proteins. Arrows indicate the intron positions in the RHL-1 gene. Highlighted residues are as follows: tyrosine-5 of H1, which is important for rapid endocytosis; serine-12 of H2, which is the major site of phosphorylation in the human ASGP receptor; serines-16 and 37 of H1, which are weakly phosphorylated; the charged residues flanking the transmembrane segment of H1 (Arg-34, Arg-40, Glu-68, Glu-69), which are involved in determining the orientation of the polypeptide in the membrane; and residues 1–8 of the ASGP-BP, which were absent in the purified protein.

Uncleaved signal-anchor sequences of membrane proteins inserted into the endoplasmic reticulum can in principle initiate the translocation of either the amino-terminal or the carboxyl-terminal polypeptide segment across the bilayer, resulting in a Type I (N-exoplasmic/C-cytoplasmic; e.g., cytochrome P-450; Sakaguchi *et al.*, 1987) or a Type II membrane protein (N-cytoplasmic/C-exoplasmic, like the ASGP receptors), respectively. The topology appears to be determined not by the hydrophobic segment but rather by the hydrophilic sequences flanking it. The cytoplasmic flanking sequences of transmembrane segments, including those of the ASGP receptors, are characteristically enriched in positively charged amino acids (the positive inside rule; Von Heijne, 1986; Von Heijne and Gavel, 1988). In addition, a striking correlation was found between the transmembrane orientation of signal-anchor sequences and the charge difference betwee the 15 carboxy-terminal and the 15 amino-terminal flanking residues (Hartmann *et al.*, 1989). This concept was tested by changing the flanking charges of the ASGP receptor subunit H1 (Fig. 4) to amino acids of opposite charge by site-directed mutagenesis (Beltzer *et al.*, 1991). Upon expression in transfected COS-7 cells, approximately half the polypeptides inserted with the inverted orientation (amino terminus exoplasmic, carboxy terminus cytoplasmic). When, in addition, the amino-terminal domain of the mutant protein was truncated, as much as 90% of the polypeptides acquired the inverted topology. Based on these results, it appears that the flanking charges are primarily responsible for orienting the signal domain within the translocation machinery. However, the translocation competence of the flanking domains is also likely to influence the ratio of the two possible orientations. In particular the amino-terminal domain, which is already folded when insertion occurs, could inhibit its own translocation. Natural proteins might have evolved to be devoid of inefficiently translocated sequences in the segments to be transferred and to contain such inhibitory features in cytoplasmic domains to ensure a unique topology.

B. Segregation into Coated Pits

The first step in receptor-mediated endocytosis is the specific clustering of receptors into clathrin-coated pits on the plasma membrane. The determinants for this process were shown for several plasma membrane receptors to be contained within their cytoplasmic domains. Deletion of this portion of the receptors for LDL, poly-Ig, EGF, transferrin, and mannose-6-phosphate resulted in a drastic reduction of the internalization rate (Lehrman *et al.*, 1985; Mostov *et al.*, 1986; Prywes *et al.*, 1986; Rothenberger *et al.*, 1987; Lobel *et al.*, 1989). In the case of the macrophage F_c receptor (RII-B2), quantitative immunoelectron microscopy studies have shown that a truncated form lacking the cytoplasmic domain was largely excluded from coated pits (Miettinen *et al.*, 1989). When the cytoplasmic domain of the chicken hepatic lectin was replaced by unrelated sequences (the cytoplasmic domain of the Na^+/K^+-ATPase or a stretch of β-globin), the internalization rate in transfected fibroblast cells was strongly reduced; when it was replaced by the cytoplasmic domain of RHL-1, efficient endocytosis was retained (Verrey et al., 1990).

1. Tyrosine Signal for Endocytosis

The cytoplasmic domains of endocytic receptors are diverse in length, primary structure, and even orientation with respect to the membrane (some are amino-, others carboxy-terminal). Analysis of natural LDL receptor mutants deficient in LDL uptake revealed that mutation of a single residue, tyrosine-807, strongly affected internalization (Davis et al., 1986). No other amino acid at this position except phenylalanine and, to a lesser degree, tryptophan, mediated efficient endocytosis of LDL (Davis et al., 1987). A general importance of tyrosine residues for internalization was suggested by the finding that insertion of a tyrosine into the short cytoplasmic domain of influenza virus hemagglutinin caused this protein to enter coated pits and to be internalized (Lazarovits and Roth, 1988). The sequence context appeared also to be important because at only one of three insertion positions tested was the tyrosine functional. By an extensive mutational analysis, a degenerate recognition element has been identified within the diverse sequences of these cytoplasmic domains. It consists of a short sequence containing at least one aromatic amino acid, typically a tyrosine, within a context that most likely assumes a turn conformation (Collawn et al., 1990; Ktistakis et al., 1990). For natural endocytic receptors, reports have been published demonstrating the importance of tyrosine-807 of the LDL receptor (Davis et al., 1986), tyrosine-734 of the polymeric immunoglobulin receptor (Breitfeld et al., 1990), tyrosine-20 of the transferrin receptor (Jing et al., 1990; Alvarez et al., 1990), and tyrosine-2362 (and to a lesser extent tyrosine-2360) of the cation-independent mannose-6-phosphate receptor (Canfield et al., 1991).

The human ASGP receptor subunit H1 was shown, independently of its association with H2, to contain all the signals necessary for internalization and recycling (Geffen et al., 1989). The only aromatic residue in the cytoplasmic tail of H1, a tyrosine in position 5, is necessary for efficient clustering into coated pits and rapid internalization. Mutation of this tyrosine to an alanine reduced the rate of constitutive endocytosis by a factor of approximately 4 (Fuhrer et al., 1991), and the fraction of H1 at the cell surface showed a corresponding increase from 50% to 80–85%. Residual internalization of this mutant still occurred at a low but significant rate, which is higher than expected for a bona fide resident plasma membrane protein. Similar levels of residual endocytosis have been reported for the tyrosine mutants of the LDL, the poly-Ig, and the transferrin receptor (Davis et al., 1986; Breitfeld et al., 1990; Jing et al., 1990). Selective inhibition of clathrin-mediated endocytosis by cytosol acidification blocked internalization of the functional hetero-oligomeric ASGP receptor as well as of wild-type H1 and of the tyrosine mutant, showing that internalization of these molecules occurs exclusively via clathrin-coated vesicles and that mutation of tyrosine-5 only partially inactivates recognition of H1 for clustering in coated pits (Fuhrer *et al.*, 1991). Direct localization of wild-type and mutant H1 in coated pits by immuno-gold electron microscopy confirmed this conclusion: the fractions of wild-type and mutant H1 detected in coated pits were 5.9% and 1.7%, respectively, and correlated well with the internalization rates of approximately 6%/min and 1.5%/min. Whereas wild-type H1 was concentrated in coated pits by a factor of 2.3 compared with uncoated plasma membrane, the mutant protein was neither concentrated in nor excluded from coated areas (Fuhrer *et al.*, 1991). This strengthens the view that the ASGP receptor subunit H1 has a residual affinity for the clathrin-coated plasma membrane domains that does not depend upon tyrosine residues (Fuhrer *et al.*, 1991).

It has been shown that the human ASGP receptor is phosphorylated at tyrosine (Fallon, 1990), a modification that has not been reported for other transport receptors. However, we found that a phenylalanine in place of the tyrosine supports fairly efficient internalization, suggesting that the aromatic nature of this residue rather than the hydroxyl group of tyrosine or a phosphate group in phospho-tyrosine is the crucial factor for clustering into clathrin coated pits (C. Fuhrer *et al.*, unpublished). Interestingly, subunit H2, although containing a phenylalanine in the homologous position of the cytoplasmic domain, unless associated with functional H1 polypeptides, does not mediate efficient internalization. This suggests that the minor subunit in the receptor complex does not contribute directly to the sorting into coated pits.

2. Serine Phosphorylation

Another feature common to most endocytic receptors is phosphorylation of cytoplasmic serine or threonine residues (Sibley *et al.*, 1987). This modification has been proposed to play an important role in regulating intracellular receptor traffic. This hypothesis is based on the observation that phorbol esters, which cause hyperphosphorylation of cell surface receptors by activating protein kinase C, have a marked effect on the kinetics of endocytosis of the EGF and the β-adrenergic receptor (Sibley *et al.*, 1987) and on the intracellular distribution of the transferrin, ASGP, and mannose-6-phosphate receptors (Klausner *et al.*, 1984; May *et al.*, 1984; Fallon and Schwartz, 1986; Braulke *et al.*, 1990).

Serine residues of the rat and human ASGP receptor have been shown to be phosphorylated under steady-state conditions (Schwartz, 1984a; T. Takahashi *et al.*, 1985). After addition of phorbol esters to HepG2 cells, surface ASGP receptors are rapidly hyperphosphorylated; coordinately, the number of surface receptors is reduced by 40–50% with a half-time of 20 minutes, and the majority of the phosphorylated receptors accumulate intracellularly (Fallon and

Schwartz, 1986, 1987, 1988). The rate of internalization and the total number of binding sites in the cell were unaffected, but the phosphorylated receptor population appeared not to be recycled. These findings suggested that phosphorylation–dephosphorylation might control receptor traffic, in particular recycling to the cell surface.

By *in vitro* mutagenesis of the cytoplasmic serines at positions 16 and 37 in H1 and at positions 12, 13, and 55 in H2 (Fig. 4) to alanines or glycines followed by transfection into COS-7 cells and labeling with [^{32}P]phosphate, the major site of phosphorylation was identified to be serine-12 of H2 (Geffen *et al.*, 1991). The two serines of H1 were found to be only poor targets for protein kinase C. Similarly, also the rat ASGP receptor is predominantly phosphorylated in its minor subunits RHL-2/3 (T. Takahashi *et al.*, 1985). Although the subunit specificity of phosphorylation appears to be conserved between the two species, the target residue, at position 12, is not.

To directly assess the importance of serine phosphorylation, mutant subunits H1 and H2 lacking any cytoplasmic serines were stably expressed in fibroblasts and analyzed with respect to receptor function. The mutant subunits formed high-affinity ligand-binding sites and supported ligand uptake and degradation at a rate indistinguishable from that of wild-type receptor subunits (I. Geffen *et al.*, 1991). This result shows unambiguously that cytoplasmic serines are not essential for receptor function and that the phosphorylation of serine residues is not important for the mechanism of intracellular receptor transport. In addition, the effect of phorbol esters on receptor distribution was found to be cell type-dependent because it was observed in HepG2 and transfected COS-7 cells but not in fibroblasts (Geffen and Spiess, 1992). A mutant receptor without cytoplasmic serines expressed in COS-7 cells is nevertheless down-regulated like the wild-type receptor, demonstrating that there is no causal relationship between the phorbol ester-induced down-regulation and receptor phosphorylation. More likely, receptor redistribution is an indirect effect of protein kinase C acting more generally on membrane traffic, as has previously been suggested by Buys *et al.* in studies on the phorbol ester effect in macrophages (1984). The ASGP receptor appears to behave similarly to the transferrin receptor, which has also been shown to cycle independently of serine phosphorylation (Rothenberger *et al.*, 1987; Davis and Meisner, 1987; McGraw *et al.*, 1988). Still, the ubiquity and specificity of phosphorylation of endocytic receptors may reflect a physiological role for this modification, which is not clear at present.

3. Interaction of the ASGP Receptor with Adaptor Proteins

Clustering of receptors into clathrin-coated pits involves direct interaction of receptors with adaptor or assembly proteins (APs). Two classes of APs, mediating the association of receptors and clathrin, have been described: AP-1 in clathrin-coated membranes budding off the *trans*-Golgi destined for endosomes and AP-

2 associated with coated pits and vesicles of the plasma membrane (Ahle *et al.*, 1988; Robinson, 1987). APs were first characterized by their ability to promote clathrin assembly into cage-like lattices (Keen *et al.*, 1979). AP-2 complexes have also been shown to be necessary for the *in vitro* formation of clathrin-coated pits on purified plasma membranes (Moore *et al.*, 1987; Mahaffey *et al.*, 1990).

The AP-2 plasma membrane complex is a tetramer composed of two ~100kDa proteins α and β and two smaller polypeptides of 50 kDa and 17 kDa. There are at least four subtypes of α polypeptides, α_{a1}, α_{a2}, α_{c1}, and α_{c2}, and two β subtypes (Kirchhausen *et al.*, 1989; Ahle *et al.*, 1988). The Golgi AP-1 complex is similarly a tetramer composed of two ~100 kDa molecules, β' and γ, and two smaller proteins of 47 kDa and 20 kDa. The ~100 kDa proteins of these complexes were called adaptins (Pearse, 1988) because they were postulated to interact with clathrin and with the cytoplasmic domains of receptors. AP complexes exhibit a characteristic tripartite structure consisting of a central core containing the amino-terminal portions of the two ~100 kDa adaptins plus the two smaller polypeptides, and two appendages made of the carboxy-terminal portions of the two adaptins (Heuser and Keen, 1988; Keen and Beck, 1989; Kirchhausen *et al.*, 1989). By limited proteolysis the ear-like projections of the AP-2 complexes are released and the assembly-promoting activity but not clathrin binding is eliminated (Keen and Beck, 1989). The β adaptin of the AP-2 complex has been purified and has been shown to bind clathrin (Ahle and Ungewickell, 1989).

That AP complexes interact not only with clathrin but also with receptors was first demonstrated for the mannose-6-phosphate receptor, which could be co-assembled into clathrin–AP cages (Pearse, 1985). Furthermore, Pearse (1988) also showed specific binding of the AP-2 complex to the immobilized cytoplasmic domain of the LDL receptor. In a similar manner, both AP-1 and AP-2 complexes bound to the immobilized cytoplasmic domain of the mannose-6-phosphate receptor (Glickman *et al.*, 1989). Mutation of the only two tyrosines within this domain of the mannose-6-phosphate receptor to nonaromatic residues selectively binding to the AP-2 fraction suggests separate recognition signals for each class of AP complexes.

In an *in vitro* assay an interaction of the ASGP receptor with APs has been demonstrated. Coated vesicle proteins and purified AP complexes were fractionated by sodium dodecyl sulfate (SDS)-polyacrylamide gel electrophoresis, transferred to nitrocellulose, and probed for receptor binding on the filter (Beltzer and M. Spiess, 1991). This assay showed that the receptor is bound by a coated vesicle protein of ~100 kDa that is also present in AP-2 complexes. Binding to this ~100 kDa protein is specific for the cytoplasmic domain of the receptor because a bacterial fusion protein consisting of the 30 amino-terminal residues of the receptor subunit H1 fused to mouse dihydrofolate reductase competed for receptor binding. A characteristic shift in electrophoretic mobility in SDS-polyacrylamide gels containing 6 M urea identified the binding protein as a β adaptin.

Because clathrin binding was shown to involve the core portion of the adaptor complex containing the amino-terminal protease-resistant domain of adaptins (Keen and Beck, 1989), it has been speculated that the carboxy-terminal adaptin portions, which form the appendages of the AP complex, might be responsible for receptor recognition. Using the blot assay, it could be shown that the amino-terminal domain of ~60 kDa, generated by partial proteolysis of AP-2 complexes, retains the complete receptor-binding activity of the intact molecule (Beltzer and Spiess, 1991). This indicates that the core portion of the AP-2 complex is responsible for the interactions both with clathrin and with membrane receptors. The function of the ear-like appendages may be to interact with other AP complexes in the coat assembly process.

C. Sorting in Polarized Cells

Surface ASGP receptors are found distributed over the basolateral (sinusoidal) plasma membrane of hepatocytes but are essentially absent from the apical (canalicular) surface domain (Wall and Hubbard, 1981; Matsuura *et al.*, 1982; Gueze *et al.*, 1982). By pulse-chase experiments and subcellular fractionation of rat hepatocytes, it was shown that in biosynthesis the ASGP receptor and other basolateral proteins are directly transported from the TGN to the basolateral domain (Bartles *et al.*, 1987). Three apical proteins also first appeared on the basolateral surface before being transported to the apical domain, suggesting that sorting of apical proteins occurs by transcytosis and is signal-mediated. In contrast, both basolateral and apical proteins were demonstrated to be directly sorted from the TGN to the respective surface domain in Madin-Darby canine kidney (MDCK) cells, an epithelial cell line that can be grown in culture as a polarized monolayer (Wandinger-Ness and Simons, 1990). When expressed in MDCK cells, H1 alone and the combination of RHL-1 and RHL-2 were specifically transported to the basolateral membrane also in this heterologous cell type (Wessels *et al.*, 1989; Graeve *et al.*, 1990). Expression of mutant forms of H1 in MDCK cells revealed the existence of a signal for basolateral transport within the first approximately 10 residues of the cytoplasmic domain which appears to be related to the signal for endocytosis (I. Geffen *et al.*, unpublished): deletion of amino acids 4–11 and mutation of the tyrosine residue at position 5 to alanine, which both strongly reduced the rate of internalization, abolished specific basolateral targeting and resulted in a nonpolarized surface expression. Mutation of tyrosine-5 to phenylalanine and deletion of residues 12–33, which still supported quite efficient endocytosis, did not affect basolateral delivery. Thus, there is a correlation between the two signals. Whether the same or overlapping signals are recognized by similar adaptor proteins in the plasma membrane for clathrin-dependent internalization and in the TGN for polarized surface delivery remains to be elucidated. The same phenomenon has also been discovered for influenza virus hemagglutinin (Brewer and Roth, 1991), the Fc receptor and lysosomal glycoprotein lgp120 (Hunziker *et al.*, 1991): The non-endocytic forms of these proteins, however, were delivered predominantly to the apical surface, suggesting the existence of a recessive apical signal within their ectodomains. The nonpolarized phenotype of endocytosis mutants of H1 might indicate that this protein is devoid of such an apical signal.

VI. Asialoglycoprotein Binding Protein of Peritoneal Macrophages

Expression of the genes for HL-1 and HL-2 resulting in functional ASGP receptors has so far been detected only in parenchymal hepatocytes. Only RHL-2/3 has been found to be expressed in nonhepatic cells, rat spermatogenic cells, and sperm (Abdullah and Kierszenbaum, 1989). Its function there is unknown. However, galactose–N-acetyl galactosamine-specific receptors that were immunologically cross-reactive with the hepatic ASGP receptor have been detected on elicited rat peritoneal macrophages (Kawasaki *et al.*, 1986). These cells demonstrated both binding and uptake of galactose-terminal ligands, with characteristics similar to those of the hepatic receptor system (Kawasaki *et al.* 1986; Kelm and Schauer, 1986; Lee *et al.*, 1988). By affinity purification a single galactose-binding protein of 42 kDa, the macrophage ASGP-binding protein (ASGP-BP), was isolated (Kawasaki *et al.*, 1986; Kelm and Schauer, 1988). It was recognized by antisera raised against the total rat hepatic receptor or against RHL-1 but not by RHL-2/3-specific antisera (Ii *et al.*, 1988). Partial protein sequencing revealed that the macrophage ASGP-BP was distinct from RHL-1 but highly homologous (Ii *et al.*, 1988). Its cDNA was cloned and showed sequence similarity through almost the entire length of the protein (Ii *et al.*, 1990). This is in contrast to other C-type carbohydrate-binding surface receptors, such as the avian hepatic lectin, the mannose receptor, or the Kupffer cell receptor, in which homology is restricted to the CRD. The macrophage ASGP-BP must therefore be considered a close relative of the hepatic ASGP receptors. Nevertheless, the CRD is particularly well conserved with 74% sequence identity with RHL-1, compared with only 41% in the rest of the sequence. The macrophage receptor is more closely related to HL-1 than to HL-2 because it shares 59% identical amino acids with RHL-1 and with H1, and only 45% with RHL-2/3 and 46% with H1. In addition, it lacks the cytoplasmic insertion of 18 codons found in the HL-2 sequences (Fig. 4). From the sequence similarities between the macrophage receptor and the rat and human hepatic receptors, it is also evident that their genes were generated from an ancestral gene before rodents and primates separated in evolution. The most striking difference between the structures of the macrophage and the hepatic receptors is a segment of 24 amino acids in the exoplasmic domain of the macrophage ASGP-BP, which does not have a counterpart in the hepatic proteins. Because this segment is located exactly at the position of intron 4 in the RHL-1 gene, it is either the product of a mutation affecting the splicing of the intron or else an additional exon.

Upon expression of the cDNA in transfected COS-1 cells, the ASGP-BP produced high-affinity binding as well as uptake of ASOR with a K_{uptake} (as estimated by Lineweaver-Burk type analysis) of 15 nM, which is similar to that of 23 nM determined for peritoneal macrophages (T. Kawasaki, personal communication). This result strongly suggests that, in contrast to the hepatic receptor, in the macrophage system a single polypeptide can form a high-affinity ASGP-binding complex. A detailed comparison of the binding affinities of various ligands for the two receptors will therefore be of great interest. It is conceivable that the additional segment in the polypeptide region connecting the CRD with the membrane anchor provides more flexibility to the galactose-binding domains in the complex and facilitates recognition of the asymmetrically arranged galactoses in desialylated N-linked oligosaccharides.

Interestingly, the amino terminus of the isolated macrophage ASGP-BP was found to correspond to codon 9 of the cDNA, indicating proteolytic removal of the first eight amino acids (Ii et al., 1988). This missing segment includes tyrosine-5, which is conserved in rat and human HL-1 (Fig. 4) and was shown to be important for efficient internalization of H1 (Fuhrer et al., 1991). Because there is no other tyrosine in the cytoplasmic domain of the macrophage ASGP-BP, it is yet to be determined whether internalization of the ASGP-BP is mediated by another signal within the cytoplasmic tail structure.

VII. Conclusions

Since its discovery, the ASGP receptor has provided a useful experimental model for a variety of important biological processes. It has been instrumental in the analysis of insertion of proteins into the ER membrane, protein degradation in the ER, transport to the basolateral plasma membrane domain in epithelial cells, and, most prominently, in the analysis of receptor-mediated endocytosis. The receptor is a classical marker for the early endocytic compartments while its ligands serve to delineate the intracellular organelles along the degradative route from the plasma membrane to lysosomes.

The availability of the cDNAs of several endocytic receptors, among which is the ASGP receptor, is the basis of recent progress in the identification of sorting signals. Because there is little sequence similarity between different receptors, the signals involved in directing membrane proteins through the intracellular compartments appear to be degenerate. By *in vitro* mutagenesis and expression in heterologous cells, however, general sorting signals are beginning to emerge: a tyrosine-containing sequence in the cytoplasmic receptor domain required for efficient clustering in coated pits is common for most receptors studied. Future efforts will be directed at further characterizing this signal and its three-dimensional structure. To understand the molecular mechanisms, research will increasingly focus on identifying the components of the sorting machineries, for example, molecules that decode sorting signals or provide specificity to vesicular transport. Newly developed cell-free systems for the analysis of individual transport steps may provide us with important insights toward an understanding of receptor-mediated endocytosis and protein sorting on a molecular level.

Although the primary sequences of the receptor subunits have been determined, the detailed structure of the receptor remains to be elucidated. As a lectin the ASGP receptor subunits are members of a large family of extracellular or transmembrane proteins sharing homologous Ca^{2+}-dependent sugar binding domains. How the individual galactose recognition domains are organized to form a high-affinity binding site for tri- and tetra-antennary oligosaccharides and how the subunits interact to form a hetero-oligomeric complex is still subject to speculation. The exact stoichiometry, the specific formation of the complex during biosynthesis, and the dependence of transport along the exocytic path on oligomerization are questions to be further investigated.

Acknowlegments

The authors thank Drs. T. Kawasaki, H. F. Lodish, and G. Lederkremer for sharing data prior to publication, Drs. G. Griffiths and Y. Henis for fruitful discussions, and Dr. B. Glick for critically reading the manuscript. The research was supported by grants from the Swiss National Science Foundation and the Incentive Award of the Helmut Horten Foundation.

References

Abdullah, M., and Kierszenbaum, A. L. (1989). *J. Cell Biol.* **108**, 367–375.
Ahle, S., Eichelsbacher, U., and Ungewickell, E. (1988). *EMBO J.* **7**, 919–930.
Ahle, S., and Ungewickell, E. (1989). *J. Biol. Chem.* **264**, 20089–20093.
Alvarez, E., Girones, N., and Davis, R. J. (1990). *Biochem. J.* **267**, 31–35.
Amara, J. F., Lederkremer, G., and Lodish, H. F. (1989). *J. Cell Biol.* **109**, 3315–3324.
Andersen, T. T., Freytag, J. W., and Hill, R. L. (1982). *J. Biol. Chem.* **257**, 8036–8041.
Anderson, R. G. W., Brown, M. S., and Goldstein, J. L. (1977). *Cell* **10**, 351–364.
Anderson, R. G. W., Goldstein, J. L., and Brown, M. S. (1976). *Proc. Natl. Acad. Sci. U.S.A.* **73**, 2434–2438.
Ashwell, G., and Harford, J. (1982). *Annu. Rev. Biochem.* **51**, 531–554.
Ashwell, G., and Morell, A. G. (1974). *Adv. Enzymol.* **41**, 99–128.
Atoda, H., Hyuga, M., and Morita, T. (1991). *J. Biol. Chem.* **266**, 14903–14911.

Baenziger, J. U., and Fiete, D. (1986). *J. Biol. Chem.* **261**, 7445–7454.

Baenziger, J. U., and Maynard, Y. (1980). *J. Biol. Chem.* **255**, 4607–4613.

Balch, W. E. (1990). *Trends Biochem. Sci.* **15**, 473–477.

Barr, F. A., Leyte, A., Mollner, S., Pfeuffer, T., Tooze, S. A., and Huttner, W. B. (1991). *FEBS Lett.* **294**, 239–243.

Bartles, J. R., Feracci, H. M., Stieger, B., and Hubbard, A. L. (1987). *J. Cell Biol.* **105**, 1241–1251.

Beltzer, J. P., and Spiess, M. (1991). *EMBO J.* **10**, 3735–3742.

Beltzer, J. P., Fiedler, K., Fuhrer, C., Geffen, I., Handschin, C., Wessels, H. P., and Spiess, M. (1991). *J. Biol. Chem.* **266**, 973–978.

Benson, B., Hawgood, S., Schilling, J., Clements, J., Damm, D., Cordell, B., and White, R. T. (1985). *Proc. Natl. Acad. Sci. U.S.A.* **82**, 6379–6383.

Bischoff, J., Libresco, S., Shia, M. A., and Lodish, H. F. (1988). *J. Cell Biol.* **106**, 1067–1074.

Bischoff, J., and Lodish, H. F. (1987). *J. Biol. Chem.* **262**, 11825–11832.

Block, M. R., Glick, B. S., Wilcox, C. A., Wieland, F. T., and Rothman, J. E. (1988). *Proc. Natl. Acad. Sci. U.S.A.* **85**, 7852–7856.

Boggaram, V., Qing, K., and Mendelson, C. R. (1988). *J. Biol. Chem.* **263**, 2939–2947.

Braiterman, L. T., Chance, S. C., Porter, W. R., Lee, Y. C., Townsend, R. R., and Hubbard, A. L. (1989). *J. Biol. Chem.* **264**, 1682–1688.

Braulke, T., Tippmer, S., Chao, H. J., and Vonfigura, K. (1990). *Eur. J. Biochem.* **189**, 609–616.

Breitfeld, P. P., Casanova, J. E., Mckinnon, W. C., and Mostov, K. E. (1990). *J. Biol. Chem.* **265**, 13750–13757.

Breitfeld, P. P., Rup, D., and Schwartz, A. L. (1984). *J. Biol. Chem.* **259**, 10414–10421.

Breitfeld, P. P., Simmons, C. F., Strous, G. J. A. M., Geuze, H. J., and Schwartz, A. L. (1985). *Int. Rev. Cytol.* **97**, 47–95.

Bretscher, M. S. (1984). *Science* **244**, 681–686.

Bretscher, M. S., and Pearse, B. M. F. (1984). *Cell* **38**, 3–4.

Brewer, C. B., and Roth, M. G. (1991). *J. Cell Biol.* **114**, 413–421.

Brown, W. R., and Kloppel, T. M. (1989). *Immunol. Invest.* **18**, 269–285.

Buys, S. S., Keogh, E. A., and Kaplan, J. (1984). *Cell* **38**, 569–576.

Canfield, W. M., Johnson, K. F., Ye, R. D., Gregory, W., and Kornfeld, S. (1991). *J. Biol. Chem.* **266**, 5682–5688.

Casciola-Rosen, L. A. F., and Hubbard, A. L. (1991). *J. Biol. Chem.* **266**, 4341–4347.

Changeux, J. P., Llinas, R. R., Purves, D., and Bloom, F. E. (1990). *In* "Functional Architecture and Dynamics of the Nicotinic Acetyl Choline Receptor: An Allosteric Ligand-Gated Ion Channel" pp. 21–168. Raven Press, New York.

Chiacchia, K. B., and Drickamer, K. (1984). *J. Biol. Chem.* **259**, 15440–15446.

Clarke, B. L., Oka, J. A., and Weigel, P. H. (1987). *J. Biol. Chem.* **262**, 17384–17392.

Collawn, J. F., Stangel, M., Kuhn, L. A., Esekogwu, V., Jing, S. Q., Trowbridge, I. S., and Tainer, J. A. (1990). *Cell* **63**, 1061–1072.

Collins, T., Williams, A., Johnston, G. I., Kim, J., Eddy, R., Shows, T., Gimbrone, M. R., and Bevilacqua, M. P. (1991). *J. Biol. Chem.* **266**, 2466–2473.

Colombo, M. I., Mayorga, L. S., Casey, P. J., and Stahl, P. D. (1992). *Science* **255**, 1695–1697.

Daniels, C. K., Schmucker, D. L., and Jones, A. L. (1989). *Hepatology* **9**, 229–234.

Dautry-Varsat, A., Ciechanover, A., and Lodish, H. F. (1983). *Proc. Natl. Acad. Sci. U.S.A.* **80**, 2258–2262.

Davis, C. G., Lehrman, M. A., Russell, D. W., Anderson, R. G. W., Brown, M. S., and Goldstein, J. L. (1986). *Cell* **45**, 15–24.

Davis, R. J., and Meisner, H. (1987). *J. Biol. Chem.* **262**, 16041–16047.

Davis, C. G., Van Driel, I. R., Russell, D. W., Brown, M. S., and Goldstein, J. L. (1987). *J. Biol. Chem.* **262**, 4075–4082.

De Caro, A. M., Bonicel, J. J., Roumi, P., De Caro, J. D., Sarles, H., and Rovery, M. (1987). *Eur. J. Biochem.* **168**, 201–207.

DiPaola, M., and Maxfield, F. R. (1984). *J. Biol. Chem.* **259**, 9163–9171.

Doege, K., Fernandez, P., Hassell, J. R., Sasaki, M., and Yamada, Y. (1986). *J. Biol. Chem.* **261**, 8108–8111.

Dowbenko, D. J., Diep, A., Taylor, B. A., Lusis, A. J., and Lasky, L. A. (1991). *Genomics* **9**, 270–277.

Dragsten, P. R., Mitchell, D. B., Covert, G., and Baker, T. (1987). *Biochim. Biophys. Acta* **926**, 270–279.

Drickamer, K. (1981). *J. Biol. Chem.* **256**, 5827–5839.

Drickamer, K. (1988). *J. Biol. Chem.* **263**, 9557–9560.

Drickamer, K., Dordal, M. S., and Reynolds, I. (1986). *J. Biol. Chem.* **261**, 6878–6887.

Drickamer, K., Mamon, J. F., Binns, G., and Leung, J. O. (1984). *J. Biol. Chem.* **259**, 770–778.

Drickamer, K., and McCreary, V. (1987). *J. Biol. Chem.* **262**, 2582–2589.

Duncan, J. R., and Kornfeld, S. (1988). *J. Cell Biol.* **106**, 617–628.

Dunn, K. W., McCraw, T. E., and Maxfield, F. R. (1989). *J. Cell Biol.* **109**, 3303–3314.

Fallon, R. J. (1990). *J. Biol. Chem.* **265**, 3401–3406.

Fallon, R. J., and Schwartz, A. L. (1986). *J. Biol. Chem.* **261**, 15081–15089.

Fallon, R. J., and Schwartz, A. L. (1987). *Mol. Pharmacol.* **32**, 348–355.

Fallon, R. J., and Schwartz, A. L. (1988). *J. Biol. Chem.* **263**, 13159–13166.

Felder, S., Miller, K., Moehren, G., Ullrich, A., Schlessinger, J., and Hopkins, C. R. (1990). *Cell* **61**, 623–634.

Fuhlendorff, J., Clemmensen, I., and Magnusson, S. (1987). *Biochemistry* **26**, 6757–6764.

Fuhrer, C., Geffen, I., and Spiess, M. (1991). *J. Cell Biol.* **114**, 423–432.

Gartung, C., Braulke, T., Hasilik, A., and Von Figura, K. (1985). *EMBO J.* **4**, 1725–1730.

Geffen, I., and Spiess, M. (1992). *FEBS Lett.* **305**, 209–212.

Geffen, I., Wessels, H. P., Roth, J., Shia, M. A., and Spiess, M. (1989). *EMBO J.* **8**, 2855–2862.

Geffen, I., Fuhrer, C., and Spiess, M. (1991). *Proc. Natl. Acad. Sci. U.S.A.* **88**, 8425–8429.

Geuze, H. J., Slot, J. W., and Schwartz, A. L. (1987). *J. Cell Biol.* **104**, 1715–1723.

Geuze, H. J., Slot, J. W., Strous, G. J., Luzio, J. P., and Schwartz, A. L. (1984a). *EMBO J.* **3**, 2677–2685.

Geuze, H. J., Slot, J. W., Strous, G. J., and Schwartz, A. L. (1983a). *Eur. J. Cell Biol.* **32**, 38–44.

Geuze, H. J., Slot, J. W., Strous, G. J. A. M., Lodish, H. F., and Schwartz, A. L. (1982). *J. Cell Biol.* **92**, 865–870.

Geuze, H. J., Slot, J. W., Strous, G. J. A. M., Lodish, H. F., and Schwartz, A. L. (1983b). *Cell* **32**, 277–287.

Geuze, H. J., Slot, J. W., Strous, G. J. A. M., Peppard, J., Von Figura, K., Hasilisk, A., and Schwartz, A. L. (1984b). *Cell* **37**, 195–204.

Gierasch, L. M. (1989). *Biochemistry* **28**, 924–930.

Giga, Y., Ikai, A., and Takahashi, K. (1987). *J. Biol. Chem.* **262**, 6197–6203.

Giorda, R., Rudert, W. A., Vavassori, C., Chambers, W. H., Hiserodt, J. C., and Trucco, M. (1990). *Science* **249**, 1298–1300.

Glickman, J. N., Conibear, E., and Pearse, B. M. F. (1989). *EMBO J.* **8**, 1041–1047.

Goda, Y., and Pfeffer, S. R. (1988). *Cell* **55**, 309–320.

Goldstein, J. L., Brown, M. S., Anderson, R. G. W., Russell, D. W., and Schneider, W. J. (1985). *Annu. Rev. Cell Biol.* **1**, 1–39.

Graeve, L., Patzak, A., Drickamer, K., and Rodriguez-Boulan, E. (1990). *J. Biol. Chem.* **265**, 1216–1224.

Griffiths, G., Back, R., and Marsh, M. (1989). *J. Cell Biol.* **109**, 2703–2720.

Griffiths, G., and Gruenberg, J. (1991). *Trends Cell Biol.* **1**, 5–9.

Griffiths, G., Hoflack, B., Simons, K., Mellman, I., and Kornfeld, S. (1988). *Cell* **52**, 329–341.

Gruenberg, J., and Howell, K. E. (1989). *Annu. Rev. Cell Biol.* **5**, 453–481.

Gruenberg, J., Griffiths, G., and Howell, K. E. (1989). *J. Cell Biol.* **108**, 1301–1316.

Halberg, D. F., Wager, R, E., Farrell, D. C., Hildreth, J. IV, Quesenberry, M. S., Loeb, J. A., Holland, E. C., and Drickamer, K. (1987). *J. Biol. Chem.* **262**, 9828–9838.

Hardy, M. R., Townsend, R. R., Parkhurst, S. M., and Lee, Y. C. (1985). *Biochemistry* **24**, 22–28.
Hartmann, E., Rapoport, T. A., and Lodish, H. F. (1989). *Proc. Natl. Acad. Sci. U.S.A.* **86**, 5786–5790.
Henis, Y. I., Katzir, Z., Shia, M. A., and Lodish, H. F. (1990). *J. Cell Biol.* **111**, 1409–1418.
Heuser, J. (1989). *J. Cell Biol.* **108**, 401–411.
Heuser, J., and Keen, J. (1988). *J. Cell Biol.* **107**, 877–886.
Hirabayashi, J., Kusunoki, T., and Kasai, K. (1991). *J. Biol. Chem.* **266**, 2320–2326.
Holland, E. C., and Drickamer, K. (1986). *J. Biol. Chem.* **261**, 1286–1292.
Holland, E. C., Leung, J. O., and Drickamer, K. (1984). *Proc. Natl. Acad. Sci. U.S.A.* **81**, 7338–7342.
Hong, W., Le, A. V., and Doyle, D. (1988). *Hepatology* **8**, 553–558.
Hopkins, C. R., Gibson, A., Shipman, M., and Miller, K. (1990). *Nature* **346**, 335–339.
Hoyle, G. W., and Hill, R. L. (1991). *J. Biol. Chem.* **266**, 1850–1857.
Hsueh, E. C., Holland, E. C., Carrera, G. M. J., and Drickamer, K. (1986). *J. Biol. Chem.* **261**, 4940–4947.
Hunziker, W., Harter, C., Matter, K., and Mellman, I. (1991). *Cell* **66**, 907–920.
Hurtley, S. M., and Helenius, A. (1989). *Annu. Rev. Cell Biol.* **5**, 277–307.
Iacopetta, B. J., Rothenberger, S., and Kühn, L. C. (1988). *Cell* **54**, 485–489.
Ii, M., Kawasaki, T., and Yamashina, I. (1988). *Biochem. Biophys. Res. Commun.* **155**, 720–725.
Ii, M., Kurata, H., Itoh, N., Yamashina, I., and Kawasaki, T. (1990). *J. Biol. Chem.* **265**, 11295–11298.
Inoue, S., Kogaki, H., Ikeda, K., Samejima, Y., and Omorisatoh, T. (1991). *J. Biol. Chem.* **266**, 1001–1007.
Ishihara, H., Hara, T., Aramaki, Y., Tsuchiya, S., and Hosoi, K. (1990). *Pharmacol. Res.* **7**, 542–546.
Jin, M., Sahagian, G. G., and Snider, M. D. (1989). *J. Biol. Chem.* **264**, 7675–7680.
Jing, S. Q., Spencer, T., Miller, K., Hopkins, C., and Trowbridge, I. S. (1990). *J. Cell Biol.* **110**, 283–294.
Kawasaki, T., and Ashwell, G. (1976). *J. Biol. Chem.* **251**, 1296–1302.
Kawasaki, T., Li, M., Kozutsumi, Y., and Yamashina, I. (1986). *Carbohydr. Res.* **151**, 197–206.
Keen, J., Willingham, M. C., and Pastan, I. (1979). *Cell* **16**, 303–312.
Keen, J. H., and Beck, K. A. (1989). *Biochem. Biophys. Res. Commun.* **158**, 17–23.
Kelm, S., and Schauer, R. (1986). *Biol. Chem. Hoppe-Seyler* **367**, 989–998.
Kelm, S., and Schauer, R. (1988). *Biol. Chem. Hoppe-Seyler* **369**, 693–704.
Kerr, M. A. (1990). *Biochem. J.* **271**, 285–296.
Kirchhausen, T., Nathanson, K. L., Natsui, W., Vaisberg, A., Chow, E. P., Burne, C., Keen, J. H., and Davis, A. E. (1989). *Proc. Natl. Acad. Sci. U.S.A.* **86**, 2612–2616.
Klausner, R. D., Harford, J., and Van Renswoude, J. (1984). *Proc. Natl. Acad. Sci. U.S.A.* **81**, 3005–3009.
Kornfeld, S. (1987). *FASEB J.* **1**, 462–468.
Kornfeld, S., and Mellman, I. (1989). *Annu. Rev. Cell Biol.* **5**, 483–525.
Ktistakis, N. T., Thomas, D., and Roth, M. G. (1990). *J. Cell Biol.* **111**, 1393–1407.
Larkin, J. M., Donzell, W. C., and Anderson, R. G. W. (1986). *J. Cell Biol.* **103**, 2619–2627.
Lazarovits, J., and Roth, M. (1988). *Cell* **53**, 743–752.
Lederkremer, G. Z., and Lodish, H. F. (1991). *J. Biol. Chem.* **266**, 1237–1244.
Lee, H., Kelm, S., Yoshino, T., and Schauer, R. (1988). *Biol. Chem. Hoppe-Seyler* **369**, 705–714.
Lee, R. T., Lin, P., and Lee, Y. C. (1984). *Biochemistry* **23**, 4255–4261.
Lee, Y. C., Townsend, R. R., Hardy, M. R., Lonngren, J., Arnarp, J., Haraldsson, M., and Lonn, H. (1983). *J. Biol. Chem.* **258**, 199–202.
Lee, Y. M., Leiby, K. R., Allar, J., Paris, K., Lerch, B., and Okarma, T. B. (1991). *J. Biol. Chem.* **266**, 2715–2723.
Lehrman, M. A., Goldstein, J. L., Brown, M. S., Russell, D. W., and Schneider, W. J. (1985). *Cell* **41**, 735–743.

Lehrman, M. A., Haltiwanger, R. S., and Hill, R. L. (1986). *J. Biol. Chem.* **261**, 7426–7432.

Leung, J. O., Holland, E. C., and Drickamer, K. (1985). *J. Biol. Chem.* **260**, 12523–12527.

Linderman, J. J., and Lauffenburger, D. A. (1988). *J. Theor. Biol.* **132**, 203–245.

Lobel, P., Fujimoto, K., Ye, R. D., Griffiths, G., and Kornfeld, S. (1989). *Cell* **57**, 787–796.

Lodish, H. F. L. (1991). *Trends Biochem. Sci.* **16**, 374–377.

Loeb, J. A., and Drickamer, K. (1987). *J. Biol. Chem.* **262**, 3022–3029.

Loeb, J. A., and Drickamer, K. (1988). *J. Biol. Chem.* **263**, 9752–9760.

Lunney, J., and Ashwell, G. (1976). *Proc. Natl. Acad. Sci. U.S.A.* **73**, 341–343.

Mahaffey, D. T., Peeler, J. S., Brodsky, F. M., and Anderson, R. (1990). *J. Biol. Chem.* **265**, 16514–16520.

Matsuura, S., Nakada, H., Sawamura, T., and Tashiro, Y. (1982). *J. Cell Biol.* **95**, 864–875.

May, W. S., Jacobs, S., and Cuatrecasas, P. (1984). *Proc. Natl. Acad. Sci. U.S.A.* **81**, 2016–2020.

McGraw, T. E., Dunn, K. W., and Maxfield, F. R. (1988). *J. Cell Biol.* **106**, 1061–1066.

McPhaul, M., and Berg, P. (1986). *Proc. Natl. Acad. Sci. U.S.A.* **83**, 8863–8867.

McPhaul, M., and Berg, P. (1987). *Mol. Cell. Biol.* **7**, 1841–1847.

Mellman, I., Fuchs, R., and Helenius, A. (1986). *Annu. Rev. Biochem.* **55**, 663–700.

Miettinen, H. M., Rose, J. K., and Mellman, I. (1989). *Cell* **58**, 317–327.

Moore, M. S., Mahaffey, D. T., Brodsky, F. M., and Anderson, R. G. W. (1987). *Science* **236**, 558–563.

Mostov, K. E., and Deitcher, D. L. (1986). *Cell* **46**, 613–621.

Mostov, K. E., and Simister, N. E. (1985). *Cell* **43**, 389–390.

Mostov, K. E., de Bruyn Kops, A., and Deitcher, D. L. (1986). *Cell* **47**, 359–364.

Mueller, S. C., and Hubbard, A. L. (1986). *J. Cell Biol.* **102**, 932–942.

Mullock, B. M., Branch, W. J., Van Schaik, M., Gilbert, L. K., and Luzio, J. P. (1989). *J. Cell Biol.* **108**, 2093–2099.

Muramoto, K., and Kamiya, H. (1986). *Biochim. Biophys. Acta* **874**, 285–295.

Murphy, R. F. (1991). *Trends Cell Biol.* **1**, 77–82.

Neefjes, J. J., Verkerk, J. M. H., Broxterman, H. J. G., Van der Marel, G. A., Van Boom, J. H., and Ploegh, H. L. (1988). *J. Cell Biol.* **107**, 79–87.

Oka, J. A., and Weigel, P. H. (1983). *J. Biol. Chem.* **258**, 10253–10262.

Oldberg, A., Antonsson, P., and Heinegard, D. (1987). *Biochem. J.* **243**, 255–259.

Ord, D. C., Ernst, T. J., Zhou, L. J., Rambaldi, A., Spertini, O., Griffin, J., and Tedder, T. F. (1990). *J. Biol. Chem.* **265**, 7760–7767.

Parton, R. G., Prydz, K., Bomsel, M., Simons, K., and Griffiths, G. (1989). *J. Cell Biol.* **109**, 3259–3257.

Pearse, B. M. F. (1985). *EMBO J.* **4**, 2457–2460.

Pearse, B. M. F. (1988). *EMBO J.* **7**, 3331–3336.

Pearse, B. M. F., and Crowther, R. A. (1987). *Annu. Rev. Biophys. Biophys. Chem.* **16**, 49–68.

Petell, J. K., and Doyle, D. (1985). *Arch. Biochem. Biophys.* **241**, 550–560.

Prywes, R., Livneh, E., Ullrich, A., and Schlessinger, J. (1986). *EMBO J.* **5**, 2179–2190.

Regoeczi, E., Chindemi, P. A., Debanne, M. T., and Hatton, W. C. (1982). *J. Biol. Chem.* **257**, 5431.

Regoeczi, E., Hatton, M. W. C., and Charlwood, P. A. (1975). *Nature* **254**, 699–701.

Reichner, J. S., Whiteheart, S. W., and Hart, G. W. (1988). *J. Biol. Chem.* **263**, 16316–16326.

Renfrew, C. A., and Hubbard, A. L. (1991). *J. Biol. Chem.* **266**, 4348–4356.

Rice, K. G., Weisz, O. A., Barthel, T., Lee, R. T., and Lee, Y. C. (1990). *J. Biol. Chem.* **265**, 18429–18434.

Robinson, A., Kaderbhai, M. A., and Austen, B. M. (1987). *Biochem. J.* **242**, 767–777.

Roederer, M., Bowser, R., and Murphy, R. F. (1987). *J. Cell. Physiol.* **131**, 200–209.

Rome, L. H. (1985). *Trends Biochem. Sci.* **10**, 151.

Roth, J., Taatjes, D. J., Lucocq, J. M., Weinstein, J., and Paulson, J. C. (1985). *Cell* **43**, 287–295.

Roth, M. G., Doyle, C., Sambrook, J., and Gething, M. J. (1986). *J. Cell Biol.* **102**, 1271–1283.

Rothenberger, S., Iacopetta, B. J., and Kühn, L. C. (1987). *Cell* **49**, 423–431.

Rothman, J. E., and Schmid, S. L. (1986). *Cell* **46**, 5–9.

Rust, K., Grosso, L., Zhang, V., Chang, D., Persson, A., Longmore, W., Cai, G. Z., and Crouch, E. (1991). *Arch. Biochem. Biophys.* **290**, 116–126.

Sai, S., Tanaka, T., Kosher, R. A., and Tanzer, M. L. (1986). *Proc. Natl. Acad. Sci. U.S.A.* **83**, 5081–5085.

Sakaguchi, M., Mihara, K., and Sato, R. (1987). *EMBO J.* **6**, 2425–2431.

Salzman, N. H., and Maxfield, F. R. (1989). *J. Cell Biol.* **109**, 2097–2104.

Sanford, J. P., Elliott, R. W., and Doyle, D. (1988). *DNA* **7**, 721–728.

Sano, K., Fisher, J., Mason, R. J., Kuroki, Y., Schilling, J., Benson, B., and Voelker, D. (1987). *Biochem. Biophys. Res Commun.* **144**, 367–374.

Sawyer, J. T., Sanford, J. P., and Doyle, D. (1988). *J. Biol. Chem.* **263**, 10534–10538.

Schlessinger, J. (1988). *Biochemistry* **27**, 3119–3123.

Schmid, S. L., Fuchs, R., Male, P., and Mellman, I. (1988). *Cell* **52**, 73–83.

Schmid, S. R., and Spiess, M. (1988). *J. Biol. Chem.* **263**, 16886–16891.

Schreiber, A. B., Libermann, T. A., Lax, I., Yarden, Y., and Schlesinger, J. (1983). *J. Biol. Chem.* **258**, 846–853.

Schwartz, A. L. (1984a). *Biochem. J.* **223**, 481–486.

Schwartz, A. L. (1984b). *C.R.C. Crit. Rev. Biochem.* **16**, 207–233.

Schwartz, A. L., Ciechanover, A., Merritt, S., and Turkewitz, A. (1986). *J. Biol. Chem.* **261**, 15225–15232.

Schwartz, A. L., Fridovich, S. E., and Lodish, H. F. (1982). *J. Biol. Chem.* **257**, 4230–4237.

Schwartz, A. L., and Rup, D. (1983). *J. Biol. Chem.* **258**, 11249–11255.

Schwartz, A. L., Steer, C. J., and Kempner, E. S. (1984). *J. Biol. Chem.* **259**, 12025–12029.

Shia, M. A., and Lodish, H. F. (1989). *Proc. Natl. Acad. Sci. U.S.A.* **86**, 1158–1162.

Sibley, D. R., Benovic, J. L., Caron, M. G., and Lefkowitz, R. J. (1987). *Cell* **48**, 913–922.

Simmons, C. F., and Schwartz, A. L. (1984). *Mol. Pharmacol.* **26**, 509–519.

Smythe, E., Pypaert, M., Lucocq, J., and Warren, G. (1989). *J. Cell Biol.* **108**, 843–853.

Snider, M. D., and Rogers, O. C. (1985). *J. Cell Biol.* **100**, 826–834.

Spiess, M. (1990). *Biochemistry*, **29**, 1009–1018.

Spiess, M., and Handschin, C. (1987). *EMBO J.* **6**, 2683–2691.

Spiess, M., and Lodish, H. F. (1985). *Proc. Natl. Acad. Sci. U.S.A.* **82**, 6465–6469.

Spiess, M., and Lodish, H. F. (1986). *Cell* **44**, 177–185.

Spiess, M., Schwartz, A. L., and Lodish, H. F. (1985). *J. Biol. Chem.* **260**, 1979–1982.

Steer, C. J., Kempner, E. S., and Ashwell, G. (1981). *J. Biol. Chem.* **256**, 5851–5856.

Steer, C. J., Weiss, P., Huber, B. E., Wirth, P. J., Thorgeirsson, S. S., and Ashwell, G. (1987). *J. Biol. Chem.* **262**, 17524–17529.

Steinman, R. M., Mellman, I. S., Muller, W. A., and Cohn, Z. A. (1983). *J. Cell Biol.* **96**, 1–27.

Stoorvogel, W., Geuze, H. J., Griffith, J. M., Schwartz, A. L., and Strous, G. J. (1989). *J. Cell Biol.* **108**, 2137–2148.

Stoorvogel, W., Geuze, H. J., and Strous, G. J. (1987). *J. Cell Biol.* **104**, 1261–1268.

Suter, U., Bastos, R., and Hofstetter, H. (1987). *Nucleic Acids Res.* **15**, 7295–7308.

Suzuki, K., Kusumoto, H., Deyashiki, Y., Nishioka, J., Maruyama, I., Zushi, M., Kawahara, S., Honda, G., Yamamoto, S., and Horiguchi, S. (1987). *EMBO J.* **6**, 1891–1897.

Suzuki, T., Takagi, T., Furukohri, T., Kawamura, K., and Nakauchi, M. (1990). *J. Biol. Chem.* **265**, 1274–1281.

Swanson, J. A., and Silverstein, S. C. (1988). *In* "Pinocytic Flow through Macrophages" (B. Pernis, S. Silverstein, and H. Vogel, eds.), pp. 15–27. Academic Press, London.

Takahashi, H., Komano, H., Kawaguchi, N., Kitamura, N., Nakanishi, S., and Natori, S. (1985). *J. Biol. Chem.* **260**, 12228–12233.

Takahashi, T., Nakada, H., Okumura, T., Sawamura, T., and Tashiro, Y. (1985). *Biochem. Biophys. Res. Commun.* **126**, 1054–1060.

Tanaka, T., Har-El, R., and Tanzer, M. L. (1988). *J. Biol. Chem.* **263**, 15831–15835.

Taylor, M. E., Brickell, P. M., Craig, R. K., and Summerfield, J. A. (1989). *Biochem. J.* **262**, 763–771.

Taylor, M. E., Conary, J. T., Lennartz, M. R., Stahl, P. D., and Drickamer, K. (1990). *J. Biol. Chem.* **265**, 12156–12162.

Thiel, S., and Reid, K. B. M. (1989). *FEBS Lett.* **250**, 78–84.

Townsend, R. R., Hardy, M. R., Wong, T. C., and Lee, Y. C. (1986). *Biochemistry* **25**, 5716–5725.

Van den Bosch, R. A., Geuze, H. J., and Strous, G. J. (1986). *Exp. Cell Res.* **162**, 231–242.

Van Deurs, B., Hansen, S. H., Petersen, O. W., Melby, E. L., and Sandvig, K. (1990). *Eur. J. Cell Biol.* **51**, 96–109.

Van Deurs, B., Sandvig, K., Petersen, O. W., Olsnes, S., Simons, K., and Griffiths, G. (1988). *J. Cell Biol.* **106**, 253–267.

Verrey, F., Gilbert, T., Mellow, T., Proulx, G., and Drickamer, K. (1990). *Cell Regul.* **1**, 471–486.

Von Heijne, G. (1986). *EMBO J.* **5**, 3021–3027.

Von Heijne, G., and Gavel, Y. (1988). *Eur. J. Biochem.* **174**, 671–678.

Wall, D. A., and Hubbard, A. (1981). *J. Cell Biol.* **90**, 687–696.

Walter, P., Gilmore, R., and Blobel, G. (1984). *Cell* **38**, 5–8.

Wandinger-Ness, A., and Simons, K. (1990). *In* "The Polarized Transport of Surface Proteins and Lipids in Epithelial Cells" (J. Hanover, and S. Steer, eds.). pp. 575–612. Cambridge University Press, Cambridge, Massachusetts.

Weigel, P. H., and Oka, J. A. (1984). *J. Biol. Chem.* **259**, 1150–1154.

Weis, W. I., Crichlow, G. V., Murthy, H., Hendrickson, W. A., and Drickamer, K. (1991a). *J. Biol. Chem.* **266**, 20678–20686.

Weis, W. I., Kahn, R., Fourme, R., Drickamer, K., and Hendrickson, W. A. (1991b). *Science* **254**, 1608–1615.

Weiss, P., and Ashwell, G. (1989). *J. Biol. Chem.* **264**, 11572–11574.

Weissleder, R., Reimer, P., Lee, A. S., Wittenberg, J., and Brady, T. J. (1990). *A.J.R.* **155**, 1161–1167.

Wessels, H. P., Geffen, I., and Spiess, M. (1989). *J. Biol. Chem.* **264**, 17–20.

White, R. T., Damm, D., Miller, J., Spratt, K., Schilling, J., Hawgood, S., Benson, B., and Cordell, B. (1985). *Nature* **317**, 361–363.

Wikström, L., and Lodish, H. F. (1991). *J. Cell Biol.* **113**, 997–1007.

Wikström, L., and Lodish, H. F. (1992). *J. Biol. Chem.* **267**, 5–8.

Wilson, J. M., Grossman, M., Wu, C. H., Chowdhury, N. R., Wu, G. Y., and Chowdhury, J. R. (1992). *J. Biol. Chem.* **267**, 963–967.

Windler, E., Greeve, J., Levkau, B., Kolb-Bachofen, V., Daerr, W., and Greten, H. (1991). *Biochem. J.* **276**, 79–87.

Wolkoff, A. W., Klausner, R. D., Ashwell, G., and Harford, J. (1984). *J. Cell Biol.* **98**, 375–381.

Yarden, Y., and Ullrich, A. (1988). *Biochemistry* **27**, 3113–3119.

Zannis, V. I., McPherson, J., Goldberger, G., Karathanasis, S. K., and Breslow, J. L. (1984). *J. Biol. Chem.* **259**, 5495–5499.

Zijderhand-Bleekemolen, J. E., Schwartz, A. L., Slot, J. W., Strous, G. J., and Geuze, H. J. (1987). *J. Cell Biol.* **104**, 1647–1654.

Zimmermann, D. R., and Ruoslahti, E. (1989). *EMBO J.* **8**, 2975–2981.

Mannose Receptor

Suzanne E. Pontow, Vladimir Kery, and Philip D. Stahl

Department of Cell Biology and Physiology, Washington University School of
Medicine, St. Louis, Missouri 63110

I. Historical Perspective

A. Discovery

The macrophage mannose receptor is a carbohydrate-binding membrane protein
that functions in receptor-mediated endocytosis and phagocytosis (Stahl, 1990).
Evidence for a novel sugar-specific receptor expressed by mononuclear phagocytes initially came from experiments on the recognition and clearance of lysosomal hydrolases *in vivo*. An observation was made that the administration of
organophosphate compounds to rodents resulted in an acute and dramatic increase in plasma β-glucuronidase (Stahl *et al.*, 1975; Mandell and Stahl, 1977).
The source of the elevated plasma β-glucuronidase in poisoned animals was
shown to be the liver hepatocyte. In rodent liver β-glucuronidase has a dual localization with perhaps 25% or more of the enzyme found in the endoplasmic
reticulum (ER) (DeDuve *et al.*, 1955; Stahl and Touster, 1971). Following
organophosphate poisoning, microsomal levels of β-glucuronidase were found
to be selectively depleted; liver lysosomal levels of the enzyme were unaffected
(Mandell and Stahl, 1977). Following injection β-glucuronidase purified either
from rat liver lysosomes or microsomes was rapidly cleared from the circulation, whereas enzyme isolated from the serum of rats poisoned with organophosphate compounds was cleared slowly (Stahl *et al.*, 1975). These differential
rates of plasma clearance, coupled with subsequent studies, led to the discovery
of the mannose receptor.

Rodent liver microsomal β-glucuronidase is tethered to ER membranes by the
protein egasyn. It has been shown that egasyn is an esterase (Novak *et al.*, 1991).
Apparently, organophosphate inhibition of the esterase activity of egasyn results
in the release of the microsomal β-glucuronidase from ER membranes. For an
unknown reason the released enzyme made its way through the secretory pathway without acquiring the mannose-phosphate marker for targeting to lysosomes

221

(unpublished observation). Instead, the oligosaccharide side chains were rapidly processed to the complex form characteristic of plasma proteins, precluding rapid clearance of the enzyme from the plasma compartment.

B. Characterization

Efforts to define the sugar-specific mechanism behind clearance of β-glucuronidase revealed that many lysosomal enzymes purified from rat liver were similarly removed from the circulation following injection (Schlesinger et al., 1976; Stahl et al., 1976a,b,c; Stockert et al., 1976). In vivo experiments showed that glycoproteins bearing terminal N-acetylglucosamine residues antagonized the clearance of injected β-glucuronidase, whereas galactose-terminated glycoproteins were ineffective. The mechanism responsible for clearance of lysosomal hydrolases was thus not the asialoglycoprotein (ASGP) receptor delineated by the pioneering work of Ashwell and Morell (1974). Rather, the experiments suggested a new recognition system. Other studies with ribonuclease B (Baynes and Wold, 1976) and with agalacto-and ahexosamino-orosomucoid identified mannose as an important determinant for in vivo clearance (Stockert et al., 1976; Lunney and Ashwell, 1976; Stahl et al., 1976b; Schlesinger et al., 1978). Subsequently, it was shown that mannan, a yeast cell wall polysaccharide rich in mannose, was a potent inhibitor of enzyme clearance (Achord et al., 1977a). With the availability of neoglycoproteins (Lee et al., 1976; Stahl et al., 1978) as sugar-specific ligands, mannose-bovine serum albumin (BSA) and L-fucose-BSA were shown to be equally effective in blocking β-glucuronidase clearance in vivo (Schlesinger et al., 1980; Shepherd et al., 1981). For this reason the receptor has been referred to as the mannose-fucose receptor (Stahl and Gordon, 1982). Because the receptor has specificity for sugars other than mannose and fucose, the term mannose receptor has been favored by most investigators. Since its discovery the list of glycoprotein ligands for the receptor has grown and includes horseradish peroxidase (Stahl et al., 1978), ricin A chain (Simmons et al., 1986), amylase (Niesen et al., 1984), tissue plasminogen activator (Smedsrod et al., 1988; Smedsrod and Einarsson, 1990), the C-terminal propeptide of Type I procollagen (Smedsrod et al., 1990), neutrophil-derived myeloperoxidase (Shepherd and Hoidal, 1990), and most lysosomal hydrolases.

C. Distribution

The tissue and cellular localization of mannose receptor activity was first delineated by in vivo clearance experiments (Schlesinger et al., 1976; Achord et al., 1977b), showing that liver and spleen were the primary sites of uptake. Further studies (Schlesinger et al., 1978) determined that the liver nonparenchymal cells were the principal sites of β-glucuronidase uptake. These results were later confirmed by Achord *et al.* (1978) using human β-glucuronidase. Careful analysis of liver nonparenchymal cells at the electron-microscopic level enabled Hubbard *et al.* (1979) to show that Kupffer cells and hepatic endothelial cells were both actively engaged in β-glucuronidase uptake. Supporting evidence for the presence of mannose receptors on hepatic endothelial cells was provided by Summerfield *et al.* (1982) and Haltiwanger and Hill (1986a,b). Magnusson and Berg (1989) demonstrated mannose-specific ligand uptake in isolated hepatic endothelial cells.

The presence of mannose receptor activity on mononuclear phagocytes was first demonstrated in experiments with alveolar macrophages (Stahl *et al.*, 1978). Subsequently, a variety of mononuclear phagocyte preparations, including resident and elicited peritoneal macrophages (Stahl and Gordon, 1982), bone marrow-derived macrophages (Shepherd *et al.*, 1985), human monocyte-derived macrophages (Shepherd *et al.*, 1982), rat liver Kupffer cells (Maynard and Baenziger, 1981), and retinal pigment epithelium (McLaughlin *et al.*, 1987; Tarnowski *et al.*, 1988) were shown to express mannose receptor activity. Most macrophage-like cell lines express low levels of the receptor with the exception of J774E, a mouse cell line that has been induced to express higher levels of mannose receptor activity (Diment *et al.*, 1987).

Exciting new evidence suggests that the mannose receptor may be expressed on cells other than those of the reticuloendothelial system. First, mannose receptor activity, assessed by the uptake of mannose-BSA and by inhibition of zymosan uptake by mannans, has been identified on murine dendritic leukocytes, potent immunostimulatory cells involved in the initiation of immune responses (C. Reis e Sousa and J. Austyn, personal communication). One potential function for this activity is the accumulation of antigens for presentation. Second, Lew and Rattazzi (1991) implicated mannose receptor activity in the mitogenic effect of lysosomal hydrolases on cultured bovine tracheal smooth muscle cells. Morphological analysis following uptake of mannose-BSA–colloidal gold conjugates confirmed the presence of a mannose recognition system on these myocytes (Lew *et al.*, 1992). The presence of the mannose receptor on cells other than macrophages suggests a broader role for the receptor than previously assumed. Further characterization of this activity may be critical to the elucidation of mannose receptor function *in vivo*.

II. Isolation of the Mannose Receptor

The mannose receptor has been isolated from several sources, including rabbit (Lennartz *et al.*, 1987a), rat (Haltiwanger and Hill, 1986b), and human (Stephenson and Shepherd, 1987) alveolar macrophages, mouse J774 cells

(Blum *et al.*, 1991b), human (Lennartz *et al.*, 1987b) and bovine (C. Newton, unpublished observations) placenta, rat liver (Townsend and Stahl, 1981), and human pigment epithelium (Shepherd *et al.*, 1991). In general, the purification protocol is a two-step procedure involving the preparation of membranes and affinity chromatography. Purified mannose receptor has been utilized in a number of ways: for the generation of specific antisera, for soluble binding assays, and for receptor structure and sequencing studies. The isolation procedure developed in our laboratory is briefly outlined next.

The human placenta is an abundant and rich source for the isolation of mannose receptor protein. The tissue is first homogenized in a buffer containing 10 mM Tris, 1.25 M NaCl, 1 mM ethylenediaminetetraacetic acid (EDTA), and protease inhibitors. From the homogenate membranes are prepared by centrifugation and washed repeatedly in solution containing EDTA and then mannose to remove soluble endogenous mannose receptor ligands. The membranes are then solubilized in loading buffer containing 10 mM Tris, 1.25 M NaCl, 15 mM CaCl$_2$ and 1% Triton X-100.

The mannose receptor-containing placental extract is combined with mannose-sepharose and incubated 4–12 hours at 4°C with continuous agitation. Sepharose coupled to D-mannose is an efficient ligand for mannose receptor isolation. The column is prepared and the flow-through fraction is collected and conserved. The affinity column is washed extensively, then eluted in buffer containing 0.2 M D-mannose. Following dialysis and concentration, purity of the isolated receptor is assessed by sodium dodecyl sulfate–polyacrylomide gel electrophoresis (SDS-PAGE). If contaminating bands are present in the preparation, the eluted fraction can be reapplied to mannose-sepharose and eluted with buffer containing EDTA.

Recent investigations in our laboratory (V. Kery, unpublished observations) have revealed that the initial flow-through fraction remains rich in mannose receptor and can be rechromatographed and eluted. A possible explanation for this phenomenon is the presence of mannose-terminal membrane proteins, which compete with the affinity column for mannose receptor binding. With repeated application of the flow-through fraction to mannose-sepharose, however, milligram quantities of mannose receptor can be obtained from a single placenta (starting weight ~500 g). Isolated mannose receptor is stored at $-70°C$ and remains active for months.

III. Sequencing, Cloning, and Expression of the Mannose Receptor in Transfected Cells

The human mannose receptor has been cloned and sequenced, revealing important features of the receptor's structure and function. Peptides derived from proteolytic cleavage of isolated placental receptor were sequenced, allowing for the screening of both placental (Taylor *et al.*, 1990) and macrophage cDNA (Ezekowitz *et al.*, 1990) libraries. Sequence data from positive clones were confirmed by comparison with the known amino acid sequences; the full-length clone consists of 5185 bases and predicts a mature protein product of 1438 amino acids and molecular weight 164,135. The placental and macrophage cDNAs were found to exhibit a single nucleotide polymorphism that does not alter the resulting amino acid sequence. In addition, the screening of total mRNA from several sources, including HepG2 and HeLa cells, myelomonocytic cell lines, peripheral blood monocytes, and alveolar macrophages revealed that only those cells expressing mannose receptor activity were positive for the receptor message.

The primary structure of the mannose receptor is indicative of a Type I transmembrane orientation and contains several distinct domains (Taylor *et al.*, 1990; Ezekowitz *et al.*, 1991). The extracytoplasmic amino-terminal portin of the receptor (residues 1–1365) can be divided into three regions, which are preceded by a cleaved hydrophobic signal sequence. Within the first 139 amino acids are six cysteine residues; this region shares no homology with other known amino acid sequences, making the prediction of function difficult. A second domain (residues 139–192) is characterized by fibronectin Type II-like repeats, suggesting a possible role in the interactions of differentiated macrophages with the extracellular matrix. Consisting of eight segments resembling the carbohydrate recognition domains (CRDs) of C-type animal lectins (Drickamer, 1988), the remaining extracellular portion of the receptor mediates ligand binding (Taylor *et al.*, 1990, 1992). The eight potential CRDs are not identical; however, each contains at least some of the invariant amino acid residues of C-type lectins. Domains 4 and 5 are the most homologous, exhibiting 34% identity and 49% similarity, and each contains the WND sequence characteristic of sugar-binding proteins. In addition, the sequences of these domains are highly homologous to those of other mannose-binding proteins.

The presence of eight potential ligand-binding sites may explain the receptor's dependence on ligand multivalency for high-affinity binding. Other C-type lectins, such as the ASGP receptor and the chicken hepatic lectin, possess a single CRD per polypeptide chain and require oligomerization prior to binding with high affinity (Drickamer, 1988). Multiple binding sites may preclude a necessity for oligomerization by the mannose receptor. Studies on the exact nature of this receptor's binding characteristics have been facilitated by the availability of the receptor clone (Taylor *et al.*, 1992) and are described in detail next.

Consistent with early studies on mannose receptor structure and glycosylation state (Lennartz *et al.*, 1989), the Asn-X-Ser/Thr consensus motif for N-linked glycosylation occurs eight times within the extracytoplasmic amino acid sequence (Taylor *et al.*, 1990). The actual number of oligosaccharide chains present on the mature receptor remains to be determined for both N- and O-linked sites. Threonine residues available for O-glycosylation are also present and, interestingly, reside between the CRDs. However, no role for O-linked sugars in the acquisition or maintenance of binding activity has been demonstrated (S. E. Pontow and P. D. Stahl, manuscript in preparation).

The putative ligand-binding domain of the mannose receptor is followed by a hydrophobic sequence (residues 1366–1393) and four positively charged amino acids (Taylor *et al.*, 1990). This configuration is typical of transmembrane segments of Type I membrane proteins. The remaining amino acids (residues 1398–1438) make up the carboxy-terminal cytoplasmic tail, which has been shown in transfection studies to be required for both endocytic and phagocytic mannose receptor functions (Ezekowitz *et al.*, 1990). This region of the receptor may be particularly important in view of its potential interactions with factors involved in both ligand internalization and signal transduction. The sequence FENTLY contained within the receptor tail is highly similar to the FXNPXY motif required by certain endocytic receptors for localization to clathrin-coated pits (Ezekowitz *et al.*, 1990; Chen *et al.*, 1990; Jing *et al.*, 1990). In addition, several Ser and Thr residues exist as potential phosphorylation sites. The ability of the mannose receptor to mediate phagocytosis (Ezekowitz *et al.*, 1991) and induce lysosomal enzyme secretion (Shepherd *et al.*, 1985; Oshumi and Lee, 1987) indicates the existence of some signal-transducing capability. However, this aspect of mannose receptor function has not been established.

The mannose receptor clone has proven functional through studies on transfected cells. In a stably transfected rat fibroblast line, the mannose receptor conferred upon the cells the ability to endocytose and degrade [^{125}I]mannose-BSA specifically (Taylor *et al.*, 1990). COS-1 cells have also been transfected transiently with cDNA encoding the receptor; these cells were capable of internalizing both soluble and particulate mannose receptor ligands (Ezekowitz *et al.*, 1990, 1991). This finding has been critical to the confirmation of mannose receptor-mediated phagocytosis and distinguishes the mannose receptor from other carbohydrate recognition systems that carry out the internalization of only soluble substrates. In addition, the functional expression of a phagocytic receptor in a nonphagocytic cell will be significant to the general study of phagocytic mechanisms.

IV. Mannose Receptor Biosynthesis

A. Synthesis and Maturation of Oligosaccharide Side Chains

As an integral membrane protein (Wileman *et al.*, 1986), the mannose receptor is synthesized in the rough endoplasmic reticulum (RER) and transported through the Golgi en route to the cell surface. The biosynthesis of the mannose receptor has been studied in cultured human monocyte-derived macrophages (Lennartz et al., 1989; S. E. Pontow and P. D. Stahl, manuscript in preparation) and in fibroblasts transfected with the human receptor gene (unpublished observations). Typically, metabolic pulse labeling followed by immunoprecipitation and SDS-PAGE–autoradiography has been employed to characterize the maturation of mannose receptor structure. The results have revealed that newly synthesized receptor undergoes a characteristic shift in electrophoretic mobility that corresponds to modifications involving both N- and O-linked oligosaccharide chains (Lennartz et al., 1989). Of the other possible modifications that membrane glycoproteins can undergo, only proteolytic processing has been ruled out. Following a short pulse of [^{35}S]Met and SDS-PAGE, isolated mannose receptor migrates as a single band of M_r 154 kDa. After a chase period of 15 minutes, a second band of M_r 162 kDa appears, and the intensity of the 154-kDa protein is decreased. By 60 minutes of chase, all labeled mannose receptor exhibits an M_r of 162 kDa.

Lennartz et al. (1989) utilized tunicamycin and a battery of glycosidases to determine the relative contribution of N-linked sugars to the observed decrease in electrophoretic mobility. Synthesis of the receptor in the presence of tunicamycin produced a 150-kDa receptor that apparently lacked both N- and O-linked oligosaccharides. Because tunicamycin acts as an acceptor for the initial glycosylation reaction of the N-linked pathway, it is possible that N-linked sugars are prerequisite to the exit of the mannose receptor from the RER and subsequent O-glycosylation. Enzymatic removal of total N-linked sugars with peptide:N-glycanase reduced the M_r of the 154-kDa band to 150 kDa, whereas the 162-kDa band shifted an apparent 8 kDa in molecular weight. Endo H, which specifically cleaves N-linked sugars of the high-mannose and hybrid types (Kornfeld and Kornfeld, 1985), produced a 150-kDa band when incubated with the receptor precursor but was unable to shift the higher molecular weight band. The development of Endo H resistance coincides kinetically with the receptor's characteristic shift in electrophoretic mobility, which serves as a marker for the transport of the receptor to the medial Golgi stacks and occurs with a half-time of approximately 45 minutes (Lennartz et al., 1989). Together, these results indicate that Asn-linked oligosaccharides account for 8 kDa of the receptor's apparent molecular weight and that these oligosaccharide chains are of the complex type.

The presence of O-linked sugars on the mannose receptor was demonstrated by binding of neuraminidase-treated mannose receptor to peanut agglutinin-agarose (Lennartz et al., 1989). The synthesis of the mannose receptor in the presence of the O-glycosylation inhibitor, phenyl-N-acetyl-galactosaminide (Kuan et al., 1989), produced a protein of M_r 156 (S. E. Pontow and P. D. Stahl, manuscript in preparation). The shift in electrophoretic mobility attributable to the presence of O-linked sugars is apparently due to charge differences resulting from terminal sialylation. Treatment of cells with either brefeldin A (Klausner et al., 1992) or monensin (Collier, 1975) prevents terminal sialylation of all oligosaccharide moieties by disrupting transport to and function of the late

Golgi, respectively. These agents produced a receptor of identical mobility to that of mannose receptor synthesized in phenyl-GalNAc-treated cells (S. E. Pontow and P. D. Stahl, manuscript in preparation). In addition, enzymatic digestion of O-linked sugars can be achieved by serial incubation with neuraminidase and O-glycanase. The mannose receptor's electrophoretic mobility is shifted approximately 4 kDa by neuraminidase treatment but is not altered further during the subsequent O-glycanase digestion (Lennartz *et al.*, 1989).

B. Maturation of Binding Activity

The kinetics with which the mannose receptor traverses the biosynthetic pathway ($\tau_{1/2}$ = 45 minutes) are similar to that of other membrane glycoproteins (Lodish, 1988), which indicates that the receptor is not retained within the ER following synthesis (Lennartz *et al.*, 1989). However, the early compartments of the biosynthetic pathway are rich in potential mannose receptor ligands because both resident and newly synthesized glycoproteins bear mannose-terminal chains (Brands *et al.*, 1985; Kornfeld and Kornfeld, 1985). Binding ligand intracellularly could potentially disrupt mannose receptor maturation and/or function at the cell surface. We have now shown that the mannose receptor is synthesized as an inactive precursor that develops binding activity following exit from the ER (Pontow *et al.*, 1990; S. E. Pontow and P. D. Stahl, manuscript in preparation).

By labeling the receptor with a short pulse of [^{35}S]Met and applying the labeled receptor to mannose-sepharose affinity columns, it is possible to follow the receptor's shift from the flow-through to the eluted fractions during a chase period. The mannose receptor develops binding activity with a half-time of approximately 40 minutes (Pontow *et al.*, 1990). Development of activity can be inhibited by treatment of cells with either carbonyl cyanide *m*-chlorophenylhydrazone (CCCP) or tunicamycin. However, maturation of binding capacity is independent of N-linked oligosaccharide processing, addition of O-linked sugars, and exposure to the *trans*-Golgi environment (S. E. Pontow and P. D. Stahl, manuscript in preparation). Although the exact nature of the activating event is not known, one candidate is the arrangement of disulfide bonds. Experiments are underway to test this and other hypotheses concerning the delayed activation of the mannose receptor.

V. Mannose Receptor in Endocytosis

The mannose receptor has provided a valuable tool for the study of receptor-mediated endocytosis, both in macrophages and in general. Work with the receptor helped to establish several principles that apply to receptor-mediated endocytosis, such as receptor recycling and the acid intracellular compartments necessary for receptor–ligand uncoupling, receptor–ligand cycling, or retroendocytosis and endosomal proteolysis. Rapid progress in these endeavors was made possible by the availability of two high-affinity ligands, β-glucuronidase and mannose-BSA. Utilized in conjunction with isolated macrophages, primary macrophage cultures or macrophage-like cell lines, these molecules have allowed a detailed characterization of mannose receptor-mediated endocytosis.

A. Binding

Receptor binding at the cell surface requires calcium and a neutral pH (Stahl *et al.*, 1980). Binding of labeled ligand can be inhibited by inclusion of calcium chelators, excess unlabeled ligand, or yeast mannan in the assay medium (Stahl *et al.*, 1978). High concentrations of D-mannose are also inhibitory, although low concentrations have been shown to enhance binding, apparently by increasing the affinity of the mannose receptor for its ligand (Hoppe and Lee, 1982). This result fits well with the recent detailing of the mannose receptor binding site, which apparently requires the occupation of several recognition domains for high-affinity binding (Taylor *et al.*, 1992).

B. Uptake and Receptor Recycling

Receptor internalization occurs through clathrin-coated pits and vesicles, independently of ligation (Tietze *et al.*, 1982). When the receptor is occupied, internalization stimulates a slow secretion of lysosomal enzymes by an unidentified pathway (Oshumi and Lee, 1987; Shepherd *et al.*, 1985). Initial experiments on the endocytosis of mannose-BSA revealed that alveolar macrophages could accumulate about 2×10^6 molecules of ligand per cell per hour. Ligand-binding studies performed at 4°C showed that these cells express about 100,000 receptors on their surfaces. Recycling of the mannose receptor between the plasma membrane and intracellular compartments was proposed to account for the difference between the amount of ligand endocytosed and the number of surface receptors (Stahl *et al.*, 1980). Experiments in which macrophages were trypsinized at 4°C indicated that an intracellular pool provided a continual supply of receptors to the cell surface. These cells retained about 80% of their mannose-specific endocytosis capacity, suggesting that 10–20% of the receptor activity was on the surface at any given time. Moreover, incubating trypsinized cells at 37°C for a few minutes resulted in nearly complete recovery of cell surface-binding capacity. The intracellular versus cell surface distribution of the mannose receptor was confirmed by permeabilization of macrophages with saponin, which exposed an intracellular pool of binding sites about four-fold greater than the cell surface pool (Wileman *et al.*, 1984). Thus, a total pool of 500,000 receptors mediated the binding and internalization of 2×10^6 ligand molecules per hour or 4 ligand molecules per receptor per hour. For a recycling receptor this computes into a complete cycle every 15 minutes. Trypsinization of alveolar macrophages at 37°C for 10–15 minutes resulted in complete loss of receptor activity, consistent with the conclusion that all of the receptors are re-cycling within this time frame (Stahl *et al.*, 1980). Because biosynthetic and turnover studies have shown that the mannose receptor has a half-life of 12–30 hours, depending on the species (Lennartz *et al.*, 1989; unpublished observa-tion), a given mannose receptor must recycle hundreds of times during an aver-age lifetime. Thus, the mannose receptor, like the transferrin receptor and the low density lipoprotein (LDL) receptor, is an efficient molecule for internalizing ligand.

C. Endosomal Acidification

The possible role of acid intracellular compartment as determinants of receptor recycling was first shown by Tietze *et al.* (1980) in experiments using chloro-quine. Briefly, these studies showed that weak bases blocked ligand accumula-tion by macrophages. The weak base did not affect internalization per se because initial rates of ligand uptake were unaffected. Rather, the subsequent re-cycling of the receptor was impaired by the drug. Importantly, the presence of ligand was not required for the weak base to produce its effect because incuba-tion of cells with weak base in the absence of ligand resulted in a dramatic but reversible loss of cell surface receptors. Although direct evidence for an acid in-tracellular prelysosomal compartment did not exist at the time, these studies pointed out the importance of acidification for receptor recycling. Subsequently, Tietze *et al.* (1982) showed that weak bases block receptor–ligand dissociation, promoting accumulation of receptors within the endosomal compartment. Un-coupling of receptor and ligand was found to require the activity of an adeno-sine triphosphate (ATP)-dependent proton pump (Wileman *et al.*, 1985). The experiments with weak bases also revealed the presence of a hitherto unde-scribed process, referred to as receptor–ligand cycling or retroendocytosis, which occurs in cells in the absence of weak bases. The molecular basis of this process, which has now been described in a number of systems (Simmons and Schwartz, 1984; Townsend *et al.*, 1984; Aulinskas *et al.*, 1981), is poorly un-derstood. It may reflect some inefficiency in the endocytic system or it may be due to parallel internalization pathways that have different intracellular itiner-aries. The latter is intriguing in relation to the recent work of Tabas *et al.* (1991) regarding the endocytosis of lipoproteins.

D. Endosomal Proteolysis

Early studies with the mannose receptor and the ligand mannose-BSA showed that degradation of this ligand was occurring well before the ligand was transferred to lysosomes, traditionally thought to be the site where proteolysis is initiated. Diment and Stahl (1986) showed that proteolysis of mannose BSA was occurring within endosomes and that such endosomes are enriched in cathepsin D. Work by others has confirmed these findings (Rijnboutt *et al.*, 1991). Endosomal proteolysis may play an important role in hormone, antigen, or toxin processing (Diment *et al.*, 1989; Blum *et al.*, 1991a). The origin of endosomal proteases and the mechanism of their targeting to the endosome remain unclear. Equally undefined is the mechanism by which the mannose receptor and other receptors protect themselves from the effects of such endosomal proteases as the receptors pass through the recycling pathway.

VI. Specificity of the Mannose Receptor

The results from early *in vivo* experiments suggested a role for oligosaccharides in plasma clearance of lysosomal hydrolases, and defined the determinants for their specific recognition (Stahl *et al.*, 1976b,c; Achord *et al.*, 1977a, 1978). Several assays developed for the study of this system *in vitro* further delineated the sugars involved in the uptake of lysosomal glycosidases and mannose/fucose/N-acetylglucosamine-terminating glycoproteins. Utilizing β-glucuronidase, isolated glycoproteins, or synthetic glycoconjugates as ligands in studies on binding and uptake by rat or rabbit alveolar macrophages revealed the following consensus for the decreasing order of affinity of this receptor system: L-Fuc = D-Man > D-GlcNAc ≥ D-Glc > D-Xyl >>> D-Gal (Stahl *et al.*, 1978; Shepherd *et al.*, 1981; Oshumi *et al.*, 1988). Similar results were obtained when adhesion of rabbit alveolar macrophages to mannose-polyacrylamide derivatized microplates (Largent *et al.*, 1984) or binding of purified human placental mannose receptor to mannan-coated microplates (Kery *et al.*, 1992) was inhibited by monosaccharides.

The question of how one receptor can mediate recognition of multiple sugars has recently been addressed by the expression of putative mannose receptor carbohydrate-binding domains in a cell-free system, followed by affinity chromatography of the truncated polypeptides (Taylor *et al.*, 1992). Of the eight potential binding sites (Fig. 1), only 4–8 showed affinity for carbohydrate, and only one, CRD 4, exhibited multiple specificities. The order of affinity for this binding domain was shown to be L-Fuc > D-GlcNAc > D-Man >> Glc >> Gal. The authors noted that binding of the receptor to glycoproteins cannot be mediated by CRD 4 alone; however, high-affinity interactions required translation of at least three CRDs.

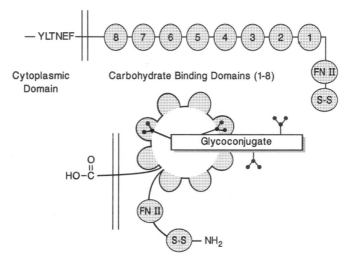

FIG. 1 Schematic representation of mannose receptor structure and ligand binding. Structural features important to mannose receptor function have been described (Taylor *et al.*, 1990, 1992) and include the following: Upper panel: The C-terminal cytoplasmic tail contains the signal sequence FENTLY for internalization through clathrin-coated pits. Eight repeats resembling carbohydrate recognition domains of C-type animal lectins comprise the majority of the extracytoplasmic portion of the receptor. The remainder consists of two distinct regions of unknown function. Lower panel: High affinity binding of ligand by the mannose receptor requires the participation of more than one CRD. In this model binding of a multivalent ligand occurs through interactions with CRDs 4 and 7.

The prerequisite of multiple recognition sites for high-affinity binding by the mannose receptor correlates with available data on receptor–ligand interactions of the mannose receptor and other carbohydrate receptors, as well. With the exception of the mannose receptor, all sugar recognition systems that have been cloned have been shown to possess a single binding site per polypeptide chain (Drickamer, 1988). These receptors require oligomerization prior to high-affinity binding. Because the mannose receptor structure contains several potential binding sites, binding of polyvalent ligands may be affected by single-receptor molecules. Studies on ligand interactions of purified mannose receptor with linear or branched mannose-oligosaccharides have revealed that the maximal chain length for binding to a single receptor molecule is 3 units and that branched oligosaccharides are better ligands than linear chains (Kery *et al.*, 1992). These results coincide with those obtained through studies on the binding domain of rat liver mannose binding protein C (Childs *et al.*, 1990), which interacts with the trimannosyl core of N-linked oligosaccharides. Additional ligands for this protein may be fucose-terminal oligosaccharides or peripheral mannose residues of high-mannose type sugars. In the binding of neoglycoproteins to alveolar

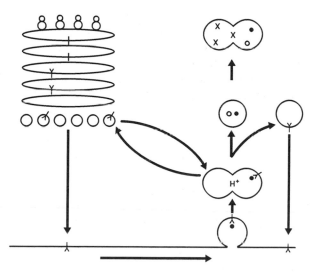

FIG. 2 Itinerary of the mannose receptor in macrophages. The mannose receptor is synthesized in the rough endoplasmic reticulum as an interactive precursor (I; Pontow *et al.*, 1990). During passage through the Golgi, oligosaccharides on the receptor are processed and the receptor is activated (Y). These two events appear to be unrelated. The active receptor passes to the cell surface where it has access to mannosylated ligands (●). Whether newly synthesized receptor is transported directly to the cell surface or to endosomes from the *trans*-Golgi network is not known. The mannose receptor is internalized, independently of ligation, through coated pits into an acidic endosomal compartment. Here receptor–ligand separation occurs and ligand proteolysis (○) is initiated by endosomal cathepsins (Blum *et al.*, 1991). In a series of fusion-dependent steps, mannosylated ligands are transferred to multivesicular bodies (Harding *et al.*, 1984) and subsequently, to the lysosomal compartment containing a full complement of hydrolases (X) where complete degradation occurs.

macrophages, sugar density was found to be an important determinant of affinity, with the tightest binding occurring when BSA was conjugated to ≥ 40 mannose residues (Hoppe and Lee, 1983).

VII. Mannose Receptor in Phagocytosis

Phagocytosis of unopsonized particles was once considered a nonspecific process but is now thought to occur through specific recognition systems, such as those for particular terminal sugar residues (Silverstein *et al.*, 1989). The mannose receptor has been implicated in the uptake of various unopsonized targets, including yeasts (Warr, 1980), such as *Candida albicans* (Karbassi, *et al.* 1987; Marodi *et al.*, 1991), zymosan (Sung *et al.*, 1983), *Leishmania promastigotes* (Blackwell *et al.*, 1985; Russell and Wilhelm, 1986; Wilson and

Pearson, 1986, 1988), and *Pneumocystis carinii* (Ezekowitz *et al.*, 1991). Several studies have shown that the ingestion of these particles can be significantly inhibited by mannose, yeast mannans, and mannose-terminal glycoconjugates. In addition, culturing macrophages on immobilized zymosan inhibits ingestion of free zymosan and of mannose-BSA (Berton and Gordon, 1983). The cDNA for the mannose receptor was expressed in COS-1 cells and shown to be capable of mediating the uptake of yeast (Ezekowitz *et al.*, 1990), and *Pneumocystis carinii* (Ezekowitz *et al.*, 1991) by these nonphagocytic cells. Assessing the precise role of the mannose receptor in the uptake of particles bearing oligosaccharide structures has been difficult, however. Macrophages express the complement receptor type three, CR3, specific for C3bi, a cleavage product of the third component of complement (Unkeless and Wright, 1988). This receptor also possesses a second binding site, which recognizes phospho-sugars and apparently contributes to the phagocytosis of carbohydrate-coated particles (Ross *et al.*, 1985). Zymosan is readily phagocytosed by monocytes, which do not express the mannose receptor (Ezekowitz *et al.*, 1983). In addition, phospho-mannans are as effective as dephosphorylated mannans at reducing zymosan uptake by macrophages (Sung *et al.*, 1983). Nearly complete inhibition of zymosan ingestion can be achieved by combining yeast mannans with a monoclonal antibody directed against the carbohydrate-binding site of CR3 (Wilson and Pearson, 1988). Another complicating factor is that human monocyte-derived macrophages have been shown to secrete enough complement components to opsonize previously unopsonized targets (Ezekowitz *et al.*, 1983). This could allow uptake via the C3bi-binding site of CR3. Thus, although participation of the mannose receptor in phagocytosis appears likely, an actual role for the receptor and contributions made by other receptors require further delineation.

The study of mannose receptor participation in the binding and ingestion of particles opens a myriad of experimental avenues. For example, if the mannose receptor can mediate both endocytosis and phagocytosis, does more than one form of the receptor exist or is a single receptor capable of generating an endocytic and a phagocytic signal? Perhaps the ligand, rather than the receptor, carries the information specifying the type of internalization required. Studies on both the CR3 (Graham *et al.*, 1989) and the FcR (Miettinen *et al.*, 1989) suggest that receptor populations can be divided into functional and structural subtypes. The relevance of these findings to the regulation of endocytic and phagocytic processes is unknown. The ability of the mannose receptor to transduce a signal is evident in the stimulation of lysosomal enzyme secretion (Oshumi and Lee, 1987) and the induction of tumor necrosis factor and macrophage cytotoxicity (Lefkowitz *et al.*, 1991) by mannose receptor ligands. The ability of the mannose receptor to stimulate secretion and respiratory burst during phagocytosis needs to be assessed.

The fate of receptors and other proteins that enter phagosomes is another aspect of phagocytosis requiring further attention. Mellman *et al.* (1983) showed that FcRs internalized during phagocytosis of IgG-coated beads were rapidly degraded and that FcR activity at the cell surface decreased accordingly. These findings suggest that participation in phagocytosis may represent a mode of down-regulation, which is dictated by the presence of ligand. However, the mannose receptor apparently cycles out of phagosomes rather than undergoing proteolysis during phagocytosis (Pitt *et al.*, 1992). This receptor is not degraded in response to entry into the endosomal apparatus either, despite the presence of proteases within early endosomes (Diment and Stahl, 1986). Consequently, the mannose receptor must be resistant to the hydrolase activity within endosomes and/or be rapidly sorted away from these protease-positive compartments.

VIII. Physiological Functions of the Mannose Receptor

Two major physiologic functions have been proposed for the mannose receptor: molecular scavenging and host defense (Stahl, 1990). The former involves the clearance of mannose-terminating glycoproteins and peptides from the plasma or the extracellular medium. The presence of mannose receptors on both liver Kupffer cells and tissue macrophages is uniquely suited to the purpose of clearing mannose-terminal ligands, which occur naturally in either the circulation or at sites of inflammation. Early *in vivo* investigations, which followed the fate of injected lysosomal hydrolases, implicated the mannose receptor in the removal of these glycoproteins from the plasma by liver Kupffer cells. Shepherd and Hoidal (1990) revealed that enzymes secreted by polymorphonuclear leukocytes are taken up via the mannose receptor of neighboring macrophages. Therefore, one consequence of mannose receptor function may be protection from extensive tissue damage, which could result from the actions of secreted enzymes during an inflammatory reaction. Other physiologic ligands include tissue plasminogen activator (Smedsrod *et al.*, 1988; Smesrod and Einarsson, 1990) and circulating C-terminal propeptide of Type I procollagen (Smesrod *et al.*, 1990).

A second role proposed for the mannose receptor is the pre-immune clearance of various microorganisms. Many pathogens are incapable of processing their N-linked oligosaccarides or express complex carbohydrates, such as mannans or glucans, and therefore bear mannose and glucose residues on their surfaces for recognition by the mannose receptor. In this capacity the mannose receptor may act as the first defense against invading organisms, functioning prior to that time when more complex forms of immune clearance can be implemented. One potential problem in defining the role for the mannose receptor in host defense is the apparent redundancy of the immune system; deficient mannose receptor activity could be compensated for by serum mannose-binding proteins (Kuhlman *et al.*, 1989) or by CR3 activity (Ezekowitz *et al.*, 1983). However, the regulation of mannose receptor expression by immunomodulators indicates that it may play a significant role in host defense.

IX. Regulation of Mannose Receptor Expression

Macrophages are stimulated to undergo a variety of metabolic changes during maturation and activation; each phase in the life of a macrophage is character-ized by a distinct set of capabilities and functions. The mannose receptor may be useful as a marker for certain macrophage states because its expression is clearly regulated by mediators of macrophage differentiation and immune func-tion, both *in vitro* and *in vivo*. For example, resident tissue macrophages and macrophages elicited by sterile inflammation express mannose receptor activity, although other properties of these two cell types differ (Imber *et al.*, 1982). Macrophages elicited through infection exhibit further changes in their capabil-ities, including a loss of mannose receptor activity. In general, differentiation and possibly priming promote mannose receptor expression, whereas macro-phage activation and exposure to certain products of inflammatory cells down-regulate its activity.

A. Macrophage Differentiation

Macrophage differentiation has profound effects on mannose receptor expres-sion. Peripheral blood monocytes, the precursor to tissue macrophages, do not express mannose receptor protein or activity when freshly isolated (Shepherd *et al.*, 1982). However, receptor synthesis commenses within hours of cell plating (Lennartz *et al.*, 1989). The cells are able to internalize soluble mannose recep-tor ligands after 2 days in either suspended or adherent culture, and this capa-bility peaks 4–5 days postplating (Shepherd *et al.*, 1982). The basis for this regulatory action remains undefined, although its reflects influences on receptor biosynthesis (Lennartz *et al.*, 1989; unpublished observations).

Factors that promote macrophage maturation have been shown to specifically up-regulate mannose receptor expression in mouse bone marrow-derived macrophages. These agents include glucocorticoids (Shepherd *et al.*, 1985; Cowan and Shepherd, 1989), prostaglandin E (Schreiber *et al.*, 1990), mono-meric IgG2a (Schreiber *et al.*, 1991a), and 1,25-dihydroxyvitamin D_3 (Clohisy *et al.*, 1987). The mechanism involves an enhanced rate of receptor synthesis at early time points; the maximum amount of mannose receptor activity expressed is not altered by the experimental treatments. It is likely that this form of regulation occurs at the level of mRNA transcription. The mannose receptor gene must therefore contain elements that allow for its regulation by various biological mediators.

B. Macrophage Activation

Agents that activate macrophages, such as IFN-γ (Harris *et al.*, 1991; S. Schreiber, personal communication), lipopolysaccharide and phorbol esters (Shepherd *et al.*, 1990), aggregated IgG (Clohisy *et al.*, 1987, 1989; Schreiber *et al.*, 1991b), or infection by bacillus Calmette-Guerin (Ezekowitz *et al.*, 1981; Imber *et al.*, 1982), or *Leishmania* (Basu *et al.*, 1991) selectively down-regulate mannose receptor expression. The end result in each case is a diminution of total available binding sites, and new protein synthesis is required for recovery of activity, at least in the case of *Leishmania* infection (Basu *et al.*, 1991). It should be noted that significant mannose receptor activity does remain in the activated macrophage, however. In the aforementioned studies, 10–30% of the mannose receptor activity was retained. In addition, Shepherd *et al.* (1983) were able to study the accessibility of parasitophorous vacuoles to β-glucuronidase entering *Leishmania*-infected macrophages via the mannose receptor. Whether this residual ability to internalize mannose receptor ligands fulfills some physiological function is unknown. The mechanism responsible for diminished mannose receptor expression in activated macrophages may differ depending on the stimulant. Murine (S. Schreiber *et al.*, submitted) and human (unpublished observation) macrophages cultured in the presence of IFN-γ synthesize little mannose receptor, as assessed by metabolic radiolabeling and immunoprecipitation. In the mouse cell line J774E, decreased mannose receptor synthesis in response to IFN-γ has been shown to reflect decreased steady-state levels of receptor mRNA (Harris *et al.*, 1991). In contrast, Shepherd *et al.* (1990) found that the irreversible loss of mannose receptor activity in response to macrophage activation by lipopolysaccharide and Phorbol 12-myristate 13-acetate (PMA) may be the result of receptor inactivation rather than diminished biosynthesis. The authors proposed two explanations. One possible scenario is that lipopolysaccharide and PMA trigger a cascade of phosphorylation events wherein the mannose receptor itself is phosphorylated, causing direct inactivation. Although it is possible that phosphorylation of the receptor or other substrates plays a role in mannose receptor regulation, the time course of inactivation by lipopolysaccharide and PMA favors the second model. This proposed mechanism is indirect and is based on the observation that macrophages treated with lipopolysaccharide and PMA exhibit increased synthesis of proteins bearing high-mannose type oligosaccharide chains, which could compete with exogenous ligand for mannose receptor binding.

Evidence for overproduction of endogenous ligands as a means of mannose receptor regulation is derived from experiments that tested the effect of glycosylation inhibitors on mannose receptor function (Arumugham and Tanzer, 1983; Chung *et al.*, 1984). Treatment of cells with swainsonine, an inhibitor of α-mannosidase II, causes the accumulation of membrane glycoproteins bearing high-mannose and hybrid carbohydrates (Elbein, 1987) and specifically down-regulates mannose receptor activity in macrophages (Arumugham and Tanzer, 1983; Chung *et al.*, 1984). Similar results are obtained using the glucosidase inhibitor, castanospermine. Chung *et al.* (1984) suggested that the swainsonine-induced impairment was due to occupation of the mannose receptor by neighboring membrane glycoproteins bearing truncated oligosaccharides. Recovery of mannose receptors cell surface activity following swainsonine treatment required about 24 hours and new protein synthesis (Chung *et al.*, 1984), although mannose receptor synthesis (S. E. Pontow and P. D. Stahl, manuscript in preparation) and turnover apparently continue at the same rate (unpublished observation).

X. Selective Delivery to Macrophages via the Mannose Receptor

The mannose receptor, unlike other receptors involved in host defense, is relatively specific to macrophages and certain macrophage-like cells. As such, this receptor offers unique opportunities for selective targeting of therapeutic and other agents to macrophages. Examples of application include the eradication of organisms that thrive within macrophages, such as *Myobacterium tuberculosis, Myobacterium leprae, Legionella,* and *Leishmania,* by targeting drugs that may be contraindicated due to systemic toxicity. Targeting of anti-inflammatory agents may be useful in allergic diseases, especially when delivering aerosolized steroids to pulmonary macrophages in children and adolescents in which systemic distribution is associated with impaired growth. The activation of macrophages within tumors via the selective delivery of activating reagents or, following the removal of tumors, the widespread activation of macrophages to prevent metastasis may have considerable merit. Finally, the selective delivery of antiviral agents to mononuclear phagocytes harboring viruses, such as HIV, may provide an excellent collaterals therapy for treatment of diseases, such as AIDS. Progress is currently underway in the development of carriers for targeting via the mannose receptor.

A. Mannosylated Liposomes

Capitalizing on the phagocytic properties of the mannose receptor, several groups have described the generation of mannose-coated liposomes. Different approaches for synthesis of these carriers include mannosylation of preformed liposomes (Yamashita *et al.*, 1991; Muller and Schuber, 1989; Van Rooijen, 1989), formation of liposomes from mannosylated phospholipids (Barratt *et al.*, 1986, 1987), or exposing a mannose-terminal glycoprotein, such as the major surface glycoprotein of *Leishmania* (Russell and Wilhelm, 1986) or BSA (Garcon *et al.*, 1988) at the liposome surface. Mannosylated liposomes have been shown to undergo mannan-inhibitable interactions with macrophages and to selectively deliver encapsulated agents. For example, mannosylated vesicles loaded with macrophage activators were shown to increase tumoricidal activity of macrophages *in vitro* and *in vivo* (Barratt *et al.*, 1987; Dumont *et al.*, 1990; Yamashita *et al.*, 1991). These treatments were effective in stimulating regression of experimental solid tumors and reducing lung metastases in mice. Umezawa and Eto (1988) reported that mannosylated liposomes efficiently crossed the blood–brain barrier in mice. Exploiting this selective passage, Huitinga *et al.* (1990) eliminated infiltrating macrophages from the brains of rats with experimental allergic encephalomyelitis by injecting mannosylated liposomes containing dichloromethylene diphosphonate. A marked suppression of clinical signs of the disease, which is characterized in part by influx of macrophages into the central nervous system, reportedly followed infusion of the liposomes. Another application of mannose receptor-targeted vesicles was devised by Garcon *et al.* (1988), who observed that production of antitoxoid antibodies following injection of mice with tetanus toxoid-loaded vesicles increased when the vesicles were coupled with mannose-BSA.

B. Soluble Ligands

A second experimental design for macrophage targeting employs the mannose receptor in mediating the endocytosis of soluble molecules. The toxic subunit of ricin toxin, A chain, is a ligand for the mannose receptor and has been shown to intoxicate macrophages *in vitro* (Simmons, 1986; Skilleter and Foxwell, 1986), an event that may be dependent upon endosomal proteolysis (Fiani, M. L., manuscript in preparation). By injecting A chain into rats, Simmons *et al.* (1987) were able to selectively deplete mannose receptor-bearing cells from the liver, as assessed by morphology and by the clearance of infused β-glucuronidase.

A second carbohydrate-based delivery system utilizes enzymatic deglycosylation to generate glycoproteins with specific terminal sugars exposed. This approach has been successfully exploited by Brady and colleagues in the clinical treatment of Gaucher disease, a prevalent sphingolipid storage disorder resulting from a deficiency of glucocerebrosidase, especially in cells of the reticuloendothelial system. Purified human placental glucocerebrosidase was injected into patients in an attempt to supplement endogenous enzyme levels, but was not effective in reducing the clinical symptoms unless mannose residues first were exposed by sequential enzymatic digestion of the oligosaccharide side chains (Barton, *et al.*, 1990; 1991). Mannose-terminating glucocerebrosidase was more efficiently recognized and internalized than the native enzyme.

A final method for targeting soluble molecules is to covalently couple drugs or other agents to mannose-BSA. Roche *et al.* (1985) successfully treated metastases in tumor-bearing mice using muranyl dipeptide, a general macrophage activator, linked to mannose-BSA. Utilizing methotrexate-conjugated mannose-BSA, Chakraborty *et al.* (1990) significantly inhibited the growth of *Leishmania donovani* within macrophages *in vitro*. Intravenous injection of the conjugate reduced by 85% the number of parasites in the spleens of infected mice.

XI. New Directions

The studies described in this chapter have centered on the isolation and characterization of the mannose receptor protein, the cell biology of its involvement in endocytosis and phagocytosis, and its use as a marker for mononuclear phagocyte differentiation. Future mannose receptor research will attempt to define a physiological role for this protein. Working hypotheses concerning mannose receptor function have assumed that the mannose receptor, as a macrophage-specific protein, is primarily involved in scavenging microorganisms and glycoproteins bearing mannose-terminal oligosaccharides. Recent studies indicate that expression of the mannose receptor may not be restricted to macrophages, however. The presence of mannose receptor activity on cells, such as osteoclasts, tracheal myocytes, endothelial cells, and dendritic cells, suggests that this may be a multifunctional receptor system. For example, mannosylated ligands or other ligands that bind uncharted regions of the receptor may generate intracellular signals that are not associated with endocytic or phagocytic processes. In addition, the mannose receptor may function in oligosaccharide-mediated cell-to-cell interactions, which, in turn, may play a role in processes, such as the immune response. The availability of the cDNA encoding the receptor, specific antibodies, and a variety of mannose receptor ligands should allow critical questions of function to be addressed in the near future.

Acknowledgment

The authors acknowledge the generous and insightful contributions made by our collaborators over the past two decades of mannose receptor research: B. Mandell, D. Achord, W. Sly, O. Touster, J. Rodman, P. Schlesinger, S. Lang, T. Doebber, R. Townsend, Y. C. Lee, C. Tietze, S. Gordon, A. Ezekowitz, R. Senior, M. Rabinovitch, V. L. Shepherd, M. Konish, T. Wileman, R. Boshans,

K. N. Chung, S. Diment, M. Zenilman, J. Heuser, M. Levy, C. Harding, B. Simmons, J. Russell, M. Lennartz, F. S. Cole, L. Mayorga, M. Dean, R. Diaz, M. Colombo, K. Drickamer, S. Schreiber, M. Fiani, J. Blum, S. Teitelbaum, A. L. Schwartz, A. Pitt, J. Lenhard, M. Koval, L. Petus, C. Adles, G. P. Li, and J. T. Conary.

References

Achord, D. T., Brot, F. E., and Sly, W. S. (1977a). *Biochem. Biophys. Res. Commun.* **77**, 409–415.

Achord, D., Brot, F., Gonzalez-Noriega, A., Sly, W., and Stahl, P. D. (1977b). *Pediatr. Res.* **11**, 816–822.

Achord, D. T., Brot, F. E., Bell, C. E., and Sly, W. S. (1978). *Cell* **15**, 269–278.

Arumugham, R. G., and Tanzer, M. L. (1983). *Biochem. Biophys. Res. Commun.* **116**, 922–930.

Ashwell, G., and Morell, A. G. (1974). *Adv. Enzymol.* **41**, 99–128.

Aulinskas, T. H., Van Der Westhuyzen, D. R., Bierman, E. L., Gevers, W., and Coetzee, G. A. (1981). *Biochim. Biophys. Acta* **664**, 255–265.

Barratt, G. M., Nolibe, D., Yapo, A., Petit, J. F., and Tenu, J. P. (1987). *Ann. Inst. Pasteur Immunol.* **138**, 437–450.

Barratt, G. M., Tenu, J. P., Yapo, A., and Petit, J. F. (1986). *Biochim. Biophys. Acta* **862**, 153–164.

Barton, N. W., Brady, R. O., Dambrosia, J. M., Di Bisceglie, A. M., Doppelt, S. H., Hill, S. C., Mankin, H. J., Murray, G. J., Parker, R. I., Argoff, C. E., Grewal, R. P., Yu, K.-T. *et al.* (1991). *N. Engl. J. Med.* **324**, 1464–1468.

Barton, N. W., Furbish, F. S., Murray, G. J., Garfield, M. and Brady, R. O. (1990). *Proc. Natl. Acad. Sci. USA* **87**, 1913–1916.

Basu, N., Sett, R., and Das, P. K. (1991). *Biochem. J.* **277**, 451–456.

Baynes, J. W., and Wold, F. (1976). *J. Biol. Chem.* **251**, 6016–6024.

Berton, G., and Gordon, S. (1983). *Immunology* **49**, 705–715.

Blackwell, J. M., Ezekowitz, R. A. B., Roberts, M. B., Channon, J. Y., Sim, R. B., and Gordon, S. (1985). *J. Exp. Med.* **162**, 324.

Blum, J. S., Fiani, M. L., and Stahl, P. D. (1991a). *J. Biol. Chem.* **266**, 22091–22095.

Blum, J. S., Stahl, P. D., Diaz, R., and Fiani, M. L. (1991b). *Carbohydr. Res.* **213**, 145–153.

Brands, R., Snider, M. D., Hino, Y., Park, S. S., Gelboin, H. V., and Rothman, J. E. (1985). *J. Cell Biol.* **101**, 1724–1732.

Chakraborty, P., Bhaduri, A. N., and Das, P. K. (1990). *Biochem. Biophys. Res. Commun.* **166**, 404–410.

Chen, W. J., Goldstein, J. L., and Brown, M. S. (1990). *J. Biol. Chem.* **265**, 3116–3123.

Childs, R. A., Feizi, T., Yuen, C. T., Drickamer, K., and Quensberry, M. S. (1990). *J. Biol. Chem.* **265**, 20770–20777.

Chung, K. N., Shepherd, V. L., and Stahl, P. D. (1984). *J. Biol. Chem.* **259**, 14637–14641.

Clohisy, D. R., Bar-Shavit, Z., Chappel, J. C., and Teitelbaum, S. L. (1987). *J. Biol. Chem.* **262**, 15922–15929.

Clohisy, D. R., Chappel, J. C., and Teitelbaum, S. L. (1989). *J. Biol. Chem.* **264**, 5370–5377.

Collier, R. J. (1975). *In* "The Specificity and Action of Animal, Bacterial and Plant Toxins, Receptors and Recognition Series B" (P. Cuatracasas, ed.), Vol 1, pp. 69–98. John Wiley and Sons, New York.

Cowan, H., and Shepherd, V. (1989). *Am. J. Respir. Dis.* **139**, A157.

DeDuve, C., Pressman, B. C., Gianetto, R., Appelmans, F., and Wattiaux, R. (1955). *Biochem. J.* **60**, 604–617.

Diment, S., and Stahl, P. D. (1986). *J. Biol. Chem.* **260**, 15311–15317.

Diment, S., Leech, M. L., and Stahl, P. D. (1987). *J. Leukocyte Biol.* **42**, 485–490.

Diment, S., Martin, K. J., and Stahl, P. D. (1989). *J. Biol. Chem.* **264**, 13403–13406.
Drickamer, K. (1988). *J. Biol. Chem.* **263**, 9557–9560.
Dumont, S., Muller, C. D., Schuber, F., and Bartholeyns, J. (1990). *Anticancer Res.* **10**, 155–160.
Elbein, A. D. (1987). *Annu. Rev. Biochem.* **56**, 497–534.
Ezekowitz, R. A. B., Austyn, J., Stahl, P. D., and Gordon, S. (1981). *J. Exp. Med.* **154**, 60–76.
Ezekowitz, R. A. B., Sastry, K., Bailly, P., and Warner, A. (1990). *J. Exp. Med.* **172**, 1785–1794.
Ezekowitz, A. B., Sim, R. B., Hill, M., and Gordon, S. (1983). *J. Exp. Med.* **159**, 244–260.
Ezekowitz, R. A. B., Williams, D. J., Koziel, H., Armstrong, M. Y. K., Warner, A., Richards, F. F., and Rose, R. M. (1991). *Nature* **351**, 155–158.
Garcon, N., Gregoriadis, G., Taylor, M., and Summerfield, J. (1988). *Immunology* **64**, 743–745.
Graham, I. L., Gresham, H. D., and Brown, E. J. (1989). *J. Immunol.* **142**, 2352–2358.
Haltiwanger, R. S., and Hill, R. L. (1986a). *J. Biol. Chem.* **261**, 15696–15702.
Haltiwanger, R. S., and Hill, R. L. (1986b). *J. Biol. Chem.* **261**, 7440–7444.
Harding, C., Levy, M. A., and Stahl, P. D. (1985). *Eur. J. Cell Biol.* **36**, 230–238.
Harris, N. S., Super, M., Chang, G., and Ezekowitz, R. A. B. (1991). *Am. Soc. Cell Biol.* Meeting abstract 1193, Boston.
Hoppe, C. A., and Lee, Y. C. (1982). *J. Biol. Chem.* **257**, 12831–12834.
Hoppe, C. A., and Lee, Y. C. (1983). *J. Biol. Chem.* **258**, 14194–14199.
Hubbard, A., Wilson, G., Ashwell, G., and Stukenbrok, H. (1979). *J. Cell Biol.* **83**, 47–64.
Huitinga, I., Van Rooijen, N., de Groot, C. J., Uitdehaag, B. M., and Dijkstra, C. D. (1990). *J. Exp. Med.* **172**, 1025–1033.
Imber, M. J., Pizzo, S. V., Johnson, W. J., and Adams, D. O. (1982). *J. Biol. Chem.* **257**, 5129–5135.
Jing, S., Spencer, T., Miller, K., Hopkins, C., and Trowbridge, I. S. (1990). *J. Cell Biol.* **110**, 283–294.
Karbassi, A., Becker, J. M., Foster, J. S., and Moore R. N. (1987). *J. Immunol.* **139**, 417–421.
Kery, V., Krepinsky, J. J. F., Warren, C. D., Capek, P., and Stahl, P. D. (1992). *Arch. Biochem. Biophys.* **298**, No. 1.
Klausner, R. D., Donaldson, J. G., and Lippincott-Schwartz, J. (1992). *J. Cell Biol.* **166**, 1071–1080.
Kornfeld, R., and Kornfeld, S. (1985). *Annu. Rev. Biochem.* **54**, 631–664.
Kuan, S. F., Byrd, J. C., Basbaum, C., and Young, Y. S. (1989). *J. Biol. Chem.* **264**, 19271–19277.
Kuhlman, M., Joiner, K., and Ezekowitz, R. A. B. (1989). *J. Exp. Med.* **169**, 1733–1745.
Largent, B. L., Walton, K. M., Hoppe, C. A., Lee, Y. C., and Schnaar, R. L. (1984). *J. Biol. Chem.* **259**, 1764–1769.
Lee, Y. C., Stowell, C., and Krantz, M. J. (1976). *Biochemistry* **15**, 3956–3962.
Lefkowitz, D. L., Mills, K., Castro, A., and Lefkowitz, S. S. (1991). *J. Leukocyte Biol.* **50**, 615–623.
Lennartz, M. R., Wileman, T. E., and Stahl, P. D. (1987). *Biochem. J.* **245**, 705–711.
Lennartz, M. R., Cole, F. S., Shepherd, V. L., Wileman, T. E., and Stahl, P. D. (1987b). *J. Biol. Chem.* **262**, 9942–9944.
Lennartz, M. R., Cole, F. S., and Stahl, P. D. (1989). *J. Biol. Chem.* **264**, 2385–2390.
Lew, D. B., and Rattazzi, M. C. (1991). *J. Clin. Invest.* **88**, 1969–1975.
Lew, D. B., Tran, N., and Rattazi, M. C. (1992). *Am. Pediatr. Soc./Soc. Pediatr. Res.* Spring Meeting.
Lodish, H. F. (1988). *J. Biol. Chem.* **263**, 2107–2110.
Lunney, J., and Ashwell, G. (1976). *Proc. Natl. Acad. Sci. U.S.A.* **73**, 341–343.
Magnusson, S., and Berg, T. (1989). *Biochem. J.* **257**, 651–656.
Mandell, B., and Stahl, P. D. (1977). *Biochem. J.* **164**, 549–556.
Marodi, L., Korchak, H. M., and Johnston, R. B. (1991). *J. Immunol.* **146**, 2783–2789.
Maynard, Y., and Baenziger, J. U. (1981). *J. Biol. Chem.* **256**, 8063–8068.
McLaughlin, B. J., Tarnowski, B. I., and Shepherd, V. L. (1987). *Prog. Clin. Biol. Res.* **247**, 243–257.
Mellman, I. S., Plutner, H., Steinman, R. M., Unkeless, J. C., and Cohn, Z. A. (1983). *J. Cell Biol.* **96**, 887–895.

Miettinen, H. M., Rose, J. K., and Mellman, I. (1989). *Cell* **58**, 317–327.

Muller, C. D., and Schuber, F. (1989). *Biochim. Biophys. Acta* **986**, 97–105.

Niesen, T. E., Alpers, D. H., Stahl, P. D., and Rosenblum, J. L. (1984). *J. Leukocyte Biol.* **36**, 307–320.

Novak, E. K., Baumann, H., Ovnic, M., and Swank, R. T. (1991). *J. Biol. Chem.* **266**, 6377–6380.

Oshumi, Y., and Lee, Y. C. (1987). *J. Biol. Chem.* **262**, 7955–7962.

Oshumi, Y., Chen, V. J., Yan, S. C., Wold, F., and Lee, Y. C. (1988). *Glucoconjugate J.* **5**, 99–106.

Pitt, A., Mayorga, L. S., Schwartz, A. L., and Stahl, P. D. (1992). *J. Biol. Chem.* **267**, 126–132.

Pontow, S. E., Blum, J. S., and Stahl, P. D. (1990). *Proceedings of the 1st Annual Midwest Regional Meeting of the ASCB*, p. 41. Chicago.

Rijnboutt, S., Kal, A. J., Geuze, H. J., Aerts, A., and Strous, G. J. (1991). *J. Biol. Chem.* **266**, 23586–23592.

Roche, A. C., Bailly, P., and Monsigny, M. (1985). *Invasion Metastasis* **5**, 218–232.

Ross, G. D., Cain, J. A., and Lachmann, P. J. (1985). *J. Immunol.* **134**, 3307–3315.

Russell, D. G., and Wilhelm, H. (1986). *J. Immunol.* **136**, 2613–2620.

Schlesinger, P., Rodman, J. S., Frey, M., Lang, S., and Stahl, P. D. (1976). *Arch. Biochem. Biophys.* **177**, 606–614.

Schlesinger, P. H., Doebber, T. W., Mandell, B. F., White, R., DeSchryver, C., Rodman, J. S., Miller, M. J., and Stahl, P. D. (1978). *Biochem. J.* **176**, 103–109.

Schlesinger, P. H., Rodman, J. S., Doebber, T. W., Stahl, P. D., Lee, Y. C., Stowell, C. P., and Kuhlenschmidt, T. B. (1980). *Biochem. J.* **192**, 597–606.

Schreiber, S., Blum, J. S., Chappel, J. C., Stenson, W. F., Stahl, P. D., Teitelbaum, S. L., and Perkins, S. L. (1990). *Cell Regul.* **1**, 403–413.

Schreiber, S., Blum, J. S., Stenson, W. F., MacDermott, R. P., Stahl, P. D., Teitelbaum, S. L., and Perkins, S. L. (1991a). *Proc. Natl. Acad. Sci. U.S.A.* **88**, 1616–1620.

Schreiber, S., Stenson, W. F., MacDermott, R. P., Chappel, J. C., Teitelbaum, S. L., and Perkins, S. L. (1991b). *J. Immunol.* **147**, 1377–1382.

Shepherd, V. L., Abdolrasulnia, R., Garrett, M., and Cowan, H. B. (1990). *J. Immunol.* **145**, 1530–1536.

Shepherd, V. L., and Hoidal, J. R. (1990). *Am. J. Respir. Cell. Mol. Biol.* **2**, 335–340.

Shepherd, V. L., Campbell, E. J., Senior, R. M., and Stahl, P. D. (1982). *J. Ret. Soc.* **32**, 423–431.

Shepherd, V. L., Konish, M. G., and Stahl, P. D. (1985). *J. Biol. Chem.* **260**, 160–164.

Shepherd, V. L., Lee, Y. C., Schlesinger, P. H., and Stahl, P. D. (1981). *Proc. Natl. Acad. Sci. U.S.A.* **78**, 1019–1022.

Shepherd, V. L., Stahl, P. D., Bernd, P., and Rabinovitch, M. (1983). *J. Exp. Med.* **157**, 1471–1482.

Shepherd, V. L., Tarnowski, B. I., and McLaughlin, B. J. (1991). *Invest. Ophthalmol. Vis. Sci.* **32**, 1779–1784.

Silverstein, S. C., Greenberg, S., Di Virgilio, F., and Steinberg, T. H. (1989). *In* "Fundamental Immunology" Second Edition. (W. P. Paul, ed.). Raven Press, New York.

Simmons, C. F., and Schwartz, A. L. (1984). *Mol. Pharmacol.* **26**, 509–519.

Simmons, B. M., Stahl, P. D., and Russell, J. H. (1986). *J. Biol. Chem.* **261**, 7912–7920.

Simmons, B., Stahl, P. D., and Russell, J. (1987). *Biochem. Biophys. Res. Commun.* **146**, 849–854.

Skilleter, D. N., and Foxwell, B. M. J. (1986). *FEBS Lett* **196**, 344–348.

Smedsrod, B., Einarsson, M., and Pertoft, H. (1988). *Thromb. Haemost.* **59**, 480–484.

Smedsrod, B., and Einarsson, M. (1990). *Thromb. Haemost.* **63**, 60–66.

Smedsrod, B., Melkko, J., Risteli, L., and Risteli, J. (1990). *Biochem. J.* **271**, 345–350.

Stahl, P. D. (1990). *Am. J. Respir. Cell. Mol. Biol.* **2**, 317–318.

Stahl, P., and Gordon, S. (1982). *J. Cell Biol.* **93**, 49–56.

Stahl, P. D., and Touster, O. (1971). *J. Biol. Chem.* **246**, 5398–5406.

Stahl, P. D., Mandell, B., Rodman, J. S., Schlesinger, P., and Lang, S. (1975). *Arch. Biochem. Biophys.* **170**, 536–546.

Stahl, P. D., Rodman, J. S., and Schlesinger, P. (1976a). *Arch. Biochem. Biophys.* **177**, 594–605.
Stahl, P. D., Rodman, J. S., Miller, M. J., and Schlesinger, P. H. (1978). *Proc. Natl. Acad. Sci. U.S.A.* **75**, 1399–1403.
Stahl, P. D., Schlesinger, P. H., Rodman, J. S., and Doebber, T. (1976b). *Nature* **264**, 86–88.
Stahl, P. D., Schlesinger, P. H., Sigardson, E., Rodman, J. S., and Lee, Y. C. (1980). *Cell* **19**, 207–215.
Stahl, P. D., Six, H., Rodman, J. S., Schlesinger, P. H., Tusliani, D. R. P., and Touster, O. (1976c). *Proc. Natl. Acad. Sci. U.S.A.* **73**, 4045–4049.
Stephenson, J. D., and Shepherd, V. L. (1987). *Biochem. Biophys. Res. Commun.* **148**, 883–889.
Stockert, R. J., Morell, G. A., and Scheinberg, I. H. (1976). *Biochem. Biophys. Res. Commun.* **68**, 988–993.
Summerfield, J. A., Vergalla, J., and Jones, E. A. (1982). *J. Clin. Invest.* **69**, 1337–1347.
Sung, S. S. J., Nelson, R. S., and Silverstein, S. C. (1983). *J. Cell Biol.* **96**, 160–166.
Tabas, I., Myers, J. N., Innerarity, T. L., Xu, X. X., Arnold, K., Boyles, J., and Maxfield, F. R. (1991). *J. Cell Biol.* **155**, 1547–1560.
Tarnowski, B. I., Shepherd, V. L., and McLaughlin, B. J. (1988). *Invest. Ophthalmol. Vis. Sci.* **29**, 742–748.
Taylor, M. E., Conary, J. T., Lennartz, M. R., Stahl, P. D., and Drickamer, K. (1990). *J. Biol. Chem.* **265**, 12156–12162.
Taylor, M. E., Bezouska, K., and Drickamer, K. (1992). *J. Biol. Chem.* **267**, 1719–1726.
Tietze, C., Schlesinger, P., and Stahl, P. (1980). *Biochem. Biophys. Res. Commun.* **93**, 1–8.
Tietze, C., Schlesinger, P., and Stahl, P. D. (1982). *J. Cell Biol.* **92**, 417–424.
Townsend, R., and Stahl, P. D. (1981). *Biochem. J.* **194**, 209–214.
Townsend, R., and Stahl, P. D. (1981). *Biochem. J.* **194**, 209–214.
Townsend, R. R., Wall, D. A., Hubbard, A. L., and Lee, Y. C. (1984). *Proc. Natl. Acad. Sci. U.S.A.* **81**, 466–470.
Umezawa, F., and Eto, Y. (1988). *Biochem. Biophys. Res. Commun.* **153**, 1038–1044.
Unkeless, J. C., and Wright, S. D. (1988). *In* "Inflammation, Basic Principles and Clinical Correlates" (J. I. Gallin, ed.), p. 343–362. Raven Press, New York.
Van Rooijen, N. (1989). *J. Immunol. Methods* **124**, 1.
Warr, G. A. (1980). *Biochem. Biophys. Res. Commun.* **93**, 737–745.
Wileman, T., Boshans, R., and Stahl, P. (1985). *J. Biol. Chem.* **260**, 7387–7393.
Wileman, T., Boshans, R. L., Schlesinger, P. H., and Stahl, P. D. (1984). *Biochem. J.* **220**, 665–675.
Wileman, T. E., Lennartz, M. R., and Stahl, P. D. (1986). *Proc. Natl. Acad. Sci. U.S.A.* **83**, 2501–2505.
Wilson, M. E., and Pearson, R. D. (1986). *J. Immunol.* **136**, 4681–4688.
Wilson, M. E., and Pearson, R. D. (1988). *Infect. Immun.* **56**, 363–369.
Yamashita, C., Sone, S., Ogura, T., and Kiwada, H. (1991). *Jpn. J. Cancer Res.* **82**, 569–576.

Index

A

N-Acetyl galactosamine, asialoglycoprotein receptor and, 182–183, 211
Acetylation, rhodopsin and, 67, 69
Acetylcholine, adenylyl cyclase-coupled β-adrenergic receptors and, 10, 12, 18–19
Acidification, mannose receptor and, 230
Actin
    Dictyostelium and, 37
    rhodopsin and, 52
Actinomycin D, adenylyl cyclase-coupled β-adrenergic receptors and, 14
Acute myeloblastic leukemia, cytokine receptors and, 129, 131, 143
Adaptation, Dictyostelium and, 37–38, 41, 45
Adaptins, asialoglycoprotein receptor and, 209–210
Adaptor proteins, asialoglycoprotein receptor and, 208–210
Adenylate cyclase, erythropoietin receptor and, 115
Adenylyl cyclase-coupled β-adrenergic receptors, 1, 27–28
    desensitization, 14–15
        arrestin proteins, 20–22
        classification, 15
        down-regulation, 26–27
        phosphorylation, 22–24
        protein kinase, 15–20
        sequestration, 24–26
    function
        G protein-coupling domains, 10–13
        gene organization, 13–14
        ligand binding domains, 8–10
        posttranslation modifications, 3–8
        subtypes, 2–3
    structure, 3
Adenylyl cyclases, Dictyostelium and, 36–38, 41, 45

Adneylate cyclase, cytokine receptors and, 149
ADP, rhodopsin and, 66
β-Adrenergic receptor
    adenylyl cyclase-coupled, see Adenylyl cyclase-coupled β-adrenergic receptors
    asialoglycoprotein receptor and, 207
    Dictyostelium and, 39, 41
    rhodopsin and, 61, 86
β-Adrenergic receptor kinase (βARK)
    desensitization, 16–20, 22, 25, 28
    Dictyostelium and, 41
    rhodopsin and, 86, 90
Affinity
    asialoglycoprotein receptor and, 182, 208, 212–213
        intracellular traffic, 200–201
        protein structure, 183, 189, 191
    cytokine receptors and, 122, 148
        lymphokines, 126–135, 137
        molecular properties, 139–144
    Dictyostelium and, 43–44
    erythropoietin receptor and, 101, 103–105, 109–110, 114
    mannose receptor and, 224, 229, 231
    nerve growth factor receptor and, 175–177
    rhodopsin and, 89
AIC2A, cytokine receptors and, 129–130, 133, 138, 142–143
AIC2B, cytokine receptors and, 130–133, 138, 140
Amebas, Dictyostelium and, 35–36
Amino acids
    adenylyl cyclase-coupled β-adrenergic receptors and
        desensitization, 17–19, 21–22, 26
        function, 3, 5, 10, 12–13
    asialoglycoprotein receptor and, 181, 212
        protein structure, 184–185, 187, 189
        sorting signals, 205–206

P

p75 nerve growth factor, 169–172, 174–177
p140, nerve growth factor receptor and, 173–177
Palmitate, rhodopsin and, 68, 71
Palmitic acid, adenylyl cyclase-coupled β-adrenergic receptors and, 7–8
Palmitoylation
  adenylyl cyclase-coupled β-adrenergic receptors and, 8
  rhodopsin and, 68–69, 74, 90
PC12 cells, nerve growth factor receptor and, 169, 171–174, 176–177
Peptides
  adenylyl cyclase-coupled β-adrenergic receptors and desensitization, 16–17, 19, 23–24
  function, 9, 12
  asialoglycoprotein receptor and, 187
  cytokine receptors and, 136
  mannose receptor and, 224, 227, 235
  rhodopsin and, 79, 84, 86–87, 90
Peripherin, rhodopsin and, 60–61
Peritoneal macrophages
  asialoglycoprotein receptor and, 182, 211–212
  mannose receptor and, 223
pH
  asialoglycoprotein receptor and
    intracellular traffic, 194–195, 198, 201
    protein structure, 183–184, 188–189
  mannose receptor and, 229
  polymeric immunoglobulin receptor and, 166
Phagocytosis, mannose receptor and, 221, 223, 226, 233–235, 238, 240
Pharmacology
  adenylyl cyclase-coupled β-adrenergic receptors and, 2–3, 8
  asialoglycoprotein receptor and, 200
  cytokine receptors and, 122, 128
  Dictyostelium and, 46
Phenotype
  asialoglycoprotein receptor and, 211
  Dictyostelium and, 45
  erythropoietin receptor and, 110
  nerve growth factor receptor and, 173
Phorbol esters
  adenylyl cyclase-coupled β-adrenergic receptors and, 24

asialoglycoprotein receptor and, 201–202, 207–208
cytokine receptors and, 148
Phosducin, rhodopsin and, 66
Phosphate
  adenylyl cyclase-coupled β-adrenergic receptors and, 18, 23, 26
  asialoglycoprotein receptor and, 207
  Dictyostelium and, 43–44
  polymeric immunoglobulin receptor and, 165
  rhodopsin and, 68, 83–84
Phosphatidylinositol
  cytokine receptors and, 148
  turnover, 148
Phosphatidylinositol-specific phospholipase C (PI-PLC), Dictyostelium and, 37
Phosphodiesterase
  Cyclic GMP, see Cyclic GMP-phosphodiesterase
  rhodopsin and, 58, 81, 89
Phospholipase C
  adenylyl cyclase-coupled β-adrenergic receptors and, 22
  Dictyostelium and, 36
  erythropoietin receptor and, 116
Phospholipids, mannose receptor and, 239
Phosphorylation
  adenylyl cyclase-coupled β-adrenergic receptors and, 28
    desensitization, 14–26
    function, 4, 8, 11
  asialoglycoprotein receptor and, 202, 207–208
  cytokine receptors and, 122
  Dictyostelium and, 37, 39, 41, 45
  erythropoietin receptor and, 114, 117
  mannose receptor and, 226, 237
  nerve growth factor receptor and, 172–173, 177
  polymeric immunoglobulin receptor and, 165
  rhodopsin and, 60, 68, 71, 82–90
Photoaffinity labeling
  adenylyl cyclase-coupled β-adrenergic receptors and, 9
  Dictyostelium and, 39
Photolysis
  asialoglycoprotein receptor and, 191
  rhodopsin and, 77–80, 86–87
Phototransduction, rhodopsin and, see Rhodopsin, phototransduction and